Mastering Digital Television:
The Complete Guide to the DTV Conversion

Mastering Digital Television:
The Complete Guide to the DTV Conversion

Jerry C. Whitaker

McGraw-Hill

New York   Chicago   San Francisco   Lisbon   London
Madrid   Mexico City   Milan   New Delhi   San Juan
Seoul   Singapore   Sydney   Toronto

The McGraw-Hill Companies

1 2 3 4 5 6 7 8 9 0 DOC/DOC 0 1 2 1 0 9 8 7 6

ISBN 0-07-147016-6

The sponsoring editor for this book was Steve Chapman. The production supervisor was Pamela A. Pelton. This book was set in Times New Roman and Helvetica by Technical Press, Morgan Hill, CA. The art director for the cover was Anthony Landi

Printed and bound by RR Donnelley.

McGraw-Hill books are available at special quantity discounts to use as premiums and sales promotions, or for use in corporate training programs. For more information, please write to the Director of Special Sales, McGraw-Hill, 2 Penn Plaza, 12th Floor, New York, NY 10121. Or contact your local bookstore.

This book is printed on acid-free paper.

For

Christy, Tara, and Lisa

You never really leave the broadcast business...

Other books in the McGraw-Hill Video/Audio Series

The McGraw-Hill Video/Audio series continues to expand with new books on a variety of topics of importance to engineers and technicians involved in professional media. Current offerings include:

Contents

Preface

The conversion from analog to digital television is proceeding at a rapid pace. In the U.S., nearly all broadcasters have their DTV stations on the air, carrying high-definition programming or in many cases multiple standard-definition programs. Progress is equally swift in other parts of the world.

The promise of DTV has bee realized. Stations around the world are broadcasting innovative programming and fashioning new business strategies made possible by this new technology. While broadcasters are the most visible early adopters of DTV, many other industries are gearing up and—indeed—are already making commercials, television programs, and motion pictures using HDTV equipment. Put side-by-side, the breadth of the HDTV applications now finding commercial success is staggering—from NASA to postproduction to medial imaging.

The launch of digital television in North America has fundamentally changed the broadcast landscape. Image quality really does count, and program producers and stations are competing on quality. Another factor influencing the trend toward higher quality is that making good pictures is easier than ever. Encoders improve with every generation, and receivers are improving as well. The experience in North America is that consumers do care about picture quality. In the current multi-source (terrestrial/cable/satellite) environment, consumers can have it all—quality and quantity.

The launch of every new technology has been fraught with growing pains. Consider what television engineers of the 1950s faced when their management decided that it was good business to convert to color. In 1954, getting a black-and-white signal on the air was an accomplishment, let alone holding the frequency and waveform tolerances sufficiently tight that the NTSC color system would look halfway decent. And then there were the receivers: big, expensive, and unstable. Round picture tubes with a long necks. The sets commandeered living rooms across America. And service these things? There were two dozen convergence adjustments alone in the RCA CTC-15 chassis, the first design that really worked—and it was released nearly a decade after the first color broadcasts began.

These problems were, in the final analysis, only details. Obstacles to be overcome. Challenges to be met.

There have been challenges with DTV, but like the transition to color decades ago, those challenges have also been met.

About the Book

Mastering Digital Television examines the technology of DTV, with particular emphasis on the system documented by the Advanced Television Systems Committee (ATSC). Within the practical constraints of limited pages in a printed book, it is impossible to cover the other digital television systems in common use—DVB and ISDB. Rather than provide a surface treatment of all systems, the author chose to focus on one, and to then examine it in detail. In this book the underlying technology of DTV is examined and implementation issues are addressed. New developmental efforts also are explained and the benefits of the work are outlined.

This publication is directed toward technical and engineering personnel involved in the design, specification, installation, and maintenance of broadcast television systems and non-broadcast professional imaging systems. The basic principles of digital television—in general—and HDTV—in particular—are discussed, with emphasis on how the underlying technologies influence the ultimate applications.

In any technical handbook, finding the information that a reader needs can be difficult. For this reason, *Mastering Digital Television* has been organized into—essentially—six separate "books." The section titles listed in the Table of Contents outline the scope of the handbook and each section is self-contained. The section introductions include a detailed table of contents of the chapters that follow. It is the goal of this approach to make the book easier to use and more useful on the job. In addition, a master subject index is provided at the end of the publication.

This book is all about fundamentals. The path to DTV is outlined in Section 1, which provides a detailed look at the developments leading to the television system that we now enjoy. On this point, please note that "The views expressed are the authors' and do not necessarily reflect the conclusions of any particular organization." Having dispensed with the disclaimer, Section 1 provides a fascinating glimpse of the long road to DTV. That road has been marked by many inspired moments and many years of plain hard work.

The chapters that follow focus on fundamental technical principles, specifically:

- **Section 2**: Fundamental Video Principles

- **Section 3**: Fundamental Audio Principles

- **Section 4**: Fundamental Transmission Principles

The goal here is provide a solid foundation of understanding in audio, video, and transmission technologies—all of which leads to an extensive review of the ATSC DTV standards in Section 5.

Implementation considerations are addressed in Section 6. The author acknowledges that this treatment does not go into great detail. The primary reason for this approach is that implementation problems for any new technology tend to be transitory in nature. For example, when the author was preparing the first "DTV Handbook" in the mid-1990s, perhaps the greatest implementation challenge was transmission tower siting and licensing. This issue, of course, is well behind us now. Other implementation issues have arisen and been resolved over the years since the DTV transition began. Still more will be identified as the affected industries continue to move forward. The intent of Section 6 is to identify some of the fundamental implementation considerations and offer some guidance on dealing with them.

The editor has made every effort to cover the subject of digital television in a comprehensive manner. Extensive references are provided at the end of each chapter to direct readers to sources of additional information.

Considerable detail has been included on the ATSC digital television system. These chapters are based on documents published by the ATSC, and the editor gratefully acknowledges this contribution.

The field of science encompassed by DTV is broad and exciting. It is an area of enormous importance to market segments of all types and—of course—to the public. It is the intent of *Mastering Digital Television* to bring these diverse concepts and technologies together in an understandable form.

Jerry C. Whitaker

Mastering Digital Television:
The Complete Guide to the DTV Conversion

The Road to Digital Television

Advances in technology and increased competition have created new opportunities and new challenges the television industry. The steady advance of technology makes it easy to lose sight of the great progress made from one decade to the next in broadcast equipment design. Only by stepping back and comparing where we are with where we have been can the true distance covered be measured.

Throughout the history of product development, design standardization has been critically important. The term "standard" envisions a means of promoting an atmosphere of interchangeability of basic hardware.

Many of the early standards relating to broadcasting in the U.S. were developed by equipment manufacturers, first under the banner of the Radio Manufacturers Association (RMA), then the Radio, Electronic and Television Manufacturers Association (RETMA), and still later the Electronic Industries Association (EIA). The Institute of Radio Engineers (the forerunner of the IEEE) was responsible for measurement standards and techniques.

The Society of Motion Picture and Television Engineers (SMPTE) has played an important role in the development of standards related to the motion picture and television industries. On or about 1915 it became obvious that the rapidly expanding motion picture industry needed to standardize the basic dimensions and tolerances of film stocks and transport mechanisms. After two unsuccessful attempts to form industry-based standardizing committees, the Society of Motion Picture Engineers (SMPE) was formed. The founding goals were to standardize the nomenclature, equipment and processes of the industry; to promote research and development; and to remain independent of, while cooperating with, its business partners.

By the late 1940s it was apparent that the future of motion pictures and television would involve sharing technologies and techniques. SMPE subsequently was expanded to SMPTE.

The NAB has traditionally been involved with standards involving audio equipment used principally by broadcasters. A disc recording pre-emphasis specification was developed around 1940 to bring order to the chaos then existing in electrical transcriptions used for music libraries and program syndication. Again, in the 1950s, NAB was instrumental in the adoption of standard audio tape recording/playback pre-emphasis curves. Tape cartridge specifications were subsequently developed during the early 1960s.

The wheels of standardization may seem to grind slowly at times, but the adoption of a standard that might have to endure for 50 years or more cannot be taken lightly. This section charts the path from the earliest work in radio and television broadcasting to the ATSC Digital Television Standard.

In This Section: _____

1.1

A Brief History of Television

1.1.1 Introduction

From humble beginnings, television has advanced to become the most effective communications medium in the history of this planet.

"Standardization at the present stage is dangerous. It is extremely difficult to change a standard, however undesirable it may prove, after the public has invested thousands of dollars of equipment. The development goes on, and will go on. There is no question that the technical difficulties will be overcome."

The writer is not addressing the problems faced by high-definition television or fiber optic delivery of video to the home. The writer is addressing the problems faced by *television*. The book containing this passage was published in April 1929. Technology changes, but the problems faced by the industry do not.

1.1.2 Television: A Revolution in Communications

The mass communications media of television is one of the most significant technical accomplishments of the 20th century. The ability of persons across the country and around the world to *see* each other, to communicate with each other, and experience each other's cultures and ideas is a monumental development. Most of us have difficulty conceiving of a world without instant visual communication to virtually any spot on earth. The technology that we enjoy today, however, required many decades to mature.

1.1.2a The Nipkow Disc

The first working device for analyzing a scene to generate electrical signals suitable for transmission was a scanning system proposed and built by Paul Nipkow in 1884. The scanner consisted of a rotating disc with a number of small holes (or *apertures*) arranged in a spiral, in front of a photo-electric cell. As the disc rotated, the spiral of 18 holes swept across the image of the scene from top to bottom in a pattern of 18 parallel horizontal lines.

The Nipkow disc was capable of about 4,000 picture "dots" (or pixels) per second. The scanning process analyzed the scene by dissecting it into picture elements. The fineness of picture detail that the system was capable of resolving was limited in the vertical and horizontal axes by the diameter of the area covered by the aperture in the disc. For reproduction of the scene, a light source controlled in intensity by the detected electrical signal was projected on a screen through a similar Nipkow disc rotated in synchronism with the pickup disc.

Despite subsequent improvements by other scientists (J. L. Baird in England and C. F. Jenkins in the United States) and in 1907 the use of Lee De Forest's vacuum-tube amplifier, the serious limitations of the mechanical approach discouraged any practical application of the Nipkow disc. The principle shortcomings were:

- Inefficiency of the optical system

- Use of rotating mechanical components

- Lack of a high-intensity light source capable of being modulated by an electrical signal at the higher frequencies required for video signal reproduction.

Nevertheless, Nipkow demonstrated a scanning process for the for the analysis of images by dissecting a complete scene into an orderly pattern of picture elements that could be transmitted by an electrical signal and reproduced as a visual image. This approach is—of course—the basis for present-day television.

Nipkow lived in Berlin, although he was of Russian birth. The U.S.S.R. claims a Russian invented television, not because of Nipkow, but another man who experimented with the Nipkow disc in 1905 in Moscow. The Germans, English and Japanese also claim their share of the fame for inventing television.

No one argues, however, that credit for the development of modern *electronic* television belongs to two men: Philo T. Farnsworth and Vladimir Zworykin. Each spent their lives perfecting this new technology.

1.1.2b   Zworykin: The Brains of RCA

A Russian immigrant, Vladimir Zworykin came to the United States after World War I and went to work for Westinghouse in Pittsburgh. During his stay at the company, 1920 until 1929, Zworykin performed some of his early experiments in television. Zworykin had left Russia for America to develop his dream: television. His conception of the first practical TV camera tube, the *Iconoscope* (1923), and his development of the *kinescope* picture tube formed the basis for subsequent advances in the field. Zworykin is credited by most historians as *the father of television*.

Zworkin's Iconoscope (from Greek for "image" and "to see") consisted of a thin aluminum-oxide film supported by a thin aluminum film and coated with a photosensitive layer of potassium hydride. With this crude camera tube and a CRT as the picture reproducer, he had the essential elements for electronic television.

Continuing his pioneering work, Zworykin developed an improved Iconoscope six years later that employed a relatively thick, 1-sided target area. He had, in the meantime, continued work on improving the quality of the CRT and presented a paper on his efforts to the Eastern Great Lakes District Convention of the Institute of Radio Engineers (IRE) on November 18, 1929. The presentation attracted the attention of another former Russian immigrant, David Sarnoff, then vice president and general manager of RCA. Sarnoff persuaded Zworykin to join RCA Victor in Camden, NJ, where he was made director of RCA's electronics research laboratory. The company provided the management and financial backing that enabled Zworykin and the RCA scientists working with him to develop television into a practical system.

Both men never forgot their first meeting. In response to Sarnoff's question, Zworykin—thinking solely in research terms—estimated that the development of television would cost $100,000. Years later, Sarnoff delighted in teasing Zworykin by telling audiences what a great salesman the inventor was. "I asked him how much it would cost to develop TV. He told me $100,000, but we spent $50 million before we got a penny back from it."

By 1931, with the Iconoscope and CRT well-developed, electronic television was ready to be launched—and Sarnoff and RCA were ready for the new industry of television.

1.1.2c Farnsworth: The Boy Wonder

Legend has it that Philo Farnsworth conceived of electronic television when he was a 15 year old high school sophomore in Rigby, Idaho, a small town about 200 miles north of Salt Lake City. Farnsworth met a financial expert by the name of George Everson in Salt Lake City when he was 19 years old and persuaded him to try and secure venture capital for an all-electronic television system.

The main concern of the financial investors whom Everson was able to persuade to put up money for this unproven young man with unorthodox ideas, was that no one else was investigating an electronic method of television. Obviously, many people were interested in capturing the control over patents of a vast new field for profit. If no one was working on this method, then Farnsworth had a clear field. If, on the other hand, other companies were working in secret without publishing their results, then Farnsworth would have little chance of receiving the patent awards and the royalty income that would surely result. Farnsworth and Everson were able to convince the financial backers that they alone were on the trail of a total electronic television system.

Farnsworth established his laboratory first in Los Angeles, and later in San Francisco at the foot of Telegraph Hill. Farnsworth was the proverbial lone basement experimenter. It was at his Green Street (San Francisco) laboratory that Farnsworth gave the first public demonstration in 1927 of the television system he had dreamed of for six years.

He was not yet 21 years of age!

Farnsworth was quick to developed the basic concepts of an electronic television system, giving him an edge on most other inventors in the race for patents. His patents included the principle of blacker-than-black synchronizing signals, linear sweep and the ratio of forward sweep to retrace time. Zworykin won a patent for the principle of field interlace.

In 1928 Farnsworth demonstrated a non-storage electronic pickup and image scanning device he called the *Image Dissector*. The detected image was generated by electrons emitted from a photocathode surface and deflected by horizontal and vertical scanning fields (applied by coils surrounding the tube) so as to cause the image to scan a small aperture. (See Figure 1.1.1.) In other words, rather than an aperture or electron beam scanning the image, the aperture was stationary and the electron image was moved across the aperture. The electrons passing through the aperture were collected to produce a signal corresponding to the charge at an element of the photocathode at a given instant.

The limitation of this invention was the extremely high light level required because of the lack of storage capability. Consequently, the Image Dissector found little use other than as a labora-

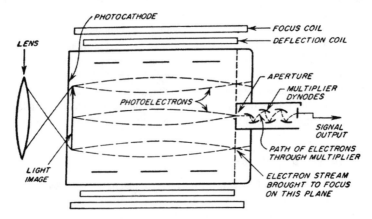

Figure 1.1.1 Operating principles of the Image Dissector. (*From* [1]. *Used with permission.*)

tory signal source. Still, in 1930, the 24-year-old Farnsworth received a patent for his Image Dissector, and in the following year entertained Zworykin at his San Francisco laboratory.

Farnsworth's original "broadcast" included the transmission of graphic images, film clips of a Dempsey/Tunney fight and scenes of Mary Pickford combing her hair (from her role in the *Taming of the Shrew*). In his early systems, Farnsworth could transmit pictures with 100- to 150-line definition at a repetition rate of 30 lines per second. This pioneering demonstration set in motion the progression of technology that would lead to commercial broadcast television a decade later.

Farnsworth held many patents for television, and through the mid-1930s remained RCA's fiercest competitor in developing new technology. Indeed, Farnsworth's thoughts seemed to be directed toward cornering patents for the field of television and protecting his ideas. In the late 1930s, fierce patent conflicts between RCA and Farnsworth flourished. They were settled in September 1939 when RCA capitulated and agreed to pay continuing royalties to Farnsworth for the use of his patents. The action ended a long period of litigation. By that time Farnsworth held an impressive list of key patents for electronic television.

Farnsworth died in 1971 and is credited only slightly for the giant television industry that he helped create.

1.1.2d Other Experimenters

Unsuccessful attempts were made to use pickup devices without storage capability, such as the Farnsworth Image Dissector, for studio applications. The most ambitious was the Allen B. DuMont Laboratories' experiments in the 1940s with an electronic *flying-spot* camera. The set in the studio was illuminated with a projected raster frame of scanning lines from a cathode-ray tube. The light from the scene was gathered by a single photo-cell to produce a video signal.

The artistic and staging limitations of the dimly-lit studio are all too obvious. Nevertheless, while useless for live pickups, it demonstrated the flying-spot principle, a technology that has been widely used for television transmission of motion picture film.

General Electric also played an early role in the development of television. In 1926, Ernst Alexanderson, an engineer at the company, developed a mechanical scanning disc for video transmission. He gave a public demonstration of the system two years later. Coupled with the GE experimental TV station, WGY (Schenectady, N.Y.), Alexanderson's system made history on September 11, 1928, by broadcasting the first dramatic program on television. It was a 40-minute play titled, *The Queen's Messenger*. The program consisted of two characters performing before three simple cameras.

There was a spirited race to see who could begin bringing television programs to the public first. In fact, the 525 line 60 Hz standards promoted in 1940 and 1941 were known as "high-definition television," as compared with some of the experimental systems of the 1930s. The original reason for the 30 frame per second rate was the simplified receiver design that it afforded. With the field scan rate the same as the power system frequency, ac line interference effects were minimized in the reproduced picture. Both Zworykin and Farnsworth were members of the committee that came up with proposed standards for a national (U.S.) system. The standard was to be in force before any receiving sets could be sold to the public.

The two men knew that to avoid flicker, it would be necessary to have a minimum of 40 complete pictures per second; this was known from the motion picture industry. Although film is exposed at 24 frames per second, the projection shutter is opened twice for each frame, giving a net effect of 48 frames per second. If 40 complete pictures per second were transmitted, even with 441 lines of horizontal segmentation (which was high-definition TV prior to WW II), the required bandwidth of the transmitted signal would have been greater than the technology of the day could handle. The *interlace* scheme was developed to overcome the technical limitations faced by 1940s technology.

1.1.2e   Pickup Tubes

Zworykin's Iconoscope television camera tube, first patented in 1923, stored an image of a scene as a mosaic pattern of electrical charges. A scanning electron-beam then released secondary electrons from the photosensitive mosaic to be read out sequentially as a video signal (Figure 1.1.2).

Although the Iconoscope provided good resolution, relatively high light levels (studio illumination of 500 or more foot-candles) was necessary. In addition, picture quality was degraded by spurious flare. This was caused by photoelectrons and secondaries, resulting from the high potential of the scanning beam, falling back at random on the storage surface. The presence of flare, and the lack of a reference black signal (because of capacitive coupling through the signal plate) resulted in a gray scale that varied with scene content. The pickup system thus required virtually continuous manual correction of video-gain and blanking levels.

Furthermore, because the light image was focused on the same side of the signal plate as the charge image, it was necessary to locate the electron gun and deflection coils off the optical axis in order to avoid obstructing the light path. Because the scanning beam was directed at the signal plate at an average angle of 45 degrees, vertical keystone correction of the horizontal scan was needed.

With all of its shortcomings, the Iconoscope was the key to the introduction of the first practical all-electronic television system. Because a cathode-ray picture-display tube (necessary to supplant the slowly-reacting modulated light source and cumbersome rotating disc of the Nip-

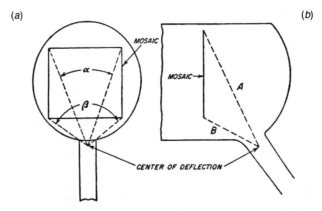

Figure 1.1.2 Iconoscope deflection geometry: (*a*) front view, (*b*) side view, where α and β are horizontal deflection angles of the scanning beam, *A* at top and *B* at bottom, of the raster. (*From* [1]. *Used with permission.*)

kow display system) had been demonstrated as early as 1905, a television system composed entirely of electronic components was then feasible.

Zworykin continued to refine the Iconoscope, demonstrating improved tubes in 1929 and 1935. His work culminated in the development of the *Image Iconoscope* in 1939, which offered greater sensitivity and overcame some of the inherent problems of the earlier devices. (See Figure 1.1.3.) In the Image Iconoscope, a thin-film transparent photocathode was deposited on the inside of the faceplate. Electrons emitted and accelerated from this surface at a potential several hundred electronvolts were directed and focused on a target storage plate by externally-applied magnetic fields. A positive charge image was formed on the storage plate, this being the equivalent of the storage mosaic in the Iconoscope. A video signal was generated by scanning the positive charge image on the storage plate with a high-velocity beam in exactly the same manner as the Iconoscope.

Both types of Iconoscopes had a light-input/video-output characteristic that compressed highlights and stretched lowlights. This less-than-unity relationship produced signals that closely matched the exponential input-voltage/output-brightness characteristics of picture display tubes, thus producing a pleasing gray-scale of photographic quality.

The *Orthicon* camera tube, introduced in 1943, was the next major development in tube technology. It eliminated many of the shortcomings of the Iconoscope through the use of low-velocity scanning. The Orthicon, so named because the scanning beam landed on the target at right angles to the charge surface, used a photoemitter composed of isolated light-sensitive granules deposited on an insulator. A similar tube, the *CPS Emitron* (so named for its cathode-potential stabilized target scanning), was developed in England (Figure 1.1.4). The CPS Emitron target was made up of precise squares of semitransparent photoemissive material deposited on the target insulator through a fine mesh. Both of these tubes produced improved-resolution pictures with good gray-scales.

In 1943, an improved Orthicon pickup device, the *Image Orthicon*, was introduced. The tube incorporated 3 important technologies to make possible studio and field operations under reduced and varied lighting conditions (Figure 1.1.5). The technologies involved were:

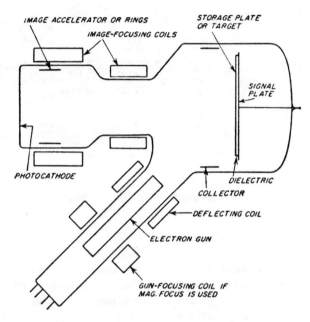

Figure 1.1.3 Schematic diagram of the image iconoscope and its associated components. (*From* [1]. *Used with permission.*)

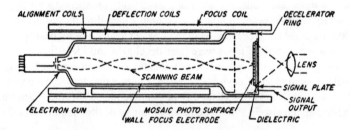

Figure 1.1.4 Diagram of the CPS emitron. (*From* [1]. *Used with permission.*)

- Imaging the charge pattern from a photosensitive surface on an electron storage target
- Modulation of the scanning beam by the image charge of the target
- Amplification of the scanning-beam modulation signal by secondary-electron emission in a multistage multiplier.

The image of the scene being televised was focused on a transparent photocathode on the inside of the faceplate of the tube. The diameter of the photocathode on a 3-inch diameter Image

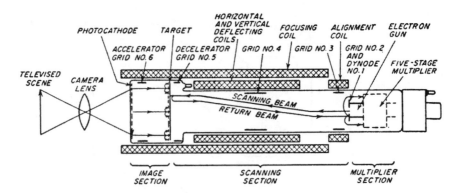

Figure 1.1.5 Schematic arrangement of the Image Orthicon. (*From* [1]. *Used with permission.*)

Orthicon was 1.6 in. (41 mm), the same as double-frame 35 mm film, a fortunate choice because it permitted the use of already-developed conventional lenses. Light from the scene caused a charge pattern of the image to be set up. Because the faceplate was at a negative voltage (approximately –450 V), electrons were emitted in proportion to scene illumination and accelerated to the target-mesh surface, which was at (nearly) zero potential. The fields from the accelerator grid and focusing coil focused the electrons on the target.

The *Vidicon* was introduced to the broadcast industry in 1950. It was the first successful television camera tube to use a photoconductive surface to derive a video signal. The photosensitive target material of the Vidicon consisted of a continuous light-sensitive film deposited on a transparent signal electrode. An antimony trisulphide photoconductor target was scanned by a low-velocity electron beam to provide an output signal directly. No intermediate electron-imaging or electron-emission processes, as in the Image Orthicon or Iconoscope, were employed.

While a variety of tubes were later developed and identified under different trade names (or by the type of photoconductor used), the name *Vidicon* became the generic classification for all such photoconductive devices. Milestones in this developmental process include introduction of the 1-in Plumbicon in 1968, the 2/3-in Plumbicon in 1974, the 1/2-in Plumbicon in 1981, and the 1/2-in Saticon also in 1981. The basic geometry of the Plumbicon is shown in Figure 1.1.6.

1.1.2f   Sold State Imaging

Solid-state imaging devices using a flat array of photosensitive diodes were proposed as early as 1964 and demonstrated publicly in 1967. The charge voltage of each sensor element was sampled in a horizontal and vertical, or X-Y, addressing pattern to produce an output voltage corresponding to a readout of the image pixels. The resolution capability of these first laboratory models did not exceed 180×180 pixels, a tenth of that required for television broadcasting applications. Nevertheless, the practicability of solid-state technology was demonstrated.

In the first solid-state camera system, a video signal was generated by sampling the charge voltages of the elements of the array directly in an X and Y scanning pattern. In the early 1970s a major improvement was achieved with the development of the *charge-coupled device* (CCD), in

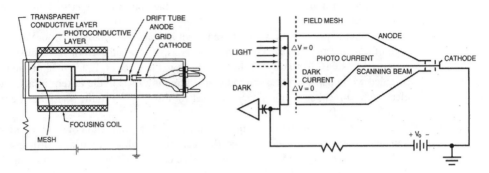

Figure 1.1.6 Schematic arrangement of the Plumbicon. (From [1]. Used with permission.)

operation a charge-transfer system. The photosensitive action of a simple photodiode was combined in one component with the charge-transfer function and metal-oxide capacitor storage capability the CCD. The photo-generated charges were transferred to metal-oxide semiconductor (MOS) capacitors in the CCD and stored for subsequent readout as signals corresponding to pixels.

Thus, rather than sampling directly the instantaneous charge on each photosensitive picture element, the charges were stored for readout either as a series of picture scanning lines in the *interline-transfer* system, or as image fields in the *frame-transfer* system.

The early CCD chips were interline-transfer devices in which vertical columns of photosensitive picture elements were alternated with vertical columns of sampling gates. The gates in turn fed registers to store the individual pixel charges. The vertical storage registers were then sampled one line at a time in a horizontal and vertical scanning pattern to provide an output video signal. This approach was used in early monochrome cameras and in three-sensor color cameras. It was also used with limited success in a single-tube color camera wherein cyan, green, and yellow stripe filters provided three component color signals for encoding as a composite signal. The interline system is of only historical interest, having been overtaken by frame-transfer technology.

Milestones in the development of CCD devices for professional applications include the introduction in 1979 by Bosch of the FDL-60 CCD-based telecine, the NEC SPC-3 CCD camera in 1983, and the RCA CCD-1 camera in 1984.

1.1.2g   Image Reproduction

From the start of commercial television in the 1940s until the emergence of color as the dominant programming medium in the mid-1960s, virtually all receivers were the direct-view monochrome type. A few large-screen projection receivers were produced, primarily for viewing in public places by small audiences. Initially the screen sizes were 10 to 12-in. diagonal.

The horizontal lines of the two fields on a receiver or monitor screen are produced by a scanning electron beam which, upon striking the back of the picture tube screen, causes the phosphor to glow. The density of the beam, and the resultant brightness of the screen, is controlled by the voltage level of a video signal applied between the controlling aperture and the cathode in the electron gun.

In the old days, viewers were advised to sit at least one foot away from the screen for every inch of screen size as measured diagonally. Thus, if you had a 10-inch screen TV set, you were supposed to sit 10-feet away. In those early days the electron beam scan of the CRT phosphor revealed with crisp sharpness the individual scanning lines in the raster. In fact, the focus of the electron beam was sometimes purposely set for a soft focus so the scan lines were not as easily seen.

All color television picture displays synthesize the reproduction of a color picture by generating light, point by point, from three fluorescent phosphors, each of a different color. This is called an *additive* system. The chroma characteristic, or hue, of each of color light source is defined as a primary color. The most useful range of reproduced colors is obtained from the use of three primaries with hues of red, green, and blue. A combination of the proper intensities of red, green and blue light will be perceived by an observer as white.

Utilizing this phenomenon of physics, color television signals were first produced by optically combining the images from three color tubes, one for each of the red, green and blue primary transmitted colors. This early *Trinescope*, as it was called by RCA, demonstrated the feasibility of color television. The approach was, however, too cumbersome and costly to be a practical solution for viewing in the home.

The problem was solved by the invention of the shadow-mask picture tube in 1953. The first successful tube used a triad assembly of electron guns to produce three beams that scanned a screen composed of groups of red, green, and blue phosphor dots. The dots were small enough not to be perceived as individual light sources at normal viewing distances. Directly behind the screen, a metal mask perforated with small holes approximately the size of each dot triad, was aligned so that each hole was behind an R-G-B dot cluster.

The three beams were aligned by *purity* magnetic fields so that the mask *shadowed* the green and blue dots from the beam driven by the red signal. Similarly, the mask shadowed the red and blue dots from the from green beam, and the red and green dots from the blue beam.

1.1.2h   Who was First?

The technology of television actually was the creation of many people in many countries over many years. It is correctly viewed as an international invention.

British developments in television saw great advances during the thirties. While Baird was conducting daily nighttime 30 line broadcasts from 1929 on two BBC MW transmitters (video and audio), EMI Laboratories in Hayes (Middlesex, England) was developing electronic television of their own. Patents were freely shared, but much development took place in the U.K. (EMI was linked to RCA from the days when the Victor Talking Machine Company of Camden and Gramophone Company of Hayes were related.)

EMI and Marconi in 1934 proposed to the Selsdon Committee on the future of U.K. Television, a system comprised of 405 line interlaced scanning; the 405 waveform included the serrated vertical sync pulse but did not include "equalizing" pulses to improve interlace. Baird proposed 240 lines 25 frames progressive scanning. On November 2, 1936, daily programming was inaugurated, with the BBC alternating weekly each transmission system. Programming began with two hours a day (not including test transmissions). The first sets were made switchable to both systems. (Fortunately each system shared the same video/audio carrier frequencies of 45 MHz and 41.5 MHz, respectively, and each used AM sound and positive modulation of the video.) EMI used iconoscope (Zworykin) camera tubes, whereas Baird, in 1936, used a flying-spot studio camera (four years earlier than DuMont), an "Intermediate Film/Flying Spot" camera of German origin, and an imported Farnsworth Image Dissector camera.

The Baird System was ordered to shut down on February, 8, 1937, and the 405 line system remained. It is estimated that 20,000 TV sets were sold to the public by September, 1939, when the BBC Television service was ordered to close for the duration of WW-II. Prices of receivers plummeted.

The Germans claimed to have begun broadcasting television to public audiences in 1935. The Farnsworth system was used in Germany to broadcast the 1936 Olympics at Berlin.

The U.S. did not come on-line with a broadcast system that sought to inform or entertain audiences until shortly before WW-II. While both Farnsworth and Zworykin had transmitters in place and operational early in their experiments, whatever programing present was incidental. The main purpose was to experiment with the new communication medium. The goal at the time was to improve the picture being transmitted until it compared reasonably well with the 35 mm photographic images available in motion picture theaters. There were, however, some pioneer TV broadcasters during the 1930s that offered entertainment and information programs to the few people who had a television receiver.

In 1933, station W9XK and radio station WSUI, broadcasting from the campus of the State University of Iowa, thrilled select midwesterners with a regular evening program of television. WSUI broadcast the audio portion on its assigned frequency of 880 kHz, and W9XK transmitted the video at 2.05 MHz with a power of 100 W. The twice-per-week program, initiating educational television, included performances by students and faculty with brief skits, lectures, and musical selections. During the early 1930s, W9XK was the only television station in the world located on a university campus, transmitting simultaneous video and audio programs.

1.1.2i   TV Grows Up

Both NBC and CBS took early leads in paving the way for commercial television. NBC, through the visionary eyes of David Sarnoff and the resources of RCA, stood ready to undertake pioneering efforts to advance the new technology. Sarnoff accurately reasoned that TV could establish an industry-wide dominance only if television set manufacturers and broadcasters were using the same standards. He knew this would only occur if the FCC adopted suitable standards and allocated the needed frequency spectrum. Toward this end, in April 1935, Sarnoff made a dramatic announcement that RCA would put millions of dollars into television development. One year later, RCA began field testing television transmission methods from a transmitter atop the Empire State Building.

In a parallel move, CBS (after several years of deliberation) was ready to make its views public. In 1937, the company announced a $2 million experimental program that consisted of field testing various TV systems. It is interesting to note that many years earlier, in 1931, CBS put an experimental TV station on the air in New York City and transmitted programs for more than a year before becoming disillusioned with the commercial aspects of the new medium.

The Allen B. DuMont Laboratories also made significant contributions to early television. While DuMont is best known for CRT development and synchronization techniques, the company's major historical contribution was its production of early electronic TV sets for the public beginning in 1939.

It was during the 1939 World's Fair in New York and the Golden Gate International Exposition in San Francisco the same year that exhibits of live and filmed television were demonstrated on a large scale for the first time. Franklin Roosevelt's World's Fair speech (April 30, 1939) marked the first use of television by a U.S. president. The public was fascinated by the new technology.

Television sets were available for sale at the New York Fair's RCA pavilion. Prices ranged from $200 to $600, a princely sum for that time. Screen sizes ranged from 5 in. to 12 in. (the "big screen" model). Because CRT technology at that time did not permit wide deflection angles, the pictures tubes were long. So long, in fact, that the devices were mounted (in the larger-sized models) vertically. A hinge-mounted mirror at the top of the receiver cabinet permitted viewing.

At the San Francisco Exposition, RCA had another large exhibit that featured live television. The models used in the demonstrations could stand the hot lights for only a limited period. The studio areas were small, hot, and suitable only for interviews and commentary. People were allowed to walk through the TV studio and stand in front of the camera for a few seconds. Friends and family members were able to watch on monitors outside the booth. It was great fun, the lines were always long, and the crowds enthusiastic. The interest caused by these first mass demonstrations of television sparked a keen interest in the commercial potential of television broadcasting. Both expositions ran for a second season in 1940, but the war had started in Europe and television development was about to grind to a halt.

1.1.2j   Transmission Standard Developed

In 1936 the Radio Manufacturers Association (RMA), the forerunner of today's Electronics Industries Association, set up a committee to recommend standards for a commercial TV broadcasting service. In December 1937 the committee advised the FCC to adopt the RCA 343-line/30-frame system that had been undergoing intensive development since 1931. The RCA system was the only one tested under both laboratory and field conditions. A majority of the RMA membership objected to the RCA system because of the belief that rapidly advancing technology would soon render this marginal system obsolete, and perhaps more importantly, would place them at a competitive disadvantage (RCA was prepared to immediately start manufacturing TV equipment and sets). Commercial development of television was put on hold.

At an FCC hearing in January 1940, a majority of the RMA was willing to embrace the RCA system, now improved to 441 lines. However, a strong dissenting minority (Zenith, Philco, and DuMont) was still able to block any action.

The result was that the National Television Systems Committee functioned essentially as a forum to investigate various options. DuMont proposed a 625-line/15-frame/4-field interlaced system. Philco advocated a 605-line/24-frame system. Zenith took the stance that it was still premature to adopt a national standard. Not until June 1941 did the FCC accept the consensus of a 525-line/30-frame (60 Hz) black and white system, which still exists today with minor modifications.

Television was formally launched in July 1941 when the FCC authorized the first two commercial TV stations to be constructed in the U.S. However, the growth of early television was ended by the licensing freeze that accompanied World War II. By the end of 1945 there were just nine commercial TV stations authorized, with six of them on the air. The first post-war full-service commercial license was issued to WNBW, the NBC-owned station in Washington, D.C.

As the number of TV stations on the air began to grow, the newest status symbol became a TV antenna on the roof of your home. Sets were expensive and not always reliable. Sometimes there was a waiting list to get one. Nobody cared; it was all very exciting—pictures through the air. People would stand in front of a department store window just to watch a test pattern.

1.1.2k   Color Standard

During early development of commercial television systems—even as early as the 1920s—it was assumed that color would be demanded by the public. Primitive field sequential systems were demonstrated in 1929. Peter Goldmark of CBS showed a field sequential (color filter wheel) system in the early 1940s and promoted it vigorously during the post war years. Despite the fact that it was incompatible with existing receivers, had limited picture size possibilities, and was mechanically noisy, the CBS system was adopted by the FCC as the national color television standard in October 1950.

The engineering community felt betrayed (CBS excepted). Monochrome TV was little more than 3 years old with a base of 10 to 15 million receivers; broadcasters and the public were faced with the specter of having much of their new, expensive equipment become obsolete. The general wisdom was that color must be an adjunct to the 525/30 monochrome system so that existing terminal equipment and receivers could accept color transmissions.

Was the decision to accept the CBS color wheel approach a political one? Not entirely, because it was based on engineering tests presented to the FCC in early 1950. Contenders were the RCA dot sequential, the CTI (Color Television Incorporated) line sequential, and the CBS field sequential systems. The all-electronic compatible approach was in its infancy and there were no suitable display devices. Thus, for a decision made in 1950 based on the available test data, the commission's move to embrace the color wheel system was reasonable. CBS, however, had no support from other broadcasters or manufacturers; indeed, the company had to purchase Hytron-Air King to produce color TV sets (which would also receive black and white NTSC). Two hundred sets were manufactured for public sale.

Programming commenced on a 5 station East Coast network on June 21, 1951, presumably to gain experience. Color receivers went on sale in September, but only 100 total were sold. The last broadcast was on October 21, 1951. The final curtain fell in November when the National Production Authority (an agency created during the Korean war) imposed a prohibition on manufacturing color sets for public use. Some cynics interpreted this action as designed to get CBS off the hook because the production of monochrome sets was not restricted.

The proponents of compatible, all electronic color systems were, meanwhile, making significant advances. RCA had demonstrated a tri-color delta-delta kinescope. Hazeltine demonstrated the constant luminance principle, as well as the "shunted monochrome" idea. GE introduced the frequency interlaced color system. Philco showed a color signal composed of wideband luminance and two color difference signals encoded by a quadrature-modulated subcarrier. These, and other manufacturers, met in June 1951 to reorganize the NTSC for the purpose of pooling their resources in the development of a compatible system. By November a system employing the basic concepts of today's NTSC color system was demonstrated.

Field tests showed certain defects, such as sound interference caused by the choice of color subcarrier. This was easily corrected by the choice of a slightly different frequency, but at the expense of lowering the frame rate to 29.97 Hz. Finally, RCA demonstrated the efficacy of unequal I and Q color difference bandwidths. Following further field tests, the proposal was forwarded to the FCC on July 22, 1953.

A major problem, however, remained: the color kinescope. It was expensive and could only be built to yield a 9 × 12 in. picture. Without the promise of an affordable large screen display, the future of color television would be uncertain. Then came the development of a method of directly applying the phosphor dots on the faceplate together with a curved shadow mask mounted directly on the faceplate. This breakthrough came from the CBS-Hytron company. The commission adopted the color standard on December 17, 1953. It is interesting to note that the *phase alternation line* (PAL) principle was tried, but practical hardware to implement the scheme would not be available until 10 years later.

After more than 50 years, NTSC still bears a remarkable aptitude for improvement and manipulation. But even with the advantage of compatibility for monochrome viewers, it took 10 years of equipment development and programming support from RCA and NBC before the general public starting buying color receivers in significant numbers.

In France, Germany and other countries, engineers began work aimed at improving upon the U.S. system. France developed the SECAM system and Germany produced the PAL system.

1.1.2I   UHF Comes of Age

The early planners of the U.S. television system thought that 13 channels would more than suffice for a given society. The original channel 1 was from 44 MHz to 50 MHz, but was latter dropped prior to any active use because of possible interference with other services. There remained 12 channels for normal use.

Bowing to pressure from various groups, the FCC revised its allocation table in 1952 to permit UHF TV broadcasting for the first time. The new band was not, however, a bed of roses. Many people went bankrupt building UHF stations because there were few receivers available to the public. UHF converters soon became popular. The first converters were so-called *matchbox types* that were good for one channel only. More expensive models mounted on top of the TV receiver and were tunable.

Finally, the commission requested the authority from Congress to require that all TV set manufacturers include UHF tuning in their receivers. President Kennedy signed the bill into law on July 10, 1962, authorizing the FCC to require television receivers "be capable of adequately receiving all frequencies allocated by the Commission to television broadcasting." In September of that year, the FCC released a Notice of Proposed Rulemaking requiring any television set manufactured after April 30, 1964, to be an all-channel set. The commission also proposed a maximum noise figure for the tuner. These moves opened the doors for significant market penetration for UHF broadcasters. Without that mandate, UHF broadcasting would have lingered in the dark ages for many years.

The klystron has been the primary means of generating high power UHF-TV signals since the introduction of UHF broadcasting. The device truly revolutionized the modern world when it was quietly developed in 1937. Indeed, the klystron may have helped save the world as we know it. And, more than 70 years after it was first operated in a Stanford University laboratory by Russell Varian and his brother Sigurd, the klystron remains irreplaceable, even in this solid-state electronic age.

1.1.2m   Birth of the Klystron

The Varian brothers were unusually bright and extremely active. The two were mechanically minded, producing one invention after another. Generally, Sigurd would think up an idea, Russell would devise a method for making it work, and Sigurd would then build the device.

Through the influence of William Hansen, a former roommate of Russell and at the time (the mid 1930s) a physics professor at Stanford, the Varians managed to get non-paying jobs as research associates in the Stanford physics lab. They had the right to consult with members of the faculty and were given the use of a small room in the physics building.

Hansen's role, apparently, was to shoot down ideas as fast as the Varians could dream them up. As the story goes, the Varians came up with 36 inventions of varying impracticality. Then they came up with idea #37. This time Hansen's eyes widened.

On June 5, 1937 Russell proposed the concept that eventually became the klystron tube. The device was supposed to amplify microwave signals. With $100 for supplies granted by Stanford, Sigurd built it.

The device was simple. A filament heated by an electric current in turn heated a cathode. A special coating on the cathode gave off electrons when it reached a sufficiently high temperature.

Negatively-charged electrons attracted by a positively-charged anode passed through the first cavity of the klystron tube. Microwaves in the cavity interacted with the electrons and passed through a narrow passage called a *drift tube*. In the drift tube, the electrons tended to bunch up: some speeded up, some slowed down. At the place in the drift tube where the bunching was most pronounced, the electrons entered a second cavity, where the stronger microwaves were excited and amplified in the process.

The first klystron device was lit up on the evening of August 19, 1937. Performance was marginal, but confirmed the theory of the device. An improved klystron was completed and tested on August 30.

The Varians published the results of their discovery in the *Journal of Applied Physics*. For reasons that have never been clear, their announcement immediately impressed British scientists working in the same field, but was almost entirely ignored by the Germans. The development of the klystron allowed British and American researchers to build smaller, more reliable radar systems. Klystron development paralleled work being done in England on the magnetron.

The successful deployment of microwave radar was accomplished by the invention to the cavity magnetron at Manchester University in the late 1930s. It was one of Britain's "Top Secrets" handed over to the Americans early in the war. The cavity magnetron was delivered to the Radiation Laboratory at MIT, where it was incorporated into later wartime radar systems. During the Battle of Britain in May 1940, British defenses depended upon longer wavelength radar (approximately 5 m), which worked but with insufficient resolution. The magnetron provided high-power microwave energy at 10 cm wavelengths, which improved detection resolution enormously. (Incidentally, the mass production of television CRTs prior to WW-II also helped radar development.)

Armed with the magnetron and klystron, British and American scientists perfected radar, a key element in winning the Battle of Britain. So valuable were the secret devices that the British decided not to put radar in planes that flew over occupied Europe lest one of them crash, and the details of the components be discovered.

After the war, the Varians—convinced of the potential for commercial value of the klystron and other devices they had conceived—established their own company. For Stanford University, the klystron represents one of its best investments: $100 in seed money and use of a small laboratory room were turned into $2.56 million in licensing fees before the patents expired in the 1970s, three major campus buildings, and hundreds of thousands of dollars in research funding.

1.1.3 Early Development of HDTV

The first developmental work on a high-definition television system began in 1968 at the technical research laboratories of Nippon Hoso Kyokai (NHK) in Tokyo [2]. In the initial effort, film transparencies were projected to determine the levels of image quality associated with different numbers of lines, aspect ratios, interlace ratios, and image dimensions. Image diagonals of 5 to 8 ft, aspect ratios of 5:3 to 2:1, and interlace ratios from 1:1 to 5:1 were used. The conclusions reached in the study were that a 1500-line image with 1.7:1 aspect ratio and a display size of 6.5 ft diagonal would have a subjective image quality improvement of more than 3.5 grades on a 7-grade scale, compared to the conventional 525-line, 4:3 image on a 25-in display. To translate these findings to television specifications, the following parameters were thus adopted:

- 2-to-1 interlaced scanning at 60 fields/s

- 0.7 Kell factor

- 0.5 to 0.7 interlace factor

- 5:3 aspect ratio

- Visual acuity of 9.3 cycles per degree of arc

Using these values, the preferred viewing distance for an 1125-line image was found to be 3.3 times the picture height. A display measuring 3 × 1.5 ft was constructed in 1972 using half mirrors to combine the images of three 26-in color tubes. Through the use of improved versions of wide-screen displays, the bandwidth required for luminance was established to be 20 MHz and, for wideband and narrowband chrominance, 7.0 and 5.5 MHz, respectively. The influence of viewing distance on sharpness also was investigated.

Additional tests were conducted to determine the effect of a wide screen on the realism of the display. In an elaborate viewing apparatus, motion picture film was projected on a spherical surface. As the aspect ratio of the image was shifted, the viewers' reactions were noted. The results showed that the realism of the presentation increased when the viewing angle was greater than 20°. The horizontal viewing angles for NTSC and PAL/SECAM were determined to be 11° and 13°, respectively. The horizontal viewing angle of the NHK system was set at 30°.

The so-called provisional standard for the NHK system was published in 1980. Because the NHK HDTV standard of 1125 lines and 60 fields/s was, obviously, incompatible with the conventional (NTSC) service used in Japan, adoption of these parameters raised a number of questions. No explanations appear in the literature, but justification for the values can be found in the situation faced by NHK prior to 1980. At that time, there was widespread support for a single worldwide standard for HDTV service. Indeed, the International Radio Consultative Committee (CCIR) had initiated work toward such a standard as early as 1972. If this were achieved, the NTSC and PAL/SECAM field rates of 59.94 and 50 Hz would prevent one (or all) of these systems from participation in such a standard. The 50 Hz rate was conceded to impose a severe limit on display brightness, and the 59.94 (vs. 60) Hz field rate posed more difficult problems in transcoding. Thus, a 60-field rate was proposed for the world standard. The choice of 1125 lines also was justified by the existing operating standards. The midpoint between 525 and 625 lines is

575 lines. Twice that number would correspond to 1150 lines for a *midpoint* HDTV system. This even number of lines could not produce alternate-line interlacing, then thought to be essential in any scanning standard. The nearest odd number having a common factor with 525 and 625 was the NHK choice: 1125 lines. The common factor—25—would make line-rate transcoding among the NHK, NTSC, and PAL/SECAM systems comparatively simple.

1.1.3a   1125/60 Equipment Development

The 1970s saw intense development of HDTV equipment at the NHK Laboratories. By 1980, when the NHK system was publicly demonstrated, the necessary camera tubes and cameras, picture tubes and projection displays, telecines, and videotape recorders were available. Also, the choices of transmission systems, signal formats, and modulation/demodulation parameters had been made. Work with digital transmission and fiber optics had begun, and a prototype tape recorder had been designed. Much of the HDTV equipment built to the 1125/60 NHK standard and brought to market by various vendors can be traced to these early developments.

In 1973, the NHK cameras used three 1-1/2-in Vidicons, then commercially available. The devices, however, lacked the necessary resolution and sensitivity. To solve this problem, NHK developed the *return-beam Saticon* (RBS), which had adequate resolution and sensitivity, but about 30 percent lag. Cameras using three RBS tubes came into production in 1975 and were used during much of the NHK HDTV system development. By 1980, another device—the *diode-gun impregnated-cathode Saticon* (DIS) tube—was ready for public demonstration. This was a 1-in. tube, having a resolution of 1600 lines (1200 lines outside the 80 percent center circle), lag of less than 1 percent, and 39 dB signal-to-noise ratio (S/N) across a 30 MHz band. Misregistration among the primary color images, about 0.1 percent of the image dimensions in the earlier cameras, was reduced to 0.03 percent in the DIS camera. When used for sporting events, the camera was fitted with a 14-× zoom lens of f/2.8 aperture. The performance of this camera established the reputation of the NHK system among industry experts, including those from the motion picture business.

The task of adapting a conventional television display to high definition began in 1973, when NHK developed a 22-in. picture tube with a shadow-mask hole pitch of 310 μm (compared with 660 μm for a standard tube) and an aspect ratio of 4:3. In 1978, a 30-in. tube with hole pitch of 340 μm and 5:3 aspect ratio was produced. This tube had a peak brightness of 30 foot-lamberts (100 cd/m^2). This was followed in 1979 by a 26- in tube with 370 μm hole pitch, 5:3 aspect ratio, and peak brightness of 45 foot-lamberts (ft-L).

The need for displays larger than those available in picture tubes led NHK to develop projection systems. A system using three CRTs with Schmidt-type focusing optics produced a 55-in. image (diagonal) on a curved screen with a peak brightness of about 30 ft-L. A larger image (67 in.) was produced by a light-valve projector, employing Schlieren optics, with a peak brightness of 100 ft-L at a screen gain of 10.

Development by NHK of telecine equipment was based on 70 mm film to assure a high reserve of resolution in the source material. The first telecine employed three Vidicons, but they had low resolution and high noise. It was decided that these problems could be overcome through the use of highly monochromatic laser beams as light sources: helium-neon at 632.8 nm for red, argon at 514.5 nm for green, and helium-cadmium at 441.6 nm for blue. To avoid variation in the laser output levels, each beam was passed though an acoustic-optical modulator with feedback control. The beams then were combined by dichroic mirrors and scanned mechanically. Horizontal scanning was provided by a 25-sided mirror, rotating at such a high speed (81,000 rpm) as to require aerostatic bearings. This speed was required to scan at 1125 lines, 30 frames/s, with 25 mirror facets. The deflected beam was passed through relay lenses to another mirror mounted on a galvanometer movement, which introduced the vertical scanning. The scanned beam then passed through relay lenses to a second mirror-polygon of 48 sides, rotating at 30 rpm, in accurate synchronism with the continuous movement of the 70 mm film.

The resolution provided by this telecine was limited by the mechanical scanning elements to 35 percent modulation transfer at 1000 lines, 2:1 aspect ratio. This level was achieved only by maintaining high precision in the faces of the horizontal scanning mirror and in the alignment of successive faces. To keep the film motion and the frame-synchronization mirror rotation in precise synchronism, an elaborate electronic feedback control was used between the respective motor drives. In all other respects, the performance was more than adequate.

The processes involved in producing a film version of a video signal were, essentially, the reverse of those employed in the telecine, the end point being exposure of the film by laser beams controlled by the R, G, and B signals. A prototype system was shown in 1971 by CBS Laboratories, Stamford, Conn. The difference lay in the power required in the beams. In the telecine, with highly sensitive phototubes reacting to the beams, power levels of approximately 10 mW were sufficient. To expose 35 mm film, power approaching 100 mW is needed. With the powers available in the mid-1970s, the laser-beam recorder was limited to the smaller area of 16 mm film. A prototype 16 mm version was constructed. In the laser recorder, the R, G, and B video signals were fed to three acoustic-optical modulators, and the resulting modulated beams were combined by dichroic mirrors into a single beam that was mechanically deflected. The scanned beam moved in synchronism with the moving film, which was exposed line by line. Color negative film typically was used in such equipment, but color duplicate film, having higher resolution and finer grain, was preferred for use with the 16 mm recorder. This technique is only of historical significance; the laser system was discarded for the *electron-beam recording* system.

1.1.3b   The 1125/60 System

In the early 1980s, NHK initiated a development program directed toward providing a high-definition television broadcasting service to the public. The video signal format was based upon the proposed 1125/60 standard published in 1980 by NHK. Experimental broadcasts were transmitted to prototype receivers in Japan over the MUSE (multiple sub-Nyquist encoding) satellite system. The 1984 Olympic Games, held in the United States and televised by NHK for viewers in Japan, was the first event of worldwide interest to be covered using high-definition television. The HDTV signals were *pan-scanned* to a 1.33 aspect ratio and transcoded to 525 lines for terrestrial transmission and reception over regular TV channels. On June 3, 1989, NHK inaugurated regular HDTV program transmissions for about an hour each day using its MS-2 satellite.

Various engineering committees within the Society of Motion Picture and Television Engineers (SMPTE), as well as other engineering committees in the United States, had closely studied the 1125/60 format since the 1980 NHK publication of the system details. Eventually, a production specification describing the 1125/60 format was proposed by SMPTE for adoption as an American National Standard. The document, SMPTE 240M, was published in April 1987 by the Society and forwarded to the American National Standards Institute (ANSI). Because of objections by several organizations, however, the document was not accepted as a national standard. One argument against the proposed standard was that it would be difficult to convert to the NTSC system, particularly if a future HDTV version of NTSC was to be compatible with the existing service.

NBC, for one, recommended that the HDTV studio standard be set at 1050 lines (525 × 2) and that the field rate of 59.94 Hz be retained, rather than the 60 Hz value of SMPTE 240M. Philips Laboratories and the David Sarnoff Laboratories concurred and based their proposed HDTV systems on the 1050-line, 59.94-field specifications.

Despite the rejection as a national standard for production equipment, SMPTE 240M remained a viable *recommended practice* for equipment built to the 1125/60 system.

The first full-scale attempt at international HDTV standardization was made by the CCIR in Dubrovnik, Yugoslavia, in May 1986. Japan and the United States pushed for a 1125-line/60 Hz production standard. The Japanese, of course, already had a working system. The 50 Hz countries, which did not have a working HDTV system of their own, demurred, asking for time to perfect and demonstrate a non-60 Hz (i.e., 50 Hz) system. Because of this objection, the CCIR took no action on the recommendation of its Study Group, voting to delay a decision until 1990, pending an examination of alternative HDTV production standards. There was strong support of the Study Group recommendations in the United States, but not among key European members of the CCIR.

The HDTV standardization fight was by no means simply a matter of North America and Japan vs. Europe. Of the 500 million receivers then in use worldwide, roughly half would feel the effect of any new frame rate. The Dubrovnik meeting focused mainly on a production standard. Television material can, of course, be produced in one standard and readily converted to another for transmission. Still, a single universal standard would avoid both the bother and degradation of the conversion process.

During this developmental period, the commercial implications of HDTV were inextricably intertwined with the technology. In many cases, commercial considerations tended to dominate thinking. The 1125/60 system was, basically, a Japanese system. The U.S. came in late and jumped on Japan's coattails, aided greatly by the fact that the two countries have identical television standards (CCIR System *M*). But the Europeans did not want to have to buy Japanese or U.S. equipment; did not want to pay any Japanese or U. S. royalties; and did not want to swallow their NIH ("not invented here") pride. This feeling also emerged in the United States during the late 1980s and early 1990s for the same reasons—with American manufacturers not wanting to be locked into the Japanese and/or European HDTV systems.

1.1.3c   European HDTV Systems

Early hopes of a single worldwide standard for high-definition television began a slow dissolve to black at the 1988 International Broadcasting Convention (IBC) in Brighton, England. Brighton was the public debut of the HDTV system developed by the European consortium *Eureka EU95*, and supporters made it clear that their system was intended to be a direct competitor of the 1125/60 system developed by NHK.

The Eureka project was launched in October 1986 (5 months after the Dubrovnik CCIR meeting) with the goal of defining a European HDTV standard of 1250-lines/50 Hz that would be compatible with existing 50 Hz receivers. EU95 brought together 30 television-related organizations, including major manufacturers, broadcasters, and universities. The Brighton showing included products and technology necessary for HDTV production, transmission, and reception.

HD-MAC (high-definition multiplexed analog component) was the transmission standard developed under the EU95 program. HD-MAC was an extension of the MAC-packet family of transmission standards.

The primary movers in EU95 were the hardware manufacturers Bosch, Philips, and Thomson. The aim of the Eureka project was to define a 50 Hz HDTV standard for submission to the plenary assembly of the CCIR in 1990. The work carried out under this effort involved defining production, transmission, recording, and projection systems that would bring high-definition pictures into viewers' homes.

Supporters of the 1125/60 system also were planning to present their standard to the CCIR in 1990 for endorsement. The entry of EU95 into the HDTV arena significantly changed the complexion of the plenary assembly. For one thing, it guaranteed that no worldwide HDTV production standard—let alone a broadcast transmission standard—would be developed.

The 1250/50 Eureka HDTV format was designed to function as the future direct-broadcast satellite transmission system to the entire Western European Community. Although some interest had been expressed in the former Eastern Bloc countries, work there was slowed by more pressing economic and political concerns.

1.1.3d   A Perspective on HDTV Standardization

HDTV production technology was seen from the very beginning as an opportunity to simplify program exchange, bringing together the production of programs for television and for the cinema [2]. Clearly, the concept of a single production standard that could serve all regions of the world *and* have application in the film community would provide benefits to both broadcasting organizations and program producers. All participants stated their preference for a single worldwide standard for HDTV studio production and international program exchange.

The work conducted in the field of studio standards showed that the task of creating a recommendation for HDTV studio production and international program exchange was made somewhat difficult by the diversity of objectives foreseen for HDTV in different parts of the world. There were differences in approach in terms of technology, support systems, and compatibility. It became clear that, for some participants, the use of HDTV for production of motion pictures and their subsequent distribution via satellites was the most immediate need. For others, there was a greater emphasis on satellite broadcasting, with a diversity of opinion on both the time scale for service introduction and the frequency bands to be used. For still others, the dominant consideration was terrestrial broadcasting services.

The proposal for a draft recommendation for an HDTV studio standard based on a 60 Hz field rate was submitted to the CCIR, initially in 1985. The proposal for a draft recommendation for an HDTV studio standard based on a 50 Hz field rate was submitted to the CCIR in 1987. Unfortunately, neither set of parameters in those drafts brought a consensus within the CCIR as a single worldwide standard. However, both had sufficient support for practical use in specific areas to encourage manufacturers to produce equipment.

Despite the lack of an agreement on HDTV, a great deal of progress was made during this effort in the area of an HDTV source standard. The specific parameters agreed upon included:

- Picture aspect ratio of 16:9

- Color rendition

- Equation for luminance

Thus, for the first time in the history of television, all countries of the world agreed on the technical definition of a basic tristimulus color system for display systems. Also agreed upon, in principle, were the digital HDTV bit-rate values for the studio interface signal, which was important in determining both the interface for HDTV transmission and the use of digital recording. All of these agreements culminated in Recommendation 709, adopted by the XVII Plenary Assembly of the CCIR in 1990 in Dusseldorf [3].

1.1.3e Digital Systems Emerge

By mid-1991, publications reporting developments in the U.S., the United Kingdom, France, the Nordic countries, and other parts of the world showed that bit-rate reduction schemes on the order of 60:1 could be applied successfully to HDTV source images. The results of this work implied that HDTV image sequences could be transmitted in a relatively narrowband channel in the range of 15 to 25 Mbits/s. Using standard proven modulation technologies, it would therefore be possible to transmit an HDTV program within the existing 6, 7, and/or 8 MHz channel bandwidths provided for in the existing VHF- and UHF-TV bands.

One outgrowth of this development was the initiation of studies into how—if at all—the existing divergent broadcasting systems could be included under a single unifying cover. Thus was born the *HDTV-6-7-8* program. HDTV-6-7-8 was based on the following set of assumptions [2]:

- First, the differences between the bandwidths of the 6, 7, and 8 MHz channels might give rise to the development of three separate HDTV scenarios that would fully utilize the bandwidth of the assigned channels. It was assumed that the 6 MHz implementation would have the potential to provide pictures of sufficiently high quality to satisfy viewers' wishes for a "new viewing experience." The addition of a 1 or 2 MHz increment in the bandwidth, therefore, would not be critical for further improvement for domestic reception. On this basis, there was a possibility of adopting, as a core system, a single 6 MHz scheme to provide a minimum service consisting of video, two multichannel sound services, and appropriate support data channels for conditional access, and—where appropriate—closed captioning, program identification, and other user-oriented services.

- Second, given the previous assumption, the 1 or 2 MHz channel bandwidth that could be saved in a number of countries might be used for transmission of a variety of additional information services, either within a 7 or 8 MHz composite digital signal or on new carriers in the band above the HDTV-6 signal. Such additional information might include narrowband TV signals that provide for an HDTV stereoscopic service, enhanced television services, multiprogram TV broadcasting, additional sound services, and/or additional data services.

- Third, it would be practical to combine audio and video signals, additional information (data), and new control/test signals into a single HDTV-6-7-8 signal, in order to avoid using a secondary audio transmitter. In combination with an appropriate header/descriptor protocol and appropriate signal processing, the number of frequency channels could be increased, and protection ratio requirements would be reduced. In the course of time, this could lead to a review of frequency plans at the national and international levels for terrestrial TV transmission networks and cable television. This scheme, therefore, could go a considerable distance toward meeting the growing demand for frequency assignments.

This digital television system offered the prospect of considerably improved sound and image quality, while appreciably improving spectrum utilization (as compared to the current analog services). It was theorized that one way of exploiting these possibilities would be to use the bit stream available in digital terrestrial or satellite broadcasting to deliver to the public a certain number of digitally compressed conventional television programs instead of a single conventional, enhanced-, or high-definition program. These digitally compressed TV signals would be accompanied by digital high-quality sound, coded conditional access information, and ancillary data channels. Furthermore, the same approach could be implemented in the transmission of multiprogram signals over existing digital satellite or terrestrial links, or cable TV networks.

Despite the intrinsic merits of the HDTV-6-7-8 system, it was quickly superseded by the Digital Video Broadcasting project in Europe and the Grand Alliance project in the United States.

HD-DIVINE

HD-DIVINE began as a developmental project in late 1991. The aim was to prove that a digital HDTV system was possible within a short time frame without an intermediate step based on analog technology. The project was known as "Terrestrial Digital HDTV," but later was renamed "HD-DIVINE" (*digital video narrowband emission*). Less than 10 months after the project started, HD-DIVINE demonstrated a digital terrestrial HDTV system at the 1992 IBC show in Amsterdam. It was a considerable success, triggering discussion in the Nordic countries on cooperation for further development.

HD-DIVINE subsequently was developed to include four conventional digital television channels—as an alternative to HDTV—contained within an 8 MHz channel. Work also was done to adapt the system to distribution via satellite and cable, as demonstrated at the Montreux Television Symposium in 1993.

Meanwhile, in the spring of 1992, a second coordinated effort began with the goal of a common system for digital television broadcasting in Europe. Under the auspices of a European Launching Group, composed of members of seven countries representing various organizations involved in the television business, the Working Group for Digital Television Broadcasting (WGDTB) defined approaches for a digital system. Two basic schemes were identified for further study, both multilayer systems that offered multiple service levels, either in a hierarchical or multicast mode. The starting point for this work was the experience with two analog TV standards, D2-MAC and HD-MAC, which had been operational for some time.

Eureka ADTT

On June 16, 1994, the leaders of the Eureka project approved the start of a new research effort targeted for the development of future television systems in Europe based on digital technologies. The Eureka 1187 *Advanced Digital Television Technologies* (ADTT) project was formed with the purpose of building upon the work of the Eureka 95 HDTV program. The effort, which was to run for 2-1/2 years, included as partners Philips, Thomson Consumer Electronics, Nokia, and English and Italian consortia. The initial objectives of Eureka ADTT were to address design issues relating to production, transmission, reception, and display equipment and their key technologies.

A prototype high-definition broadcast system was developed and tested, based on specifications from—and in close consultation with—the European Project for Digital Video Broadcasting. The Eureka ADTT project also was charged with exploring basic technologies and the development of key components for products, such as advanced digital receivers, recording devices, optical systems, and multimedia hardware and software.

The underlying purpose of the Eureka ADTT effort was to translate, insofar as possible, the work done on the analog-based Eureka 95 effort to the all-digital systems that emerged from 1991 to 1993.

1.1.3f Digital Video Broadcasting

The European DVB project began in the closing months of 1990. Experimental European projects such as SPECTRE showed that the digital video-compression system known as *motion-compensated hybrid discrete cosine transform coding* was highly effective in reducing the transmission capacity required for digital television [4]. Until then, digital TV broadcasting was considered impractical to implement.

In the U.S., the first proposals for digital terrestrial HDTV were made. In Europe, Swedish Television suggested that fellow broadcasters form a concerted pan-European platform to develop digital terrestrial HDTV. During 1991, broadcasters and consumer equipment manufacturers discussed how this could be achieved. Broadcasters, consumer electronics manufacturers, and regulatory bodies agreed to come together to discuss the formation of a pan-European group that would oversee the development of digital television in Europe—the European Launching Group (ELG). Over the course of about a year, the ELG expanded to include the major European media interest groups—both public and private—consumer electronics manufacturers, and common carriers.

The program officially began in September 1993, and the European Launching Group became the DVB (*Digital Video Broadcasting*) Project. Developmental work in digital television, already under way in Europe, then moved forward under this new umbrella. Meanwhile, a parallel activity, the *Working Group on Digital Television*, prepared a study of the prospects and possibilities for digital terrestrial television in Europe.

By 1999, a bit of a watershed year for digital television in general, the Digital Video Broadcasting Project had grown to a consortium of over 200 broadcasters, manufacturers, network operators, and regulatory bodies in more than 30 countries worldwide. Numerous broadcast services using DVB standards were operational in Europe, North and South America, Africa, Asia, and Australia.

At the '99 NAB Convention in Las Vegas, mobile and fixed demonstrations of the DVB system were made using a variety of equipment in various typical situations. Because mobile reception is the most challenging environment for television, the mobile system received a good deal of attention. DVB organizers used the demonstrations to point out the strengths of their chosen modulation method, the multicarrier *coded orthogonal frequency division multiplexing* (COFDM) technique.

1.1.3g Involvement of the Film Industry

From the moment it was introduced, HDTV was the subject of various aims and claims. Among them was the value of video techniques for motion picture production, making it possible to bypass the use of film in part or completely. Production and editing were said to be enhanced, resulting in reduced costs to the producer. However, the motion picture industry was in no hurry to discard film, the medium that had served it well for the better part of a century. Film quality continues to improve, and film is unquestionably *the* universal production standard. Period.

Although HDTV has made inroads in motion picture editing and special effects production, Hollywood has not rushed to hop on board the HDTV express.

Nevertheless, it is certainly true that the film industry has embraced elements of high-definition imaging. As early as 1989, Eastman Kodak unveiled the results of a long-range program to develop an *electronic-intermediate* (EI) digital video postproduction system. The company introduced the concept of an HDTV system intended primarily for use by large-budget feature film producers to provide new, creative dimensions for special effects without incurring the quality compromises of normal edited film masters.

In the system, original camera negative 35 mm film input to the electronic-intermediate system was transferred to a digital frame store at a rate substantially slower than real time, initially about 1 frame/s. Sequences could be displayed a frame at a time for unfettered image manipulation and compositing. This system was an electronic implementation of the time-standing-still format in which film directors and editors have been trained to exercise their creativity.

Kodak established a consortium of manufacturers and software developers to design and produce elements of the EI system. Although the announced application was limited strictly to the creation of artistic high-resolution special effects on film, it was hoped that the EI system eventually would lead to the development of a means for electronic real-time distribution of theatrical films.

1.1.3h   Political Considerations

In an ideal world, the design and direction of a new production or transmission standard would be determined by the technical merits of the proposed system. However, as demonstrated in the past by the controversy surrounding the NTSC monochrome and color standards, technical considerations are not always the first priority. During the late 1980s, when concern over foreign competition was at near-frenzy levels in the U.S. and Europe, the ramifications of HDTV moved beyond just technology and marketing, and into the realm of politics. In fact, the political questions raised by the push for HDTV promised to be far more difficult to resolve than the technical ones.

The most curious development in the battle over HDTV was the interest politicians took in the technology. In early 1989, while chairing the House Telecommunications subcommittee, Representative Ed Markey (D-Massachusetts) invited comments from interested parties on the topic of high-definition television. Markey's subcommittee conducted two days of hearings in February 1989. There was no shortage of sources of input, including:

- The *American Electronics Association*, which suggested revised antitrust laws, patent policy changes, expanded exports of high-tech products, and government funding of research on high-definition television.

- *Citizens for a Sound Economy*, which argued for a relaxation of antitrust laws.

- *Committee to Preserve American Television*, which encouraged strict enforcement of trade laws and consideration of government research and development funds for joint projects involving semiconductors and advanced display devices.

- *Maximum Service Telecasters*, which suggested tax credits, antitrust exemptions, and low-interest loans as ways of encouraging U.S. development of terrestrial HDTV broadcasting.

- *Committee of Corporate Telecommunications Users*, which suggested the creation of a "Technology Corporation of America" to devise an open architecture for the production and transmission of HDTV and other services.

Representative Don Ritter (R-Pennsylvania) expressed serious concern over the role that U.S.-based companies would play—or more to the point, might not play—in the development of high-definition television. Ritter believed it was vital for America to have a piece of the HDTV manufacturing pie. Similar sentiments were echoed by numerous other lawmakers.

The Pentagon, meanwhile, expressed strong interest in high-definition technology for two reasons: direct military applications and the negative effects that the lack of domestic HDTV expertise could have on the American electronics industry. The Department of Defense, accordingly, allocated money for HDTV research.

Although many of the concerns relating to HDTV that were voiced during the late 1980s now seem rather baseless and even hysterical, this was the atmosphere that drove the pioneering work on the technology. The long, twisting road to HDTV proved once again that the political implications of a new technology may be far more daunting than the engineering issues.

1.1.3i   Terminology

During the development of HDTV, a procession of terms was used to describe performance levels between conventional NTSC and "real" HDTV (1125/60-format quality). Systems were classified in one or more of the following categories:

- *Conventional systems*: The NTSC, PAL, and SECAM systems as standardized prior to the development of advanced systems.

- *Improved-definition television* (IDTV) systems: Conventional systems modified to offer improved vertical and/or horizontal definition, also known as *advanced television* (ATV) or *enhanced-definition television* (EDTV) systems.

- *Advanced systems*: In the broad sense, all systems other than conventional ones. In the narrow sense, all systems other than conventional and "true" HDTV.

- *High-definition television* (HDTV) systems: Systems having vertical and horizontal resolutions approximately twice those of conventional systems.

- *Simulcast systems*: Systems transmitting conventional NTSC, PAL, or SECAM on existing channels and HDTV of the same program on one or more additional channels.

- *Production systems*: Systems intended for use in the production of programs, but not necessarily in their distribution.

- *Distribution systems*: Terrestrial broadcast, cable, satellite, videocassette, and videodisc methods of bringing programs to the viewing audience.

1.1.4   References

1.   Fink, Donald (ed.): *Television Engineering Handbook*, McGraw-Hill, New York, N.Y., 1957.

2. Krivocheev, Mark I., and S. N. Baron: "The First Twenty Years of HDTV: 1972–1992," *SMPTE Journal*, SMPTE, White Plains, N.Y., pg. 913, October 1993.

3. CCIR Document PLEN/69-E (Rev. 1), "Minutes of the Third Plenary Meeting," pp. 2–4, May 29, 1990.

4. Based on information supplied by the DVB Project on its Web site: http://www.dvb.com.

1.1.5   Bibliography

Appelquist, P.: "The HD-Divine Project: A Scandinavian Terrestrial HDTV System," *1993 NAB HDTV World Conference Proceedings*, National Association of Broadcasters, Washington, D.C., pg. 118, 1993.

Baron, Stanley: "International Standards for Digital Terrestrial Television Broadcast: How the ITU Achieved a Single-Decoder World," *Proceedings of the 1997 BEC*, National Association of Broadcasters, Washington, D.C., pp. 150–161, 1997.

Battison, John: "Making History," *Broadcast Engineering*, Intertec Publishing, Overland Park, Kan., June 1986.

Benson, K. B., and D. G. Fink: *HDTV: Advanced Television for the 1990s*, McGraw-Hill, New York, NY, 1990.

Benson, K. B., and Jerry C. Whitaker (eds.): *Television Engineering Handbook*, rev. ed., McGraw-Hill, New York, NY, 1992.

Benson, K. B., and J. C. Whitaker (eds.): *Television and Audio Handbook for Engineers and Technicians*, McGraw-Hill, New York, NY, 1989.

CCIR Report 801-3: "The Present State of High-Definition Television," June 1989.

"Dr. Vladimir K. Zworkin: 1889–1982," *Electronic Servicing and Technology*, Intertec Publishing, Overland Park, Kan., October 1982.

Lincoln, Donald: "TV in the Bay Area as Viewed from KPIX," *Broadcast Engineering*, Intertec Publishing, Overland Park, Kan., May 1979.

McCroskey, Donald: "Setting Standards for the Future," *Broadcast Engineering*, Intertec Publishing, Overland Park, Kan., May 1989.

Pank, Bob (ed.): *The Digital Fact Book*, 9th ed., Quantel Ltd, Newbury, England, 1998.

Reimers, U. H.: "The European Perspective for Digital Terrestrial Television, Part 1: Conclusions of the Working Group on Digital Terrestrial Television Broadcasting," *1993 NAB HDTV World Conference Proceedings*, National Association of Broadcasters, Washington, D.C., 1993.

Schow, Edison: "A Review of Television Systems and the Systems for Recording Television," *Sound and Video Contractor*, Intertec Publishing, Overland Park, Kan., May 1989.

Schubin, Mark, "From Tiny Tubes to Giant Screens," *Video Review*, April 1989.

SMPTE and EBU, "Task Force for Harmonized Standards for the Exchange of Program Material as Bitstreams," *SMPTE Journal*, SMPTE, White Plains, N.Y., pp. 605–815, July 1998.

"Television Pioneering," *Broadcast Engineering*, Intertec Publishing, Overland Park, Kan., May 1979.

"Varian Associates: An Early History," Varian publication, Varian Associates, Palo Alto, Calif.

Whitaker, Jerry C.: *Electronic Displays: Technology, Design, and Applications*, McGraw-Hill, New York, N.Y., 1994.

1.2

Digital Television in the U.S.

1.2.1 Introduction

Although HDTV production equipment had been available since 1984, standardization for broadcast service was slowed by lack of agreement on how the public could best be served. The primary consideration was whether to adopt a system compatible with NTSC or a simulcast system requiring additional transmission spectrum and equipment.

1.2.2 Advisory Committee on Advanced Television Service

On November 17, 1987, at the request of 58 U.S. broadcasters, the FCC initiated a rulemaking on advanced television (ATV) services and established a blue ribbon Advisory Committee on Advanced Television Service (ACATS) for the purpose of recommending a broadcast standard. Former FCC Chairman Richard E. Wiley was appointed to chair ACATS. At that time, it was generally believed that HDTV could not be broadcast using 6 MHz terrestrial broadcasting channels. Broadcasting organizations were concerned that alternative media would be used to deliver HDTV to the viewing public, placing terrestrial broadcasters at a severe disadvantage. The FCC agreed that this was a subject of importance and initiated a proceeding (MM Docket No. 87-268) to consider the technical and public policy issues of ATV.

The first interim report of the ACATS, filed on June 16, 1988, was based primarily on the work of the Planning Subcommittee. The report noted that proposals to implement improvements in the existing NTSC television standard ranged from simply enhancing the current standard all the way to providing full-quality HDTV. The spectrum requirements for the proposals fell into three categories: 6 MHz, 9 MHz, and 12 MHz. Advocates of a 12 MHz approach suggested using two channels in one of two ways:

- An existing NTSC-compatible channel supplemented by a 6 MHz *augmentation channel* (either contiguous or noncontiguous)

- An existing NTSC-compatible channel, unchanged, and a separate 6 MHz channel containing an independent non-NTSC-compatible HDTV signal

It was pointed out that both of these methods would be "compatible" in the sense that existing TV receivers could continue to be serviced by an NTSC signal.

The first interim report stated that, "based on current bandwidth-compression techniques, it appears that full HDTV will require greater spectrum than 6 MHz." The report went on to say that the Advisory Committee believed that efforts should focus on establishing, at least ultimately, an HDTV standard for terrestrial broadcasting. The report concluded that one advantage to simulcasting was that at some point in the future—after the NTSC standard and NTSC-equipped receivers were retired—part of the spectrum being utilized might be reemployed for other purposes. On the basis of preliminary engineering studies, the Advisory Committee stated that it believed sufficient spectrum capacity in the current television allocations table might be available to allow all existing stations to provide ATV through either an augmentation or simulcast approach.

1.2.2a   The Process Begins

With this launch, the economic, political, and technical implications of HDTV caused a frenzy of activity in technical circles around the world; proponents came forward to offer their ideas. The Advanced Television Test Center (ATTC) was set up to consider the proposals and evaluate their practicality. In the first round of tests, 21 proposed methods of transmitting some form of ATV signals (from 15 different organizations) were considered. The ATTC work was difficult for a number of reasons, but primarily because the 21 systems were in various stages of readiness. Most, if not all, were undergoing continual refinement. Only a few systems existed as real, live black boxes, with "inputs" and "outputs". Computer simulation made up the bulk of what was demonstrated in the first rounds. The ATTC efforts promised, incidentally, to be as much a test of computer simulation as a test of hardware. Of the 21 initial proposals submitted to ACATS in September 1988, only six actually were completed in hardware and tested.

The race begun, engineering teams at various companies began assembling the elements of an ATV service. One of the first was the Advanced Compatible Television (ACTV) system, developed by the Sarnoff Research Center. On April 20, 1989, a short ACTV program segment was transmitted from the center, in New Jersey, to New York for broadcast over a WNBC-TV evening news program. The goal was to demonstrate the NTSC compatibility of ACTV. Consisting of two companion systems, the scheme was developed to comply with the FCC's *tentative decision* of September 1988, which required an HDTV broadcast standard to be compatible with NTSC receivers.

The basic signal, ACTV-I, was intended to provide a wide-screen picture with improved picture and sound quality on new HDTV receivers, while being compatible with NTSC receivers on a single 6 MHz channel. A second signal, ACTV-II, would provide full HDTV service on a second augmentation channel when such additional spectrum might be available.

In the second interim report of the ACATS (April 26, 1989), the committee suggested that its life be extended from November 1989 to November 1991. It also suggested that the FCC should be in a position to establish a single terrestrial ATV standard sometime in 1992. The Advisory Committee noted that work was ongoing in defining tests to be performed on proponent systems. An issue was raised relating to subjective tests and whether source material required for testing should be produced in only one format and transcoded into the formats used by different systems to be tested, or whether source material should be produced in all required formats. The Advisory Committee also sought guidance from the FCC on the minimum number of audio channels that an ATV system would be expected to provide.

The large number of system proponents, and delays in developing hardware, made it impossible to meet the aggressive timeline set for this process. It was assumed by experts at the time that consumers would be able to purchase HDTV, or at least advanced television, sets for home use by early 1992.

The FCC's tentative decision on compatibility, although not unexpected, laid a false set of ground rules for the early transmission system proponents. The requirement also raised the question of the availability of frequency spectrum to accommodate the added information of the ATV signal. Most if not all of the proposed ATV systems required total bandwidths of one, one and a half, or two standard TV channels (6 MHz, 9 MHz, or 12 MHz). In some cases, the added spectrum space that carried the ATV information beyond the basic 6 MHz did not have to be contiguous with the main channel.

Any additional use of the present VHF- and UHF-TV broadcast bands would have to take into account co-channel and adjacent-channel interference protection. At UHF, an additional important unanswered question was the effect of the UHF "taboo channels" on the availability of extra frequency space for ATV signals. These "taboos" were restrictions on the use of certain UHF channels because of the imperfect response of then-existing TV receivers to unwanted signals, such as those on image frequencies, or those caused by local oscillator radiation and front-end intermodulation.

The mobile radio industry had been a longtime combatant with broadcasters over the limited available spectrum. Land mobile had been asking for additional spectrum for years, saying it was needed for public safety and other worthwhile purposes. At that time, therefore, the chances of the FCC allocating additional spectrum to television broadcasters in the face of land mobile demands were not thought to be very good.

In any event, the FCC informally indicated that it intended to select a simulcast standard for HDTV broadcasting in the United States using existing television band spectrum and would not consider any augmentation-channel proposals.

1.2.2b   System Testing: Round 2

With new groundwork clearly laid by the FCC, the second round of serious system testing was ready to begin. Concurrent with the study of the various system proposals, the ACATS began in late 1990 to evaluate the means for transmission of seven proposed formats for the purpose of determining their suitability as the U.S. standard for VHF and UHF terrestrial broadcasting. The initial round of tests were scheduled for completion by September 30, 1992.

The FCC announced on March 21, 1990 that it favored a technical approach in which high-definition programs would be broadcast on existing 6 MHz VHF and UHF channels separate from the 6 MHz channels used for conventional (NTSC) program transmission, but the commission did not specifically address the expected bandwidth requirements for HDTV. However, the implication was that only a single channel would be allocated for transmission of an HDTV signal. It followed that this limitation to a 6 MHz channel would require the use of video-compression techniques. In addition, it was stated that no authorization would be given for any enhanced TV system, so as not to detract from development of full high-definition television. The spring of 1993 was suggested by the FCC as the time for a final decision on the selection of an HDTV broadcasting system.

Under the simulcast policy, broadcasters would be required to transmit NTSC simultaneously on one channel of the VHF and UHF spectra and the chosen HDTV standard on another 6 MHz TV broadcast channel. This approach was similar to that followed by the British in their intro-duction of color television, which required monochrome programming to continue on VHF for about 20 years after 625/50 PAL color broadcasting on UHF was introduced. Standards convert-ers working between 625-line PAL color and 405-line monochrome provided the program input for the simultaneous black-and-white network transmitters. The British policy obviously bene-fited the owners of old monochrome receivers who did not wish to invest in new color receivers; it also permitted program production and receiver sales for the new standard to develop at a rate compatible with industry capabilities. All television transmission in Great Britain now is on UHF, with the VHF channels reassigned to other radio services.

For the development of HDTV, the obvious advantage of simulcasting to viewers with NTSC receivers is that they may continue to receive all television broadcasts in either the current 525-line standard or the new HDTV standard—albeit the latter without the benefit of wide-screen and double resolution—without having to purchase a dual-standard receiver or a new HDTV receiver. Although it was not defined by the FCC, it was presumed that the HDTV channels also would carry the programs available only in the NTSC standard. Ideally for the viewer, these pro-grams would be converted to the HDTV transmission standard from the narrower 1.33 aspect ratio and at the lower resolution of the 525-line format. A less desirable solution would be to carry programs available only in the NTSC standard without conversion to HDTV and require HDTV receivers to be capable of switching automatically between standards. A third choice would be not to carry NTSC-only programs on the HDTV channel and to require HDTV receiv-ers to be both HDTV/NTSC channel and format switchable.

The development of HDTV involved exhaustive study of how to improve the existing NTSC system. It also meant the application of new techniques and the refinement of many others, including:

- Receiver enhancements, such as higher horizontal and vertical resolution, digital processing, and implementation of large displays.

- Transmission enhancements, including new camera technologies, image enhancement, adap-tive prefilter encoding, digital recording, and advanced signal distribution.

- Signal compression for direct satellite broadcasting.

- Relay of auxiliary signals within conventional TV channels.

- Allocation and optimization of transmission frequency assignments.

Concurrently, an extensive study was undertaken concerning the different characteristics of the major systems of program distribution: terrestrial broadcasting, cable distribution by wire or fiber optics, satellite broadcasting, and magnetic and optical recorders. The major purposes of this study were to determine how the wide video baseband of HDTV could be accommodated in each system, and whether a single HDTV standard could embrace the needs of all systems. This work not only provided many of the prerequisites of HDTV, but by advancing the state of the conventional art, it established a higher standard against which the HDTV industry must com-pete.

In the third interim report (March 21, 1990), the Advisory Committee approved the proposed test plans and agreed that complete systems, including audio, would be required for testing. It also was agreed that proposed systems must be precertified by June 1, 1990. That date, naturally, became quite significant to all proponents. The pace of work was accelerated even further.

It is noteworthy that the first all-digital proposal was submitted shortly before the June deadline. The third interim report also stated that psychophysical tests of ATV systems would be conducted and that the Planning Subcommittee, through its Working Party 3 (PS/WP3), would undertake the development of preliminary ATV channel allotment plans and assignment options.

In the fourth interim report (April 1, 1991), the Advisory Committee noted changes in proponents and proposed systems. Most significant was that several all-digital systems had been proposed. Testing of proponent systems was scheduled to begin later that year. Changes had been required in the test procedures because of the introduction of the all-digital systems. It was reported also that the System Standards Working Party had defined a process for recommending an ATV system, and that PS/WP3 was working toward the goal of providing essentially all existing broadcasters with a simulcast channel whose coverage characteristics were equivalent to NTSC service.

By the time the fifth interim report was issued (March 24, 1992), there were five proponent systems, all simulcast—one analog and four all-digital. The Planning Subcommittee reported that it had reconstituted its Working Party 4 to study issues related to harmonizing an ATV broadcast transmission standard with other advanced imaging and transmission schemes that would be used in television and nonbroadcast applications. The Systems Subcommittee reported that its Working Party 2 had developed procedures for field-testing an ATV system. It was noted that the intent of the Advisory Committee was to field-test only the system recommended to the FCC by the Advisory Committee based on the laboratory tests. It also was reported that the Systems Subcommittee Working Party 4 had developed a process for recommending an ATV system and had agreed to a list of 10 selection criteria.

Hundreds of companies and organizations worked together within the numerous sub-committees, working parties, advisory groups, and special panels of ACATS during the 8-year existence of the organization. The ACATS process became a model for international industry-government cooperation. Among its accomplishments was the development of a competitive process by which proponents of systems were required to build prototype hardware that would then be thoroughly tested. This process sparked innovation and entrepreneurial initiative.

1.2.2c   Formation of the Grand Alliance

Although the FCC had said in the spring of 1990 that it would determine whether all-digital technology was feasible for a terrestrial HDTV transmission standard, most observers viewed that technology as being many years in the future. Later the same year, however, General Instrument became the first proponent to announce an all-digital system. Later, all-digital systems were announced by MIT, the Philips-Thomson-Sarnoff consortium, and Zenith-AT&T.

The FCC anticipated the need for interoperability of the HDTV standard with other media. Initially, the focus was on interoperability with cable television and satellite delivery; both were crucial to any broadcast standard. But the value of interoperability with computer and telecommunications applications became increasingly apparent with the advent of all-digital systems.

Proponents later incorporated packetized transmission, headers and descriptors, and composite-coded surround sound in their subsystems. (The Philips-Thomson-Sarnoff consortium was the first to do so.) These features maximized the interoperability of HDTV with computer and telecommunications systems. The introduction of all-digital systems had made such interoperability a reality.

The all-digital systems set the stage for another important step, which was taken in February 1992, when the ACATS recommended that the new standard include a flexible, adaptive data-allocation capability (and that the audio also be upgraded from stereo to surround sound). Following testing, the Advisory Committee decided in February 1993 to limit further consideration only to those proponents that had built all-digital systems: two systems proposed by General Instrument and MIT; one proposed by Zenith and AT&T; and one proposed by Sarnoff, Philips, and Thomson. The Advisory Committee further decided that although all of the digital systems provided impressive results, no single system could be proposed to the FCC as the U.S. HDTV standard at that time. The committee ordered a round of supplementary tests to evaluate improvements to the individual systems.

At its February 1993 meeting, the Advisory Committee also adopted a resolution encouraging the digital HDTV groups to try to find a way to merge the four remaining all-digital systems. The committee recognized the merits of being able to combine the best features of those systems into a single "best of the best" system. With this encouragement, negotiations between the parties heated up, and on May 24, the seven companies involved announced formation of the Digital HDTV Grand Alliance.

By the spring of 1994, significant progress had been made toward the final HDTV system proposal. Teams of engineers and researchers had finished building the subsystems that would be integrated into the complete HDTV prototype system for testing later in the year. The subsystems—scanning formats, digital video compression, packetized data, audio, and modulation—all had been approved by the ACATS. Key features and specifications for the system included:

- Support of two fundamental arrays of pixels (picture elements): 1920 × 1080 and 1280 x 720. Each of these pixel formats supported a wide-screen 16:9 aspect ratio and square pixels, important for computer interoperability. Frame rates of 60, 30, and 24 Hz were supported, yielding a total of six different possible scanning formats—two different pixel arrays, each having three frame rates. The 60 and 30 Hz frame rates were important for video source material and 24 Hz for film. A key feature of the system was the Grand Alliance's commitment to using progressive scanning, also widely used in computer displays. Entertainment television traditionally had used interlaced scanning, which was efficient but subject to various unwanted artifacts. Of the six video formats, progressive scanning was used in all three 720-line formats and in the 30 and 24 Hz 1080-line formats. The sixth video format was a 60 Hz 1080-line scheme. It was neither technically or economically feasible to initially provide this as a progressive format, although it was a longer-term goal for the Grand Alliance. The 1080-line, 60-Hz format was handled in the initial standard by using interlaced rather than progressive scanning.

- Video compression: Utilizing the MPEG-2 (Moving Picture Experts Group)-proposed international standard allowed HDTV receivers to interoperate with MPEG-2 and MPEG-1 computer, multimedia, and other media applications.

- Packetized data transport: Also based on MPEG-2, this feature provided for the flexible transmission of virtually any combination of video, audio, and data.

- Compact-disc-quality digital audio: This feature was provided in the form of the 5.1-channel Dolby AC-3 surround sound system.

- 8-VSB (8-level vestigial sideband): The modulation system selected for transmission provided maximum coverage area for terrestrial digital broadcasting.

The Grand Alliance format employed principles that made it a highly interoperable system. It was designed with a layered digital architecture that was compatible with the international Open Systems Interconnect (OSI) model of data communications that forms the basis of virtually all modern digital systems. This compatibility allowed the system to interface with other systems at any layer, and it permitted many different applications to make use of various layers of the HDTV architecture. Each individual layer of the system was designed to be interoperable with other systems at corresponding layers.

Because of the interoperability of the system between entertainment television and computer and telecommunications technologies, the Grand Alliance HDTV standard was expected to play a major role in the establishment of the *national information infrastructure* (NII). It was postulated that digital HDTV could be an engine that helped drive deployment of the NII by advancing the development of receivers with high-resolution displays and creating a high-data-rate path to the home for a multitude of entertainment, education, and information services.

1.2.2d   Testing the Grand Alliance System

Field tests of the 8-VSB digital transmission subsystem began on April 11, 1994, under the auspices of the ACATS. The 8-VSB transmission scheme, developed by Zenith, had been selected for use in the Grand Alliance system two months earlier, following comparative laboratory tests at the ATTC. The field tests, held at a site near Charlotte, North Carolina, were conducted on channel 53 at a maximum effective radiated power of 500 kW (peak NTSC visual) and on channel 6 at an ERP of 10 kW (peak NTSC visual).

The tests, developed by the Working Party on System Testing, included measurements at approximately 200 receiving sites. Evaluations were based solely on a pseudorandom data signal as the input source; pictures and audio were not transmitted. The 8-VSB measurements included carrier-to-noise ratio (C/N), error rate, and *margin tests*, performed by adding noise to the received signal until an agreed-upon threshold of performance error rate occurred and noting the difference between the C/N and the C/N without added noise. Testing at the Charlotte facility lasted for about 3 months, under the direction of the Public Broadcasting System (PBS).

In 1995, extensive follow-up tests were conducted, including:

- Laboratory tests at the Advanced Television Test Center in Alexandria, Virginia

- Lab tests at Cable Television Laboratories, Inc. (CableLabs) of Boulder, Colorado

- Subjective viewer testing at the Advanced Television Evaluation Laboratory in Ottawa, Canada

- Continued field testing in Charlotte, North Carolina, by PBS, the Association for Maximum Service Television (MSTV), and CableLabs

The laboratory and field tests evaluated the Grand Alliance system's four principal subsystems: scanning formats, video and audio compression, transport, and transmission. Test results showed that:

- Each of the proposed HDTV scanning formats exceeded targets established for static and dynamic luminance and chrominance resolution.

- Video-compression testing, using 26 different HDTV sequences, showed that the Grand Alliance MPEG-2 compression algorithm was "clearly superior" to the four original ATV systems in both the 1080 interlaced- and 720 progressive-scanning modes. Significantly, the testing also showed little or no deterioration of the image quality while transmitting 3 Mbits/s of ancillary data.

- The 5.1-channel digital surround sound audio subsystem of the Grand Alliance system, known as Dolby AC-3, performed better than specifications in multichannel audio testing and met the expectations in long-form entertainment listening tests.

- The packetized data transport subsystem performed well when tested to evaluate the switching between compressed data streams, robustness of headers and descriptors, and interoperability between the compression and transport layers. Additional testing also demonstrated the feasibility of carrying the ATV transport stream on an *asynchronous transfer mode* (ATM) telecommunications network.

Field and laboratory testing of the 8-VSB digital transmission subsystem reinforced test results achieved in the summer of 1994 in Charlotte. Testing for spectrum utilization and transmission robustness again proved that the Grand Alliance system would provide broadcasters significantly better transmission performance than the current analog transmission system, ensuring HDTV service "in many instances where NTSC service is unacceptable." Extensive testing on cable systems and fiber optic links of the 16-VSB subsystem also showed superior results.

The final technical report, approved on November 28, 1995, by the Advisory Committee, concluded that—based on intensive laboratory and field testing—the Grand Alliance digital television system was superior to any known alternative system in the world, better than any of the four original digital HDTV systems, and had surpassed the performance objectives of the ACATS.

Marking one of the last steps in an 8-year process to establish a U.S. ATV broadcasting standard, the 25-member blue-ribbon ACATS panel recommended the new standard to the FCC on November 28, 1995. Richard E. Wiley, ACATS chairman commented, "This is a landmark day for many communications industries and, especially, for American television viewers."

1.2.2e   The Home Stretch for the Grand Alliance

With the technical aspects of the Grand Alliance HDTV system firmly in place, work proceeded to step through the necessary regulatory issues. Primary among these efforts was the establishment of a table of DTV assignments, a task that brought with it a number of significant concerns on the part of television broadcasters. Questions raised at the time involved whether a station's DTV signal should be equivalent to its present NTSC signal, and if so, how this should be accomplished.

Approval of the DTV standard by the FCC was a 3-step process:

- A *notice of proposed rulemaking* (NPRM) on policy matters, issued on August 9, 1995.

- Official acceptance of the Grand Alliance system. On May 9, 1996, the commission voted to propose that a single digital TV standard be mandated for over-the-air terrestrial broadcast of digital television. The standard chosen was that documented under the auspices of the Advanced Television Systems Committee (ATSC).

- Final acceptance of a table of assignments for DTV service.

During the comment period for the NPRM on Advanced Television Service (MM Docket 87-268), a number of points of view were expressed. Some of the more troublesome—from the standpoint of timely approval of the Grand Alliance standard, at least—came from the computer industry. Among the points raised were:

- Interlaced scanning. Some computer interests wanted to ban the 1920 × 1080 interlaced format.

- Square pixels. Computer interests recommended banning the use of formats that did not incorporate square pixels.

- 60 Hz frame rate. Computer interests recommended a frame rate of 72 Hz and banning of 60 Hz.

Meanwhile, certain interests in the motion picture industry rejected the 16:9 (1.78:1) widescreen aspect ratio in favor of a 2:1 aspect ratio.

One by one, these objections were dealt with. Negotiations between the two primary groups in this battle—broadcasters and the computer industry—resulted in a compromise that urged the FCC to adopt a standard that does not specify a single video format for digital television, but instead lets the various industries and companies choose formats they think will best suit consumers. The lack of a mandated video format set the stage for a lively competition between set makers and personal computer (PC) manufacturers, who were expected to woo consumers by combining sharp pictures with features peculiar to computers.

By early December, a more-or-less unified front had again been forged, clearing the way for final action by the FCC. With approval in hand, broadcasters then faced the demands of the commission's timetable for implementation, which included the following key points:

- By late 1998, 26 TV stations in the country's largest cities—representing about 30 percent of U.S. television households—would begin broadcasting the ATSC DTV system.

- By mid-1999, the initial group would expand to 40; by 2000, it would expand to 120 stations.

- By 2006, every TV station in the country would be broadcasting a digital signal or risk losing its FCC license.

Spirited debates about the wisdom of this plan—and whether such a plan even could be accomplished—then ensued.

1.2.2f   Digital Broadcasting Begins

If HDTV truly was going to be the "next big thing," then it was only fitting to launch it with a bang. The ATSC system received just such a sendoff, playing to excited audiences from coast to coast during the launch of Space Shuttle mission STS-95.

The first nationwide broadcast of a high-definition television program using the ATSC DTV system, complete with promos and commercials, aired on October 29, 1998. The live HDTV broadcast of Senator John Glenn's historic return to space was transmitted by ABC, CBS, Fox, NBC, and PBS affiliates from coast to coast.

The feed was available free for any broadcaster who could receive the signal. The affiliates and other providers transmitted the broadcast to viewing sites in Washington, D.C., New York, Atlanta, Chicago, Los Angeles, and 15 other cities. Audiences in those cities watched the launch on new digital receivers and projectors during special events at museums, retail stores, broadcast stations, and other locations. Many of the stations moved their on-air dates ahead of schedule in order to show the Glenn launch broadcast. The largest scheduled viewing site was the Smithsonian's National Air and Space Museum in Washington, D.C., where viewers watched the launch on an IMAX theatre screen and four new digital receivers.

Beyond the technical details was an even more important story insofar as HDTV production is concerned. All of the cameras used in the coverage provided an HD signal except for one NASA pool feed of the launch control center at the Kennedy Space Center, which was upconverted NTSC. On occasion, the director would switch to the launch center feed, providing a dramatic "A/B" comparison of high-definition versus standard-definition. The difference was startling. It easily convinced the industry observers present at the Air and Space Museum of the compelling power of the HDTV image.

The second production issue to come into focus during the broadcast was the editorial value of the wide aspect ratio of HDTV. At one point in the coverage, the program anchor described to the television audience how the Shuttle was fueled the night before. In describing the process, the camera pulled back from the launch pad shot to reveal a fuel storage tank off to one side of the pad. As the reporter continued to explain the procedure, the camera continued to pull back to reveal a second fuel storage tank on the other side of the launch pad. Thanks in no small part to the increased resolution of HDTV and—of course—the 16:9 aspect ratio, the television audience was able to see the entire launch area in a single image. Such a shot would have been wholly impossible with standard-definition imaging.

The STS-95 mission marked a milestone in space, and a milestone in television.

1.2.3   Continuing Work on the ATSC Standard

The documentation by the Advanced Television Systems Committee of the DTV standard in 1995, and the FCC's subsequent adoption of the major elements of the standard into the FCC Rules in 1996, represented landmark achievements in the history of broadcast television. While these events represented the culmination of the ATSC's work in one sense, they also marked the beginning of a whole new effort to take the DTV standard as developed and turn it into a functioning, and profitable, system that would be embraced by both industry and consumers alike. To that end, the ATSC organized three separate technical groups, each focusing on a different aspect of DTV deployment:

- The *Technology Group on Distribution* (T3), which had as its mission the development and recommendation of voluntary, international technical standards for the distribution of television programs to the public using advanced imaging technology.

- *DTV Implementation Subcommittee* (IS), established to investigate and report on the requirements for implementation of advanced television. The subcommittee evaluated technical requirements, operational impacts, preferred operating methods, time frames, and cost impacts of implementation issues.

- *Applications Subcommittee* (AS), which considered business opportunities that might be enabled by digital television standards. Based upon this analysis, the Applications Subcommittee made recommendations to the ATSC Board of Directors regarding development of voluntary standards for digital television.

1.2.3a A Surprising Challenge

As of the 1999 NAB Convention, over 50 stations were transmitting DTV signals. It was no surprise, then, that a proposal by one broadcaster—Sinclair Broadcast Group—that the chosen modulation method for the ATSC system be reevaluated, created shock waves throughout the convention, and indeed, throughout the industry. Sinclair's argument was that early production model DTV receivers performed poorly—in their estimation—relative to the NTSC service, and that the end result would be less reliance by the consumer on over-the-air signals and more reliance on cable television signals, thus putting broadcasters at a disadvantage. Sinclair suggested that COFDM, the method chosen by the European DVB consortium for terrestrial transmission, might perform better than 8-VSB. Sinclair went on to conduct side-by-side tests of COFDM and 8-VSB later in the year.

It is important to note that in the early days of the Grand Alliance program, COFDM was tested against 8-VSB. It was, in fact, tested several times. On the basis of technical merit, 8-VSB was chosen by the Grand Alliance, and ultimately endorsed by the FCC.

The Sinclair-sponsored tests did, indeed, take place during the Summer of 1999. The reviews from stations, networks, and industry groups were mixed. In general, observers stated that the tests did show the benefits of COFDM for operation with inside receive antennas in areas that experience considerable multipath. Multipath tolerance, of course, is one of the strengths of the COFDM system. Whether any significant progress was made to change industry minds, however, was unclear.

The ATSC made its position on the Sinclair tests crystal clear, however, in a statement released on July 2, 1999, calling the effort "unwarranted and irresponsible." The statement continued, "It is unwarranted because a growing body of evidence supports the performance of the VSB transmission system, and there is no clear evidence that COFDM is better. It is irresponsible because it seriously understates the impact of a change." The position statement concluded that, "Any decision to revisit the transmission standard would cause years of delay."

There were a number of other organizations and companies that weighed-in on the Sinclair demonstrations. One that received a fair amount of attention was a position paper issued by the Harris Corporation, a major broadcast equipment supplied with perhaps fewer axes to grind than most commenters. The Harris paper, released on August 10, 1999, warned that, "reopening the DTV standard debate would imperil the transition to digital broadcasting." The statement continued, "The burden on those seeking to reopen the DTV standards debate should be extremely high, given the lengthy delay, high cost, and crippling uncertainty necessarily entailed in even considering a switch to COFDM. The related industries have produced a variety of competing studio, transmission, and consumer equipment and 166 stations to date have either initiated DTV broadcasts or purchased transmission equipment. The Sinclair demonstration falls far short of warranting the delay, disruption, and confusion that would accompany reopening this debate."

The Harris position paper raised—rather convincingly—the financial implications inherent in a change to COFDM from 8-VSB, if equivalent NTSC coverage is to be maintained. "The structure of the 8-VSB signal minimizes both co-channel and adjacent channel interference to NTSC signals. By contrast, even at the same power levels, the structure of the COFDM signal will cause perceptibly more interference with existing NTSC services. This interference will be exacerbated if COFDM power levels are increased, as would be required to replicate NTSC coverage. Under any scenario, a change to COFDM would necessitate the Commission adopting a new table of allotments."

The ITU also found itself being drawn into the modulation wars of Summer 1999. A report issued on May 11 by an ITU Radiocommunication Study Group presented an objective, scientifically valid comparison of the modulation schemes under a variety of conditions. The report, titled "Guide for the Use of Digital Television Terrestrial Broadcasting Systems Based on Performance Comparison of ATSC 8-VSB and DVB-T COFDM Transmission Systems," provided considerable detail of the relative merits of each system. The report, interestingly enough, was used by both the COFDM and 8-VSB camps to support their positions.

In essence, the report concluded that the answer to the question of which system is better is: "it depends." According to the document, "Generally speaking, each system has its unique advantages and disadvantages. The ATSC 8-VSB system is more robust in an *additive white Gaussian noise* (AWGN) channel, has a higher spectrum efficiency, a lower peak-to-average power ratio, and is more robust to impulse noise and phase noise. It also has comparable performance to DVB-T on low level ghost ensembles and analog TV interference into DTV. Therefore, the ATSC 8-VSB system could be more advantageous for *multi-frequency network* (MFN) implementation and for providing HDTV service within a 6 MHz channel."

"The DVB-T COFDM system has performance advantages with respect to high level (up to 0 dB), long delay static and dynamic multipath distortion. It could be advantageous for services requiring large scale *single frequency network* (SFN) (8k mode) or mobile reception (2k mode). However, it should be pointed out that large scale SFN, mobile reception, and HDTV service cannot be achieved concurrently with any existing DTTB system over any channel spacing, whether 6, 7, or 8 MHz."

The ITU report concluded, "DTV implementation is still in its early stage. The first few generations of receivers might not function as well as anticipated. However, with the technical advances, both DTTB systems will accomplish performance improvements and provide a much improved television service."

"The final choice of a DTV modulation system is based on how well the two systems can meet the particular requirements or priorities of each country, as well as other non-technical (but critical) factors, such as geographical, economical, and political connections with surrounding countries and regions. Each country needs to clearly establish their needs, then investigates the available information on the performances of different systems to make the best choice."

Petition Filed

Sinclair formally put the question before the FCC on October 8, 1999, in a petition requesting that the Commission allow COFDM transmissions *in addition to* 8-VSB for DTV terrestrial broadcasting. If granted, the petition would have, in effect, required receiver manufacturers to develop and market dual mode receivers, something that receiver manufacturers were not predisposed to do. It was estimated that when the Sinclair petition was filed, over 5,000 DTV receivers had been sold to consumers.

Petition Rejected

On February 4, 2000, the FCC released a letter denying a Petition for Expedited Rulemaking, filed by Sinclair requesting that the Commission modify its rules to allow broadcasters to transmit DTV signals using COFDM modulation in addition to the current 8-VSB modulation standard. The Commission said that numerous studies supported the conclusion that NTSC replication is attainable under the 8-VSB standard. It said that the concerns raised in the Sinclair petition had "done no more than to demonstrate a shortcoming of early DTV receiver implementation." The Commission pointed out that receiver manufacturers were aware of problems cited by Sinclair and were aggressively taking steps to resolve multipath problems exhibited in some first-generation TV sets.

The Commission noted that the FCC Office of Engineering and Technology had analyzed the relative merits of the two standards, and concluded that the benefits of changing the DTV transmission standard to COFDM would not outweigh the costs of making such a revision. The Commission reiterated its view that allowing more than one standard could result in compatibility problems that could cause consumers and licensees to postpone purchasing DTV equipment and lead to significant delays in the implementation and provision of DTV services to the public. It said that development of a COFDM standard would result in a multi-year effort, rather than the "unrealistic" 120 days suggested in the Sinclair petition.

At the same time it dismissed the petition, the Commission recognized the importance of the issues raised by Sinclair. The Commission, stated, however, that the issue of the adequacy of the DTV standard is more appropriately addressed in the context of its biennial review of the entire DTV transition effort.

1.2.3b   FCC Reviews the Digital Television Conversion Process

On March 8, 2000, the FCC began its first periodic review of the progress of conversion of the U.S. television system from analog technology to DTV. In a Notice of Proposed Rulemaking (NPRM 00-83), the Commission invited public comment on a number of issues that it said required resolution to insure continued progress in the DTV conversion and to eliminate potential sources of delay. The Commission said its goal was to insure that the DTV transition went smoothly for American consumers, broadcasters, and other interested parties. This periodic review followed through on the conclusion, adopted by the Commission as part of its DTV construction schedule and service rules in the May 1997 5th Report and Order, that it should undertake a periodic review every two years until the cessation of analog service to help the Commission insure that the DTV conversion fully served the public interest.

In the NPRM, the Commission noted that broadcast stations were facing relatively few technical problems in building digital facilities, and that problems encountered by some stations with tower availability and/or local zoning issues did not seem to be widespread. However, it asked for comment on whether broadcasters were able to secure necessary tower locations and construction resources, and to what extent any zoning disputes, private negotiations with tower owners, and the availability of tower construction resources affected the DTV transition.

In the NPRM, the Commission asked for comments on whether to adopt a requirement that DTV licensees replicate their NTSC service area, and whether a replication requirement should be based on the population or the area served by the station. The Commission noted that several licensees had sought authority to move their DTV station to a more central location in their market—or toward a larger market—or had asked to change their DTV allotment—including their assumed transmitter site and/or technical facilities—and it asked for comments on the effect that these situations had on the general replication requirements. In addition, the Commission asked for comments on a proposed requirement that DTV stations' principal community be served by a stronger signal level than that specified for the general DTV service contour.

The Commission asked for comments on what date stations with both their NTSC and DTV channels within the DTV *core* (channels 2–51) would have to choose the channel they intend to keep following the transition. It said that with the target date for the end of the DTV transition set for December 31, 2006, it would be reasonable for stations to identify the DTV channels they will be using not later than 2004. It asked for comment on whether this date represented the proper balance between the goals of allowing DTV stations enough time to gain experience with DTV operation, and allowing stations that must move enough time to plan for their DTV channel conversion.

The Commission also invited comments on DTV application processing procedures, including whether to establish DTV application cut-off procedures, how to resolve conflicts between DTV applications to implement "initial" allotments, and the order of priority between DTV and NTSC applications. The Commission said it was seeking comment on whether to adopt a cut-off procedure for DTV area-expansion applications to minimize the number of mutually exclusive applications and to facilitate applicants' planning, and how to resolve any mutual exclusive applications that arise.

In the NPRM, the Commission noted that concerns had arisen regarding the 8-VSB DTV transmission standard. It invited comment on the status of this standard, including information on any additional studies conducted regarding NTSC replication using the 8-VSB standard. It specifically asked for comments on progress that was being made to improve indoor DTV reception, and manufacturers' efforts to implement DTV design or chip improvements.

The Commission noted certain industry agreements relating to cable compatibility, but said these agreements did not cover labeling of digital receivers, and asked whether a failure to reach agreement on the labeling issue would hinder the DTV transition. The Commission also asked for comments on the extent to which a lack of agreement on copy protection technology licensing and related issues would also hinder the DTV transition. Noting that some broadcasters had recommended that the Commission address over-the-air signal reception by setting receiver standards, the Commission asked for comments on whether the FCC had authority to set minimum performance levels for DTV receivers, whether it would be desirable to do so, and, if so, how such requirements should be structured.

In this "periodic review," the Commission said it was not asking for comment on issues that were the subject of separate proceedings—such as the issue of digital broadcast signal carriage on cable systems—or requests for reconsideration of already-decided issues—such as eligibility issues, certain issues relating to public television, and channel allotment or change requests. The Commission also said it was too early in the transition process to address other issues, including reconsidering the flexible approach to ancillary or supplementary services and the application of the simulcast requirement.

Perhaps most importantly—from the standpoint of broadcasters at least—the Commission said it would be inappropriate to review the 2006 target date for complete conversion to DTV because of Congressional action in the Balanced Budget Act of 1997, which confirmed December 31, 2006 as the target transition completion date and established procedures and for stations seeking an extension of that date.

ATSC Comments Filed

The Advanced Television Systems Committee filed comments on the FCC's DTV review, as expected. What surprised many, however, was the apparent change of public posture with respect to the 8-VSB-vs.-COFDM issue. The ATSC filing stated, "The ATSC fully endorses and supports the Commission's DTV transmission standard, based on the ATSC DTV standard, and we urge the Commission to take all appropriate action to support and promote the rapid transition to digital television. Even so, the ATSC continues to seek ways to improve its standards and the implementation thereof in order to better meet the existing and evolving service requirements of broadcasters. To this end, and also in response to the specific concerns of some broadcasters regarding RF system performance (modulation, transmission, and reception), the Executive Committee of the ATSC (has) formed a task force on RF system performance."

This task force was charged with examining a variety of technical issues that had been raised regarding the theoretical and realized performance of the DTV RF system—including receivers—and based on its findings, to make recommendations to the ATSC Executive Committee as to what, if any, technical initiatives should be undertaken by the ATSC.

At the time of the filing (May, 2000), the task force had identified three areas of focus:

- 8-VSB performance

- Broadcaster requirements

- Field test methodology

The task force was charged with evaluating the extent to which 8-VSB enabled broadcasters to achieve reliable DTV reception within their current NTSC service areas. The group then was to assess the current performance of 8-VSB, including 8-VSB receiving equipment, as well as expected improvements. In addition, the task force was to look into potential backward-compatible enhancements to VSB, i.e., enhancements that could be implemented without affecting the operation of existing receivers that are not capable of delivering the enhanced features.

The task force also hoped to recommend potential actions that could be taken by the ATSC and its member organizations that might hasten improvements in the performance of the DTV RF system.

1.2.3c   A Turning Point is Reached

January 2001 marked a turning point for the ATSC DTV System. The FCC affirmed its support—for the third time—of the 8-VSB modulation system and a closely watched comparative study of 8-VSB and COFDM concluded that, given the character of the U.S. broadcasting system, 8-VSB was superior to COFDM. The study, managed jointly by the National Association of Broadcasters (NAB) and Maximum Service Television (MSTV), was broad in scope and technical depth. With the support of 30 major broadcast organizations and the oversight of technical committees consisting of some 25 engineers representing all major technical viewpoints, the broadcasting industry conducted a comprehensive, objective, and expedited series of studies and tests to determine whether COFDM should be added to the current 8-VSB standard. Conclusions of the NAB/MSTV study, released on January 15, 2001, were as follows: "We conclude that there is insufficient evidence to add COFDM and we therefore reaffirm our endorsement of the VSB standard. We also conclude that there is an urgent need for swift and dramatic improvement in the performance of the present U.S. digital television system. We therefore will take all necessary steps to promote the rapid improvement of VSB technologies and other enhancements to digital television..."

A few days later, the FCC issued a *Report and Order and a Further Notice of Proposed Rulemaking* (FNRPM) in its first periodic review of the DTV transition. In the Report and Order, the Commission affirmed the 8-VSB modulation system for the DTV transmission standard, concluding that there was no reason to revisit its decision, thereby denying a request to allow use of an alternative DTV modulation system (COFDM).

Reactions came quickly from industry leaders. Robert Graves, Chairman of the ATSC, stated: "The decisions this week by the NAB and MSTV, and by the FCC, put an end to the debate surrounding the ATSC/VSB transmission system used for terrestrial DTV broadcasts in the U.S. With the transmission system debate behind us, with the impressive array of DTV and HDTV products now available at ever lower prices, and with the increasing amount of compelling DTV content becoming available from broadcasters, the outlook for DTV services in the U.S. is bright."

The NAB/MSTV report noted the need for improvements in receivers and called for a thorough investigation of ways to enhance the 8-VSB system. Graves addressed that issue, saying: "Significant improvements have already been made in the performance of VSB receivers, and further improvements are in the pipeline from a variety of manufacturers. Moreover, work is under way within the ATSC to further enhance the ATSC/VSB standard by adding more robust transmission modes that address emerging DTV applications…"

The consumer electronics industry also responded to the NAB/MSTV report and the FCC action. Consumer Electronics Association (CEA) President and CEO Gary Shapiro said: "The debate over the modulation standard is over. All broadcasters who are seriously committed to the DTV transition should now recommit to the transition by continuing the buildout and producing more high-definition and digital content. The results of the joint NAB/MSTV testing confirm what we have long believed: the Federal Communications Commission (FCC)-approved 8-VSB modulation system is the best choice for broadcasting digital television in the United States. These results also reaffirm that the FCC made the correct decision in initially selecting, and again re-endorsing 8-VSB last year.

A week after these pronouncements, the ATSC made good on its promise to investigate ways to enhance the 8-VSB transmission system to improve reception in fixed indoor, portable, and mobile applications. The ATSC Specialist Group on RF Transmission (T3/S9) issued a Request for Proposal (RFP) for enhancements aimed at addressing emerging DTV applications. The RFP asked for proposals for complementary specifications that will be considered as potential enhancements of the ATSC DTV Standard. Laboratory and field tests of the proposals began in the Spring of 2002. Prominent among the specialists participating in the work of T3/S9 were engineers from Sinclair.

With this flurry of activity, momentum returned to the DTV transition. As of Summer 2002, there were 422 stations transmitting DTV signals in 128 markets, according to an NAB survey.

Receiver Performance Issues

In the January 2001 Report and Order, the FCC denied requests to set performance standards for digital receivers, expressing concern that the effect of setting such standards would be to stifle innovation and limit performance to current capabilities. The Commission said, however, it would continue to monitor receiver issues.

The Commission also issued a Further Notice of Proposed Rulemaking to consider whether to require some TV sets to have the capability to demodulate and decode over-the-air DTV signals in addition to displaying the existing analog TV signals. In raising this issue, the Commission recognized broadcasters' concerns that DTV receivers are not yet available in sufficient volume to support a rapid transition to an all-digital broadcast television service. It asked whether a requirement to include DTV reception capability in certain television sets could help to develop the production volumes needed to bring DTV receiver prices down quickly to where they would be more attractive to consumers and could help to promote a more rapid development of high DTV set penetration.

The Commission asked for comments on how best to implement DTV reception capability requirements, if it were to decide to adopt them, adding that it recognized the cost considerations associated with such requirements. The Commission suggested that one approach to minimizing the impact on both consumers and manufacturers would be to impose any requirement first on a percentage of large screen televisions, such as 32 in and larger, because these are typically higher priced units where the cost of DTV components would be a smaller percentage of the cost of the receiver. In addition, it asked whether any requirement should be phased in over time such that manufacturers would increase each year the percentage of units of a designated screen size or larger that are manufactured with DTV receive capability. The FCC noted that separate set-top DTV receivers could also be included in meeting the reception capability requirements.

The Commission additionally requested comment on whether it should require any digital television receivers that cannot receive over-the-air digital broadcast signals to carry a label informing consumers of this limitation. This issue concerns receivers that are intended for use only with cable television or broadcast satellite service. The Commission indicated that while it expects consumers will continue to expect all digital television receivers to be able to receive over-the-air broadcast signals, it suggested that where receivers not able to receive such signals are marketed, consumers should be so notified prior to purchase.

1.2.4   FCC Acts on DTV Receiver Requirements

In what many broadcasters believed to be a key element in facilitating the roll-out of DTV service in the U.S., the FCC adopted a plan on August 8, 2002, that required off-air digital TV tuners on nearly all new TV sets by 2007. By enacting a five-year rollout schedule that started with larger, more expensive TV sets, the FCC said it was minimizing the costs for equipment manufacturers and consumers.

In its order, the FCC said DTV receivers were a necessary element of broadcast television service in the same way that analog TV receivers had been since the inception of analog television service. The Commission said that its jurisdiction was established by the 1962 All Channel Receiver Act (ACRA), which provided the FCC with the "authority to require" that television sets "be capable of adequately receiving all frequencies" allocated by the FCC for "television broadcasting." The Commission concluded that the authority provided under the ACRA applied to all devices used to receive broadcast television service, not just those used to receive analog signals. The FCC said its plan reflected and accounted for the following market forces:

- Including DTV reception capability in new television receivers would require the redesign of product lines

- Prices were declining and would decline even faster as economies of scale were achieved and production efficiencies were realized over time

- Prices of large TV sets had been declining at a rate of $100 to $800 per year, and so the additional cost of the DTV tuner could be partially or completely offset by the general price decline

The FCC said the plan would ensure that new TV receivers include a DTV tuner on a schedule as close as economically feasible to the December 31, 2006, target completion date for the DTV transition that was set forth in the Communications Act by Congress.

Specifically, the Second Report and Order and Second Memorandum Opinion and Order adopted by the Commission required that all television receivers with screen sizes greater than 13-in and all television receiving equipment, such as videocassette recorders (VCRs) and digital versatile disk (DVD) players/recorders, were required to include DTV reception capability after July 1, 2007, according to the following schedule:

- Receivers with screen sizes 36-in and above: 50 percent of units must include DTV tuners effective July 1, 2004, and 100 percent of such units must include DTV tuners effective July 1, 2005.

- Receivers with screen sizes 25- to 35-in: 50 percent of units must include DTV tuners effective July 1, 2005, and 100 percent of such units must include DTV tuners effective July 1, 2006.

- Receivers with screen sizes 13- to 24-in: 100 percent of all units must include DTV tuners effective July 1, 2007.

- TV interface devices, VCRs, DVD players/recorders, and similar units that receive broadcast television signals: 100 percent of all units must include DTV tuners effective July 1, 2007.

In its action, the FCC also declined to adopt labeling requirements for TV sets that were not able to receive any over-the-air broadcast signals. The FCC stated that it is unclear when, or if, such products would become commercially available or how they would be marketed. The Commission said it would monitor the state of the marketplace and take additional steps if necessary to protect consumers' interests.

Later action by the Commission accelerated the some of the DTV receiver deadlines as a way of providing additional momentum to the DTV transition.

1.2.5   Cable and CE Companies Reach Key Agreement on DTV Transition Issues

On December 20, 2002, a the DTV transition saw a major step forward with agreement between consumer electronics equipment manufacturers and cable television system operators regarding "plug-and-play" cable/receiver interoperability. The agreement, which set the stage for a national "plug-and-play" standard between digital television products and digital cable systems, was expected to help speed the transition from analog to digital television and establish much-needed marketplace, technical, and regulatory certainty for the cable and consumer electronics (CE) industries. The terms of the agreement were embodied in a Memorandum of Understanding agreed to by 14 consumer electronics companies, representing the majority of HDTV sales in the United States, and seven major cable operators, representing more than 75 percent of all cable subscribers.

The MSO/CE manufacturer agreement consisted of a package that included: 1) voluntary commitments by the signatories and 2) proposals for rules to be adopted by the Federal Communications Commission (FCC) that, once adopted, would resolve other issues.

Specifically, the joint recommendations included:

- A set of technical standards for cable systems and "cable-ready" DTV products (and testing procedures to assure compatibility)

- Proposed regulatory framework for support of digital TV receivers, digital recorders with secure interfaces, and other devices on cable systems

- Draft security technology license to ensure that high-value content could be transferred securely in a home network by consumers

- Specified "encoding rules" to resolve then-pending copyright-based concerns about home recording and viewing

In addition to proposing regulations that would assure consumer access to plug-and-play digital TV over cable, the parties agreed to begin working together immediately on standards for future interactive digital cable TV products.

Negotiations were conducted directly among company representatives and facilitated by their respective trade associations, the Consumer Electronics Association (CEA) and the National Cable Telecommunications Association (NCTA).

The parties expressed the expectation that the agreement would encourage the development and distribution of high-quality digital content. A key element of the agreement related to secure digital interfaces that protect consumers' home recording rights along with copyright owners' rights to secure their digital content. Major cable system operators agreed to support recordable IEEE 1394 connections on high-definition set-top boxes. In turn, digital TV manufacturers agreed to support FCC labeling regulations that specify Digital Visual Interface (DVI)/High-bandwidth Digital Content Protection (HDCP) (or HDMI/HDCP) display interfaces with copy protection controls in future cable-ready HDTV products. Moreover, the agreement established "rules of the road" on home recording rights and proposed copy protection rules for digital content that were based on existing law and studio-CE agreements and which were applicable to all multichannel video programming distributors.

The signatories said the agreement would provide certainty and marketplace stability, to the benefit of consumer electronics and cable companies. In addition to specifying the required digital inputs and connectors on the new HDTV sets and converter boxes, the agreement established a path for a joint process to:

- Develop technical specifications for interactive services

- Standardize and streamline technical processes to enable devices to inter-operate with each other and allow those devices to be used by consumers regardless of the cable system by which they are served

- Create a self-certification process for manufacturers to ensure compatibility between digital television products and digital cable systems

- Help cable operators establish a strong presence in the retail market

With this agreement in place, it seemed that the only remaining marketplace/regulatory issue remaining to be resolved was that of "DTV must-carry," which focused on the desire by broadcasters, and the reluctance of cable operators, to carry both NTSC and DTV signals of stations within a given market on cable systems. A second element of this debate related to whether the cable system would be required to convey the entire 19.4 Mbits/s ATSC DTV stream, or simply the "main program channel," the definition of which appeared somewhat vague.

1.2.6   DTV Transition Accelerates

Thanks in large part to the efforts of broadcasters and receiver manufacturers, the DTV transition continued to gain momentum through the early part of the decade, until—at the end of 2005—well over 1,500 digital TV stations were on the air in more than 200 markets. Receiver manufacturers continued to improve their products, with considerable advancements in performance and reduced prices at retail. Work within ATSC continued to refine the standards already in place and develop new specifications as needs dictated. The FCC, meanwhile, kept-up the pressure on all parties to move the digital transition forward, while in Congress lawmakers were discussing the merits of establishing a firm shut-off date for analog broadcasting.

1.2.6a   Enhanced VSB

The initial outcome of the ATSC T3/S9 work described previously was a system known as "Enhanced-VSB" (E-VSB), an optional mode that involves the transmission of a backward-compatible signal within the standard 8-VSB symbol stream that can be received at a lower carrier-to-noise ratio than conventional 8-VSB. The E-VSB mode allows broadcasters to trade-off some of their data capacity for additional robustness. With an E-VSB transmission, some of the approximately 19.4 Mbps data is allocated to the robust mode and some is allocated to the normal 8-VSB mode. However, the amount of delivered data (payload) is reduced for the robust mode because part of the payload is traded for additional forward error correction (FEC) bits to correct bit errors that occur with reception under weaker signal conditions (resulting in up to a 6 dB improvement).

Additional techniques to extend the capabilities of the ATSC physical layer (transmission system) were proposed and examined.

1.2.6b   Data Broadcasting

Data broadcasting was long assumed to hold considerable opportunities for digital television. Recognizing this interest, the ATSC developed a suite of data broadcast standards to enable a wide variety of data services, which may be related to one or more video programs being broadcast or stand-alone services. Potential applications range from streaming audio, video, or text services to private data delivery services. Data broadcast receivers may include personal computers, televisions, set-top boxes, or other devices.

1.2.6c   Interactive Television

In August 2002 the ATSC and the Cable Television Laboratories (CableLabs) began a multi-year effort to forge a common platform for interactive television services. The Advanced Common Appliation Platform (ACAP) was the result of this landmark harmonization effort between the ATSC DTV Application Software Environment (DASE) Standard and the CableLabs Open Cable Application Platform (OCAP) specification.

In essence, ACAP makes it appear to interactive programming content that it is running on a single platform, the so-called common receiver. This common receiver contains a well-defined architecture, execution model, syntax, and semantics. As a "middleware" specification for interactive applications, ACAP gives content and application authors assurance that their programs and data will be received and run uniformly on all brands and models of receivers.

The term interactive television (ITV) is broad and includes a vast array of applications, including:

- Customized news, weather and traffic

- Stock market data, including personal investment portfolio performance in real-time

- Enhanced sports scores and statistics on a user-selective basis

- Games associated with program

- On-line real-time purchase of everything from groceries to software without leaving home

- Video on demand (VOD)

ACAP was intended to provide consumers with advanced interactive services while providing content providers, broadcasters, cable and satellite operators, and consumer electronics manufacturers with the technical details necessary to develop interoperable services and products.

1.2.6d Future Improvements

The basic design of the ATSC DTV system both anticipates and facilitates continuing improvements in performance in many ways [1]:

- By specifying the transmitted picture formats, camera resolutions and frame rates in production can continue to improve, and display processing technology can continue to improve the overall quality delivered to the viewer.

- By specifying the MPEG decoder syntax, encoders can continue to improve the quality of ATSC signals (or further reduce bit rates to allow the introduction of additional services) as processing power increases.

- Through the flexibility of the MPEG transport protocol, the ATSC standard can deliver multiple simultaneous services, and new services can be added without disrupting the installed base of receivers.

- By specifying the transmitted VSB signal, receivers can continue to improve their ability to reject interference and deal with transmission impairments such as multipath.

By the middle of the decade, it appeared that all of the elements were in place for a relatively smooth transition from the analog NTSC service to DTV.

1.2.7 References

[1]   Richer, M., G. Reitmeier, T. Gurley, G. Jones, J. Whitaker, and R. Rast: "The ATSC Digital Television System, IEEE Proceedings, IEEE, New York, N.Y., January 2006.

1.2.8   Bibliography

ATSC: "Comments of The Advanced Television Systems Committee, MM Docket No. 00-39," ATSC, Washington, D.C., May, 2000.

"ATV System Recommendation," *1993 NAB HDTV World Conference Proceedings*, National Association of Broadcasters, Washington, D.C., pp. 253–258, 1993.

Federal Communications Commission: Notice of Proposed Rule Making 00-83, FCC, Washington, D.C., March 8, 2000.

Hopkins, R.: "Advanced Television Systems," *IEEE Transactions on Consumer Electronics*, vol. 34, pp. 1–15, February 1988.

"Sinclair Seeks a Second Method to Transmit Digital-TV Signals," *Wall Street Journal*, Dow Jones, New York, N.Y., October 7, 1999.

Fundamental Video Principles

Visible light is a form of electromagnetic radiation whose wavelengths fall into the relatively narrow band of frequencies to which the *human visual system* (HVS) responds: the range from approximately 380 nm to 780 nm. These wavelengths of light are readily measurable. The perception of color, however, is a complicated subject. Color is a phenomenon of physics, physiology, and psychology. The perception of color depends on factors such as the surrounding colors, the light source illuminating the object, individual variations in the HVS, and previous experiences with an object or its color.

Colorimetry is the branch of color science that seeks to measure and quantify color in this broader sense. The foundation of much of modern colorimetry is the CIE system developed by the Commission Internationale de l'Eclairage (International Commission on Illumination). The CIE colorimetric system consists of a series of essential standards, measurement procedures, and computational methods necessary to make colorimetry a useful tool for science and industry.

Virtually all applications of video communications deal with enormous amounts of data. Because of this, compression is an integral part of most modern digital media systems. In fact, compression is essential to all digital television systems.

Compression systems employ a combination of processing techniques that take advantage of the capabilities—and limitations—of the HVS. In order to properly understand video compression, the fundamental principles of human visit must be understood as well.

In This Section:

2.1

Principles of Color Vision

Alan R. Robertson, Joseph F. Fisher[1]

2.1.1 Introduction

The sensation of color is evoked by a physical stimulus consisting of electromagnetic radiation in the visible spectrum. The stimulus associated with a given object is defined by its *spectral concentration of radiance* $L_e(\lambda)$

$$L_e(\lambda) = \frac{1}{\pi} E_e(\lambda) R(\lambda)$$

$$(2.1.1)$$

Where:
$E_e(\lambda)$ = the *spectral irradiance*
$R(\lambda)$ = the *spectral reflectance factor*

The *spectral reflectance* $p(\lambda)$ is sometimes used instead of $R(\lambda)$. Because $p(\lambda)$ is a measure of the total flux reflected by the object, whereas $R(\lambda)$ is a measure of the flux reflected in a specified direction, the use of $p(\lambda)$ implies that the object reflects uniformly in all directions. For most objects this is approximately true, but for some, such as mirrors, it is not.

2.1.2 Trichromatic Theory

Color vision is a complicated process. Full details of the mechanisms are not entirely understood. However, it is generally believed, on the basis of strong physiological evidence, that the first stage is the absorption of the stimulus by light-sensitive elements in the retina. These light-sensitive elements, known as *cones*, form three classes, each having a different spectral sensitivity distribution. The exact spectral sensitivities are not known, but they are broad and overlap considerably. An estimate of the three classes of spectral sensitivity is given in Figure 2.1.1.

1. From *Standard Handbook of Video and Television Engineering*, 4th. ed., Jerry C. Whitaker (ed.), McGraw-Hill, New York, N.Y., 2003. Used with permission.

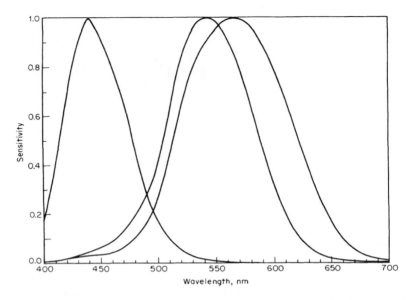

Figure 2.1.1 Spectral sensitivities of the three types of cones in the human retina. The curves have been normalized so that each is unity at its peak. (*After* [1, 2].)

It is clear from this *trichromacy* of color vision that many different physical stimuli can evoke the same sensation of color. All that is required for two stimuli to be equivalent is that they each cause the same number of quanta to be absorbed by any given class of cone. In such cases, the neural impulses—and thus the color sensations—generated by the two stimuli will be the same. The visual system and the brain cannot differentiate between the two stimuli even though they are physically different. Such equivalent stimuli are known as *metamers* and the phenomenon as *metamerism*. Metamerism is fundamental to the science of colorimetry; without it color video reproduction as we know it could not exist. The stimulus produced by a video display is almost always a metamer of the original object and not a physical (spectral) match.

2.1.2a   Color Matching

Because of the phenomenon of trichromacy, it is possible to match any color stimulus by a mixture of three primary stimuli. There is no unique set of primaries; any three stimuli will suffice as long as none of them can be matched by a mixture of the other two. In certain cases it is not possible to match a given stimulus with positive amounts of each of the three primaries, but a match is always possible if the primaries may be used in a negative sense.

Experimental measurements in color matching are carried out with an instrument called a *colorimeter*. This device provides a split visual field and a viewing eyepiece, as illustrated in Figure 2.1.2. The two halves of the visual field are split by a line and are arranged so that the mixture of three primary stimuli appears in one-half of the field. The amounts of the three primaries can be individually controlled so that a wide range of colors can be produced in this half of the

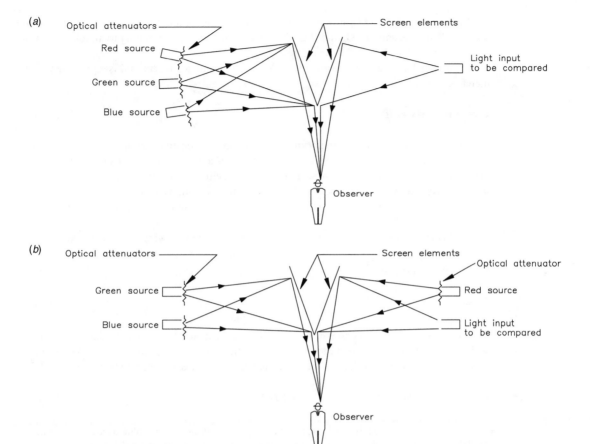

Figure 2.1.2 Tristimulus color matching instruments: (*a*) conventional colorimeter, (*b*) addition of a primary color to perform the match.

field. The other half of the field accepts light from the sample to be matched. The amounts of the primaries are adjusted until the two halves of the field match. The amounts of the primaries are then recorded. For those cases where negative values of one or more of the primaries are needed to secure a match, the instrument is arranged to transfer any of the primaries to the other half of the field. The amount of a primary inserted in this manner is recorded as negative.

The operation of color matching may be expressed by the *match equation*

$$\mathbf{C} \equiv R\mathbf{R} + G\mathbf{G} + B\mathbf{B} \tag{2.1.2}$$

This equation is read as follows: Stimulus **C** is matched by R units of primary stimulus **R** mixed with G units of primary stimulus **G** and B units of primary stimulus **B**. The quantities R, G, and B are called *tristimulus values* and provide a convenient way of describing the stimulus **C**. All the different physical stimuli that look the same as **C** will have the same three tristimulus values R, G, and B.

It is common practice (followed in this publication) to denote the primary stimuli by using boldface letters (usually **R**, **G**, and **B** or **X**, **Y**, and **Z**) and the corresponding tristimulus values by italic letters R, G, and B or X, Y, and Z, respectively.

The case in which a negative amount of one of the primaries is required is represented by the match equation

$$\mathbf{C} \equiv -R\mathbf{R} + G\mathbf{G} + B\mathbf{B} \tag{2.1.3}$$

This equation assumes that the red primary is required in a negative amount.

The extent to which negative values of the primaries are required depends upon the nature of the primaries, but no set of real physical primaries can eliminate the requirement entirely. Experimental investigations have shown that in most practical situations, color matches obey the algebraic rules of additivity and linearity. These rules, as they apply to colorimetry, are known as *Grassmann's laws* [3]

To illustrate this point, assume two stimuli defined by the following match equations:

$$\mathbf{C}_1 \equiv R_1\mathbf{R} + G_1\mathbf{G} + B_1\mathbf{B} \tag{2.1.4}$$

$$\mathbf{C}_2 \equiv R_2\mathbf{R} + G_2\mathbf{G} + B_2\mathbf{B} \tag{2.1.5}$$

If $\mathbf{C}_1$ is added to $\mathbf{C}_2$ in one half of a colorimeter field, and the resultant mixture is matched with the same three primaries in the other half of the field, the amounts of the primaries will be given by the sums of the values in the individual equations. The match equation will be

$$\mathbf{C}_1 + \mathbf{C}_2 \equiv (R_1 + R_2)\mathbf{R} + (G_1 + G_2)\mathbf{G} + (B_1 + B_2)\mathbf{B} \tag{2.1.6}$$

In this discussion, the symbols **R**, **G**, and **B** signify red, green, and blue as a set of primaries. The meaning of these color names must be specified exactly before the colorimetric expressions have precise scientific meaning. Such specification may be given in terms of three relative spectral-power-distribution curves, one for each primary. Similarly, the amounts of each primary must be specified in terms of some unit, such as watts or lumens.

The concept of matching the color of a stimulus by a mixture of three primary stimuli is, of course, the basis of color video reproduction. The three primaries are the three colored phosphors, and the additive mixture is performed in the observer's eye because of the eye's inability to resolve the small phosphor dots or stripes from one another.

2.1.2b   Color-Matching Functions

In general, a color stimulus is composed of a mixture of radiations of different wavelengths in the visible spectrum. One important consequence of Grassmann's laws is that if the tristimulus values R, G, and B of a monochromatic (single-wavelength) stimulus of unit radiance are known at each wavelength, the tristimulus values of any stimulus can be calculated by summation. Thus, if the tristimulus values of the spectrum are denoted by

$$\bar{r}(\lambda), \bar{g}(\lambda), \bar{b}(\lambda)$$

(2.1.7)

per unit radiance, then the tristimulus values of a stimulus with a spectral concentration of radiance $L_e(\lambda)$ are given by

$$R = \int_{380}^{780} L_e(\lambda)\bar{r}(\lambda)d\lambda \qquad (2.1.8)$$

$$G = \int_{380}^{780} L_e(\lambda)\bar{g}(\lambda)d\lambda \qquad (2.1.9)$$

$$B = \int_{380}^{780} L_e(\lambda)\bar{b}(\lambda)d\lambda \qquad (2.1.10)$$

If a set of primaries were selected and used with a colorimeter, all selections of color mixture could be set up and the appropriate matches made by an observer. A disadvantage of this method is that any selected observer can be expected to have color vision that differs from the "average vision" of many observers. The color matches made might not be satisfactory to the majority of individuals with normal vision.

For this reason, it is desirable to adopt a set of universal data which is prepared by averaging the results of color-matching experiments made by a number of individuals who have normal vision. A spread of the readings taken in these color matches would indicate the variation to be expected among normal individuals. Averaging the results would give a reliable set of spectral tristimulus values. Many experimenters have conducted such psychophysical experiments. The results are in good agreement and have been used as the basis for industry standards.

The curves are now generally known as *color-matching functions*, although in older literature the terms *color-mixing functions* and *distribution coefficients* were sometimes used. The set of color-matching functions used by the CIE in 1931 as a basis for an international standard are shown in Figure 2.1.3 in terms of a particular set of real primaries, **R**, **G**, and **B**.

2.1.2c   Luminance Relationships

Luminances are, by definition, additive quantities. Thus, the luminance of a stimulus with tristimulus values R, G, and B is given by

$$L = L_R R + L_G G + L_B B$$

(2.1.11)

Where:
L_R = luminance unit amount of the **R** primary
L_G = luminance unit amount of the **G** primary

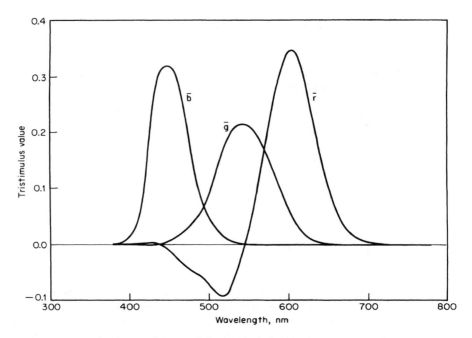

Figure 2.1.3 Color-matching functions of the CIE standard observer based on matching stimuli of wavelengths 700.0, 546.1, and 435.8 nm, with units adjusted to be equal for a match to an equienergy stimulus.

L_B = luminance unit amount of the **B** primary

In the special case of a monochromatic stimulus of unit radiance, the luminance is given by

$$L = L_R \bar{r}(\lambda) + L_G \bar{g}(\lambda) + L_B \bar{b}(\lambda)$$

(2.1.12)

This luminance condition is also given by

$$L = K_m V(\lambda)$$

(2.1.13)

Where:
K_m = 683 lm/W
$V(\lambda)$ = the spectral luminous efficiency function

It follows that $V(\lambda)$ must be a linear combination of the color-matching functions

$$K_m V(\lambda) = L_R \bar{r}(\lambda) + L_G \bar{g}(\lambda) + L_B \bar{b}(\lambda)$$

(2.1.14)

2.1.2d Vision Abnormality

In the previous discussion of color matching, the term "normal" was deliberately used to exclude those individuals (about 8 percent of males and 0.5 percent of females) whose color vision differs from the majority of the population. These people are usually called *color-blind*, although very few (about 0.003 percent of the total population) can see no color at all. About 2.5 percent of males require only two primaries to make color matches. Most of these can distinguish yellows from blues but confuse reds and greens. The remaining 5.5 percent require three primaries, but their matches are different from the majority and their ability to detect small color differences is usually less [2, 4].

2.1.2e Color Representation

Tristimulus values provide a convenient method of measuring a stimulus. Any two stimuli with identical tristimulus values will appear identical under given viewing conditions. However, the actual appearance of the stimuli (whether they are, for example, red, blue, light, or dark) depends on a number of other factors, including:

- The size of the stimuli

- The nature of other stimuli in the field of view

- The nature of other stimuli viewed prior to the present ones

Color appearance cannot be predicted simply from the tristimulus values. Current knowledge of the human color-vision system is far from complete, and much remains to be learned before color appearance can be predicted adequately. However, the idea that the first stage is the absorption of radiation (light) by three classes of cone is accepted by most vision scientists and correlates well with the concept and experimental results of tristimulus colorimetry. Furthermore, tristimulus values—and quantities derived from them—do provide a useful and orderly way of representing color stimuli and illustrating the relationships between them.

It is possible to describe the appearance of a color stimulus in words based upon a person's perception of it. The *trichromatic theory* of color leads to the expectation that this perception will have three dimensions or attributes. Everyday experience confirms this. One set of terms for these three attributes is:

- *Hue*—the attribute according to which an area appears to be similar to one, or to proportions of two, of the perceived colors red, orange, yellow, green, blue, and purple.

- *Brightness*—the attribute according to which an area appears to be emitting, transmitting, or reflecting more or less light.

- *Saturation*—the attribute according to which an area appears to exhibit more or less chromatic color.

A perceived color from a self-emitting object, therefore, is typically described by its hue, brightness, and saturation. For reflecting objects, two other attributes are often used:

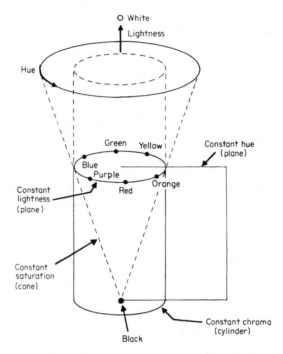

Figure 2.1.4 A geometrical model of perceptual color space for reflecting objects.

- *Lightness*—the degree of brightness judged in proportion to the brightness of a similarly illuminated area that appears to be white.

- *Perceived chroma*—the degree of colorfulness judged in proportion to a similarly illuminated area that appears to be white.

Reflecting objects, thus, may be described by hue, brightness, and saturation or by hue, lightness, and perceived chroma. The perceptual color space formed by these attributes may be represented by a geometrical model, as illustrated in Figure 2.1.4. The *achromatic* colors (black, gray, white) are represented by points on the vertical axis, with lightness increasing along this axis. All colors of the same lightness lie on the same horizontal plane. Within such a plane, the various hues are arranged in a circle with a gradual progression from red through orange, yellow, green, blue, purple, and back to red. Saturation and perceived chroma both increase from the center of the circle outward along a radius but in different ways depending on the lightness. All colors of the same saturation lie on a conical surface, whereas all colors of the same perceived chroma lie on a cylindrical surface.

If two colors have equal saturation but different lightness, the darker one will have less perceived chroma because perceived chroma is judged relative to a white area. Conversely, if two colors have equal perceived chroma, the darker one will have greater saturation because saturation is judged relative to the brightness—or, in this case, lightness—of the color itself [4].

2.1.2f   Munsell System

It is possible to construct a color chart based on the set of color attributes described in the previous section. One such chart, devised by A. H. Munsell, is known as the *Munsell system*. The names used for the attributes by Munsell are *hue*, *chroma*, and *value* [5–7], corresponding, respectively, to the terms hue, perceived chroma, and lightness. In the Munsell system, the colors are arranged in a circle in the following order:

- Red (*R*)
- Red-purple (*RP*)
- Purple (*P*)
- Purple-blue (*PB*)
- Blue (*B*)
- Blue-green (*BG*)
- Green (*G*)
- Green-yellow (*GY*)
- Yellow (*Y*)
- Yellow-red (*YR*)

The Munsell system chart defines a hue circle having 10 hues. To give a finer hue division, each of the 10 hue intervals is further subdivided into 10 parts. For example, there are 10 red subhues, which are referred to as 1*R* to 10*R*.

The attribute of chroma is described by having hue circles of various radii. The greater the radius, the greater is the chroma.

The attribute of *value* is divided into 10 steps, from zero (perfect black or zero reflectance factor) to 10 (perfect white or 100 percent reflectance factor). At each value level there is a set of hue circles of different chromas, the lightness of all the colors in the set being equal.

A color is specified in the Munsell system by stating in turn (1) hue, (2) value, and (3) chroma. Thus, the color specified as 6*RP* 4/8 has a red-purple hue of 6*RP*, a value of 4, and a chroma of 8. Not all chromas or values can be duplicated with available pigments.

This arrangement of colors in steps of hue, value, and chroma was originally carried out by Munsell using his artistic eye as a judge of the correct classification. Later, a committee of the Optical Society of America [8] made extensive visual studies that resulted in slight modifications to Munsell's original arrangement. The committee's judgment is perpetuated in the form of a book of paper swatches colored with printer's ink and marked with the corresponding Munsell notation [9]. A set of chips arranged in the form of a color tree can also be obtained.

2.1.2g   Other Color-Order Systems

In addition to the Munsell system, there are a number of other color-order systems [3]. The three scales of the various systems, and the spacing of samples along the scales, are chosen by different criteria. In some systems the scales and spacing are determined by a systematic mixture of dyes or pigments, or by systematic variation of parameters in a printing process. In others, they are based on the rules of additive mixture, as in a tristimulus colorimeter. A third class of color-order systems (which includes the Munsell system) is based on visual perceptions. Within each class the exact rules by which the colors are ordered vary significantly from one to another.

2.1.2h Color Triangle

The *color triangle* is an alternative method of classifying and specifying colors. This method was originated by Newton and used extensively by Maxwell. It is a method of representing the matching and mixing of stimuli and is derived from the tristimulus values discussed previously.

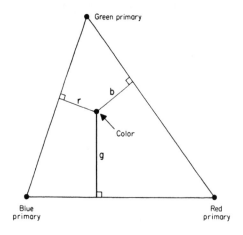

The color triangle displays a given color stimulus in terms of the relative tristimulus values; that is, the relative amounts of three primaries needed to match it. Thus, the color triangle displays only the quality of the stimulus and not its quantity. In one form, the stimulus is represented by a point chosen so that the perpendicular distances to each of the three sides are proportional to the tristimulus values. The triangle need not be equilateral, although the triangles used by Newton and Maxwell were of this type. The method is illustrated in Figure 2.1.5.

Figure 2.1.5 The color triangle, showing the use of trilinear coordinates. The amounts of the three primaries needed to match a given color are proportional to *r, g,* and *b.*

This method of display is equivalent to the use of *trilinear coordinates*, which form a well-known coordinate system in analytical geometry. In this representation, the three primaries appear one at each of the three vertices of the triangle, because two of the trilinear coordinates vanish at each vertex.

It is more common, however, to use a right-angled triangle as shown in Figure 2.1.6. The quantities *r*, *g*, and *b*, called *chromaticity coordinates*, are calculated using the following equations:

$$r = \frac{R}{R+G+B}, \quad g = \frac{G}{R+G+B}, \quad b = \frac{B}{R+G+B}$$

$$(2.1.15)$$

The chromaticity coordinates are plotted with *r* as abscissa and *g* as ordinate. Because $r + g + b = 1$, it is not necessary to plot *b* because it can be derived by $b = 1 - r - g$. This (r, g) diagram, and others like it, are known as *chromaticity diagrams*.

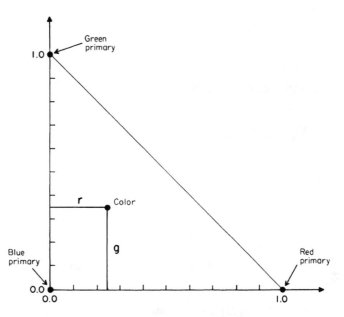

Figure 2.1.6 A chromaticity diagram. The amounts of the three primaries needed to match a given color are proportional to r, g, and b ($= 1 - r - g$).

2.1.2i   Center of Gravity Law

The chromaticity diagram is a useful way of representing additive color mixture. Consider two stimuli $\mathbf{C}_1$ and $\mathbf{C}_2$:

$$\mathbf{C}_1 \equiv R_1\mathbf{R} + G_1\mathbf{G} + B_1\mathbf{B} \tag{2.1.16}$$

$$\mathbf{C}_2 \equiv R_2\mathbf{R} + G_2\mathbf{G} + B_2\mathbf{B} \tag{2.1.17}$$

As explained previously, primary stimulus $\mathbf{C}_1$ is matched by R_1 units of primary stimulus $\mathbf{R}$, mixed with G_1 units of primary stimulus $\mathbf{G}$, and B_1 units of primary stimulus $\mathbf{B}$. In a similar manner, stimulus $\mathbf{C}_2$ is matched by R_2, G_2, and B_2 units of the same primaries. It follows that the chromaticity coordinates are

$$r_1 = \frac{R_1}{R_1 + G_1 + B_1}, \quad g_1 = \frac{G_1}{R_1 + G_1 + B_1}, \quad b_1 = \frac{B_1}{R_1 + G_1 + B_1} \tag{2.1.18}$$

The sum of $R_1 + G_1 + B_1 = T_1$, the total tristimulus value. Then

$$r_1 = \frac{R_1}{T_1}, \quad g_1 = \frac{G_1}{T_1}, \quad b_1 = \frac{B_1}{T_1}$$

$$(2.1.19)$$

After reordering the equations it follows that

$$R_1 = r_1 T_1, \quad G_1 = g_1 T_1, \quad B_1 = b_1 T_1$$

$$(2.1.20)$$

In a similar manner

$$R_2 = r_2 T_2, \quad G_2 = g_2 T_2, \quad B_2 = b_2 T_2$$

$$(2.1.21)$$

In terms of chromaticity coordinates (r, g, and b), the equations for $\mathbf{C}_1$ and $\mathbf{C}_2$ may be written as

$$\mathbf{C}_1 = (r_1 T_1)\mathbf{R} + (g_1 T_1)\mathbf{G} + (b_1 T_1)\mathbf{B} \qquad (2.1.22)$$

$$\mathbf{C}_2 = (r_2 T_2)\mathbf{R} + (g_2 T_2)\mathbf{G} + (b_2 T_2)\mathbf{B} \qquad (2.1.23)$$

Thus, by Grassmann's laws, the stimulus $\mathbf{C}$ formed by mixing $\mathbf{C}_1$ and $\mathbf{C}_2$ is

$$\mathbf{C} = R\mathbf{R} + G\mathbf{G} + B\mathbf{B} \qquad (2.1.24)$$

Where:
$R = r_1 T_1 + r_2 T_2$
$G = g_1 T_1 + g_2 T_2$
$B = b_1 T_1 + b_2 T_2$

The chromaticity coordinates of the mixture, therefore, are

$$r = \frac{r_1 T_1 + r_2 T_2}{T_1 + T_2}, \quad g = \frac{g_1 T_1 + g_2 T_2}{T_1 + T_2}, \quad b = \frac{b_1 T_1 + b_2 T_2}{T_1 + T_2}$$

$$(2.1.25)$$

The interpretation of this math in the chromaticity diagram is simply that the mixture lies on the straight lining joining the two components and divides it in the ratio T_2/T_1. This concept, illustrated in Figure 2.1.7, is known as the *center of gravity law* because of the analogy with the center of gravity of weights T_1 and T_2 placed at the points representing $\mathbf{C}_1$ and $\mathbf{C}_2$.

2.1.2j   Alychne

As stated previously, the luminance of a stimulus with tristimulus values R, G, and B is

$$L = L_R R + L_G G + L_B B$$

(2.1.26)

If this expression is divided by $R + G + B$ and if $L = 0$, then

$$0 = L_R r + L_G g + L_B b$$

(2.1.27)

The foregoing is the equation of a straight line in the chromaticity diagram. It is the line along which colors of zero luminance would lie if they could exist. The line is called the *alychne*.

The alychne is illustrated in Figure 2.1.8, which is a chromaticity diagram based on monochromatic primaries of wavelengths 700.0, 546.1, and 435.8 nm with their units normalized so that equal amounts are required to match a stimulus in which the spectral concentration of radiant power per unit wavelength is constant throughout the visible spectrum (this stimulus is called the *equienergy stimulus*). The alychne lies wholly outside the triangle of primaries, as indeed it must, for no positive combination of real primaries can possibly have zero luminance.

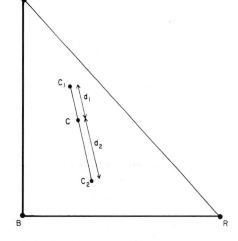

Figure 2.1.7 The center of gravity law in the chromaticity diagram. The additive mixture of color stimuli represented by C_1 and C_2 lies at C, whose location on the straight line $C_1 C_2$ is given by $d_1 T_1 = d_2 T_2$, where T_1 and T_2 are the total tristimulus values of the component stimuli.

2.1.2k   Spectrum Locus

Because all color stimuli are mixtures of radiant energy of different wavelengths, it is interesting to plot, in a chromaticity diagram, the points representing monochromatic stimuli (stimuli consisting of a single wavelength). These can be calculated from the color-matching functions

$$r(\lambda) = \frac{\bar{r}(\lambda)}{\bar{r}(\lambda) + \bar{g}(\lambda) + \bar{b}(\lambda)}, \quad g(\lambda) = \frac{\bar{g}(\lambda)}{\bar{r}(\lambda) + \bar{g}(\lambda) + \bar{b}(\lambda)}$$

(2.1.28)

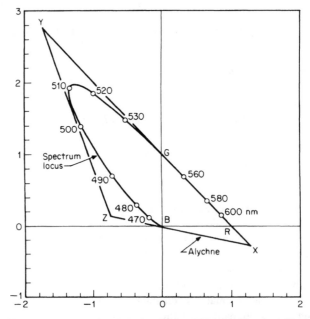

Figure 2.1.8 The spectrum locus and alychne of the CIE 1931 Standard Observer plotted in a chromaticity diagram based on matching stimuli of wavelengths 700.0, 546.1, and 435.8 nm. The locations of the CIE primary stimuli *X, Y,* and *Z* are shown.

When these spectral chromaticity coordinates are plotted, as shown in Figure 2.1.8, they lie along a horseshoe-shaped curve called the *spectrum locus*. The extremities of the curve correspond to the extremities of the visible spectrum—approximately 380 nm for the blue end and 780 nm for the red end. The straight line joining the extremities is called the *purple boundary* and is the locus of the most saturated purples obtainable.

Because all color stimuli are combinations of spectral stimuli, it is apparent from the center of gravity law that all real color stimuli must lie on, or inside, the spectrum locus.

It is an experimental fact that the part of the spectrum locus lying between 560 and 780 nm is substantially a straight line. This means that broad-band colors in the yellow-orange-red region can give rise to colors of high saturation.

2.1.2l   Subjective and Objective Quantities

It is important to distinguish clearly between perceptual (subjective) terms and psychophysical (objective) terms. Perceptual terms relate to attributes of sensations of light and color. They indicate subjective magnitudes of visual responses. Examples are hue, saturation, brightness, and lightness.

Psychophysical terms relate to objective measures of physical variables which identify stimuli that produce equal visual responses under specified viewing conditions. Examples include tristimulus values, luminance, and chromaticity coordinates.

Table 2.1.1 Perceptual Terms and their Psychophysical Correlates

Perceptual (Subjective)	Psychophysical (Objective)
Hue	Dominant wavelength
Saturation	Excitation purity
Brightness	Luminance
Lightness	Luminous reflectance or luminous transmittance

Psychophysical terms are usually chosen so that they correlate in an approximate way with particular perceptual terms. Examples of some of these correlations are given in Table 2.1.1.

2.1.3   References

1.   Smith, V. C., and J. Pokorny: "Spectral Sensitivity of the Foveal Cone Pigments Between 400 and 500 nm," *Vision Res.*, vol. 15, pp. 161–171, 1975.

2.   Boynton, R.M.: *Human Color Vision*, Holt, New York, N.Y., p. 404, 1979.

3.   Judd, D. B., and G. Wyszencki: *Color in Business, Science, and Industry,*. 3rd ed., Wiley, New York, N.Y., pp. 44-45, 1975.

4.   Wyszecki, G., and W. S. Stiles: *Color Science*, 2nd ed., Wiley, New York, N.Y., 1982.

5.   Nickerson, D.: "History of the Munsell Color System, Company and Foundation, I," *Color Res. Appl.,* vol. 1, pp. 7–10, 1976.

6.   Nickerson, D.: "History of the Munsell Color System, Company and Foundation, II: Its Scientific Application," *Color Res. Appl.*, vol. 1, pp. 69–77, 1976.

7.   Nickerson, D.: "History of the Munsell Color System, Company and Foundation, III," *Color Res. Appl.,* vol. 1, pp. 121–130, 1976.

8.   Newhall, S. M., D. Nickerson, and D. B. Judd: "Final Report of the OSA Subcommittee on the Spacing of the Munsell Colors," *Journal of the Optical Society of America*, vol. 33, pp. 385–418, 1943.

9.   *Munsell Book of Color*, Munsell Color Co., 2441 No. Calvert Street, Baltimore, MD 21218.

10.   Wright, W. D.: "A Redetermination of the Trichromatic Coefficients of the Spectral Colours," *Trans. Opt. Soc.*, vol. 30, pp. 141–164, 1928–1929.

11.   Guild, J.: "The Colorimetric Properties of the Spectrum," *Phil. Trans. Roy. Soc. A.*, vol. 230, pp. 149–187, 1931.

12.   Foley, James D., et al.: *Computer Graphics: Principles and Practice*, 2nd ed., Addison-Wesley, Reading, Mass., pp. 584–592, 1991.

13.   Smith, A. R.: "Color Gamut Transform Pairs," *SIGGRAPH 78*, 12–19, 1978.

2.1.4 Bibliography

"Colorimetry," Publication no. 15, Commission Internationale de l'Eclairage, Paris, 1971.

Kaufman, J. E. (ed.): *IES Lighting Handbook—1981 Reference Volume*, Illuminating Engineering Society of North America, New York, N.Y., 1981.

Tektronix application note #21W-7165: "Colorimetry and Television Camera Color Measurement," Tektronix, Beaverton, Ore., 1992.

Wright, W. D.: *The Measurement of Colour*, 4th ed., Adam Hilger, London, 1969.

The CIE Color System

Alan R. Robertson, Joseph F. Fisher, Jerry C. Whitaker[1]

2.2.1 Introduction

In 1931, the International Commission on Illumination (known by the initials CIE for its French name, Commission Internationale de l'Eclairage) defined a set of color-matching functions and a coordinate system that have remained the predominant, international, standard method of specifying color to this day.

The CIE system deals with the three fundamental aspects of color experience:

- The object

- The light source that illuminates the object

- The observer

The light source is important because color appearance varies considerably in different lighting conditions. Generally, the *observer* is the person viewing the color, but it can also be a camera. Working from experiments conducted in the late 1920s, the CIE derived a set of color matching functions (the *Standard Observer*) that mathematically describe the sensitivity of the average human eye with normal color vision.

2.2.2 The CIE System

The color-matching functions of the initial CIE effort were based on experimental data from many observers measured by Wright [1] and Guild [2]. Wright and Guild used different sets of primaries, but the results were transformed to a single set, namely, monochromatic stimuli of wavelengths 700.0, 546.1, and 435.8 nm. The units of the stimuli were chosen so that equal amounts were needed to match an *equienergy stimulus* (constant radiant power per unit wavelength throughout the visible spectrum). Figure 2.2.1 shows the spectrum locus in the (r, g) chromaticity diagram based on these color-matching functions.

---

1. From *Standard Handbook of Video and Television Engineering*, 4th. ed., Jerry C. Whitaker (ed.), McGraw-Hill, New York, N.Y., 2003. Used with permission.

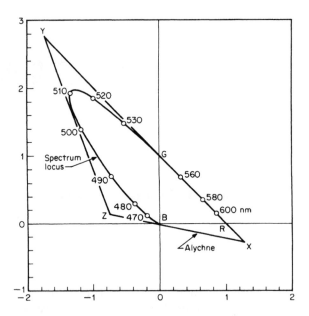

Figure 2.2.1 The spectrum locus and alychne of the CIE 1931 Standard Observer plotted in a chromaticity diagram based on matching stimuli of wavelengths 700.0, 546.1, and 435.8 nm. The locations of the CIE primary stimuli *X, Y,* and *Z* are shown.

At the same time it adopted these color-matching functions as a standard, the CIE also introduced and standardized a new set of primaries involving some ingenious concepts. The set of real physical primaries were replaced by a new set of imaginary nonphysical primaries with special characteristics. These new primaries are referred to as **X**, **Y**, and **Z**, and the corresponding tristimulus values as X, Y, and Z. The chromaticity coordinates of **X**, **Y**, and **Z** in the **RGB** system are shown in Figure 2.2.1. Primaries **X** and **Z** lie on the alychne and hence have zero luminance. All the luminance in a mixture of these three primaries is contributed by **Y**.

This convenient property depends only on the decision to locate **X** and **Z** on the alychne. It still leaves a wide choice of locations for all three primaries. The actual locations chosen by the CIE (illustrated in Figure 2.2.1) were based on the following additional considerations:

- The spectrum locus lies entirely within the triangle **XYZ**. This means that negative amounts of the primaries are never needed to match real colors. The color-matching functions $\bar{x}(\lambda), \bar{y}(\lambda), \bar{z}(\lambda)$ (shown in Figure 2.2.2) are therefore all positive at all wavelengths.

- The line $Z = 0$ (the line from **X** to **Y**) lies along the straight portion of the spectrum locus. Z is effectively zero for spectral colors with wavelengths greater than about 560 nm.

- The line $X = 0$ (the line from **Y** to **Z**) was chosen to minimize (approximately) the area of the **XYZ** triangle outside the spectrum locus. This choice led to a bimodal shape for the $\bar{x}(\lambda)$ color-matching function because the spectrum locus curves away from the line $X = 0$ at low wavelengths. See Figure 2.2.1. A different choice of $X = 0$ (tangential to the spectrum locus at

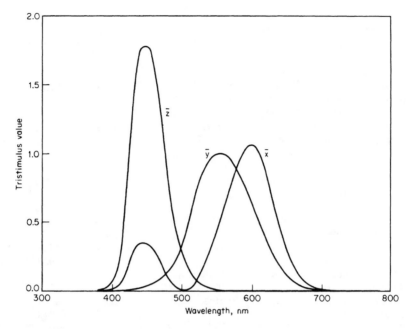

Figure 2.2.2 CIE 1931 color-matching functions.

about 450 nm) would have eliminated the secondary lobe of $\bar{x}(\lambda)$ but would have pushed **Y** much further from the spectrum locus.

- The units of **X**, **Y**, and **Z** were chosen so that the tristimulus values X, Y, and Z would be equal to each other for an equienergy stimulus.

This coordinate system and the set of color-matching functions that go with it are known as the CIE 1931 Standard Observer.

The color-matching data on which the 1931 Standard Observer is based were obtained with a visual field subtending 2° at the eye. Because of the slight nonuniformities of the retina, color-matching functions for larger fields are slightly different. In 1964, this prompted the CIE to recommend a second Standard Observer, known as the CIE 1964 Supplementary Standard Observer, for use in colorimetric calculations when the field size is greater than 4°.

2.2.2a   Color-Matching Functions

The color-matching functions of the CIE Standard Observer, shown in Figure 2.2.2, are listed in Table 2.2.1. They are used to calculate tristimulus specifications of color stimuli and to determine whether two physically different stimuli will match each other. Such calculated matches represent the results of the average of many observers, but may not represent an exact match for any single real observer. For most purposes, this restriction is not important; the match of an average observer is all that is required.

Table 2.2.1 CIE Colorimetric Data (1931 Standard Observer)

Wave-length (mm)	Trichromatic Coefficients		Distribution Coefficients, Equal-energy Stimulus			Energy Distributions for Standard Illuminants			
	r	g	$\bar{r}$	$\bar{g}$	$\bar{b}$	E_A	E_B	E_C	E_{D65}
380	0.0272	−0.0115	0.0000	0.0000	0.0012	9.80	22.40	33.00	49.98
390	0.0263	−0.0114	0.0001	0.0000	0.0036	12.09	31.30	47.40	54.65
400	0.0247	−0.0112	0.0003	0.0001	0.0121	14.71	41.30	63.30	82.75
410	0.0225	−0.0109	0.0008	−0.0004	0.0371	17.68	52.10	80.60	91.49
420	0.0181	−0.0094	0.0021	−0.0011	0.1154	20.99	63.20	98.10	93.43
430	0.0088	−0.0048	0.0022	−0.0012	0.2477	24.67	73.10	112.40	86.68
440	−0.0084	0.0048	−0.0026	0.0015	0.3123	28.70	80.80	121.50	104.86
450	−0.0390	0.0218	0.0121	0.0068	0.3167	33.09	85.40	124.00	117.01
460	0.0909	0.0517	−0.0261	0.0149	0.2982	37.81	88.30	123.10	117.81
470	−0.1821	0.1175	−0.0393	0.0254	0.2299	42.87	92.00	123.80	114.86
480	−0.3667	0.2906	−0.0494	0.0391	0.1449	48.24	95.20	123.90	115.92
490	−0.7150	0.6996	−0.0581	0.0569	0.0826	53.91	96.50	120.70	108.81
500	1.1685	1.3905	0.0717	0.0854	0.0478	59.86	94.20	112.10	109.35
510	−1.3371	1.9318	−0.0890	0.1286	0.0270	66.06	90.70	102.30	107.80
520	−0.9830	1.8534	−0.0926	0.1747	0.0122	72.50	89.50	96.90	104.79
530	−0.5159	1.4761	0.0710	0.2032	0.0055	79.13	92.20	98.00	107.69
540	0.1707	1.1628	0.0315	0.2147	0.0015	85.95	96.90	102.10	104.41
550	0.0974	0.9051	0.0228	0.2118	−0.0006	92.91	101.00	105.20	104.05
560	0.3164	0.6881	0.0906	0.1970	−0.0013	100.00	102.80	105.30	100.00
570	0.4973	0.5067	0.1677	0.1709	−0.0014	107.18	102.60	102.30	96.33
580	0.6449	0.3579	0.2543	0.1361	0.0011	114.44	101.00	97.80	95.79
590	0.7617	0.2402	0.3093	0.0975	−0.0008	121.73	99.20	93.20	88.69
600	0.8475	0,1537	0.3443	0.0625	−0.0005	129.04	98.00	89.70	90.01
610	0.9059	0.0494	0.3397	0.0356	0.0003	136.35	98.50	88.40	89.60
620	0.9425	0.0580	0.2971	0.0183	−0.0002	143.62	99.70	88.10	87.70
630	0.9649	0.0354	0.2268	0.0083	−0.0001	150.84	101.00	88.00	83.29
640	0.9797	0.0205	0.1597	0.0033	0.000	157.98	102.20	87.80	83.70
650	0.9888	0.0113	0.1017	0.0012	0.0000	165.03	103.90	88.20	80.03
660	0.9940	0.0061	0.0593	0.0004	0.0000	171.96	105.00	87.90	80.21
670	0.9966	0.0035	0.0315	0.0001	0.0000	178.77	104.90	86.30	82.28
680	0.9984	0.0016	0.0169	0.0000	0.0000	185.43	103.90	84.00	78.28
690	0.9996	0.0004	0.0082	0.0000	0.0000	191.93	101.60	80.20	69.72
700	1.0000	0.0000	0.0041	0.0000	0.0000	198.26	99.10	76.30	71.61
710	1.0000	0.0000	0.0021	0.0000	0.0000	204.41	96.20	72.40	74.15
720	1.0000	0.0000	0.0011	0.0000	0.0000	210.36	92.90	68.30	61.60
730	1.0000	0.0000	0.0005	0.0000	0.0000	216.12	89.40	64.40	69.89
740	1.0000	0.0000	0.0003	0.0000	0.0000	221.67	86.90	61.50	75.09
750	1.0000	0.0000	0.0001	0.0000	0.0000	227.00	85.20	59.20	63.59
760	1.0000	0.0000	0.0001	0.0000	0.0000	232.12	84.70	58.10	46.42
770	1.0000	0.0000	0.0000	0.0000	0.0000	237.01	85.40	58.20	66.81
780	1.0000	0.0000	0.0000	0.0000	0.0000	241.68	87.00	59.10	63.38

Table 2.2.1 (continued)

Wave length (mm)	Trichromatic Coefficients		Distribution Coefficients, Equal-energy Stimulus			Distribution Coefficients Weighted by Illuminant C		
	x	y	$\bar{x}$	$\bar{y}$	$\bar{z}$	$E_C\bar{x}$	$E_C\bar{y}$	$E_C\bar{z}$
380	0.1741	0.0050	0.0014	0.0000	0.0065	0.0036	0.0000	0.0164
390	0.1738	0.0049	0.0042	0.0001	0.0201	0.0183	0.0004	0.0870
400	0.1733	0,0048	0.0143	0.0004	0.0679	0.0841	0.0021	0.3992
410	0.1726	0.0048	0.0435	0.0012	0.2074	0.3180	0.0087	1.5159
420	0.1714	0.0051	0.1344	0.0040	0.6456	1.2623	0.0378	6.0646
430	0.1689	0.0069	0.2839	0.0116	1.3856	2.9913	0.1225	14.6019
440	0.1644	0.0109	0.3483	0.0230	1.7471	3.9741	0.2613	19.9357
450	0.1566	0.0177	0.3362	0.0380	1.7721	3.9191	0.4432	20.6551
460	0.1440	0.0297	0.2908	0.0600	1.6692	3.3668	0.6920	19.3235
470	0.1241	0.0578	0.1954	0.0910	1.2876	2.2878	1.0605	15.0550
480	0.0913	0.1327	0.0956	0.1390	0.8130	1.1038	1.6129	9.4220
490	0.0454	0.2950	0.0320	0.2080	0.4652	0.3639	2.3591	5.2789
500	0.0082	0.5384	0.0049	0.3230	0.2720	0.0511	3.4077	2.8717
510	0.0139	0.7502	0.0093	0.5030	0.1582	0.0898	4.8412	1.5181
520	0.0743	0.8338	0.0633	0.7100	0.0782	0.5752	6.4491	0.7140
530	0.1547	0.8059	0.1655	0.8620	0.0422	1.5206	7.9357	0.3871
540	0.2296	0.7543	0.2904	0.9540	0.0203	2.7858	9.1470	0.1956
550	0.3016	0.6923	0.4334	0.9950	0.0087	4.2833	9.8343	0.0860
560	0.3731	0.6245	0.5945	0.9950	0.0039	5.8782	9.8387	0.0381
570	0.4441	0.5547	0.7621	0.9520	0.0021	7.3230	9.1476	0.0202
580	0.5125	0.4866	0.9163	0.8700	0.0017	8.4141	7.9897	0.0147
590	0.5752	0.4242	1.0263	0.7570	0.0011	8.9878	6.6283	0.0101
600	0.6270	0.3725	1.0622	0.6310	0.0008	8.9536	5.3157	0.0067
610	0.6658	0.3340	1.0026	0.5030	0.0003	8.3294	4.1788	0.0029
620	0.6915	0.3083	0.8544	0.3810	0.0002	7.0604	3.1485	0.0012
630	0.7079	0.2920	0.6424	0.2650	0.0000	5,3212	2.1948	0.0000
640	0,7190	0.2809	0.4479	0.1750	0.0000	3.6882	1.4411	0.0000
650	0.7260	0.2740	0.2835	0.1070	0.0000	2.3531	0.8876	0.0000
660	0.7300	0.2700	0.1649	0,0610	0.0000	1.3589	0.5028	0.0000
670	0.7320	0.2680	0.0874	0.0320	0.0000	0.7113	0.2606	0.0000
680	0.7334	0.2666	0.0468	0.0170	0.0000	0.3657	0.1329	0.0000
690	0.7344	0.2656	0.0227	0.0082	0.0000	0.1721	0.0621	0.0000
700	0.7347	0.2653	0.0114	0.0041	0.0000	0.0806	0.0290	0.0000
710	0.7347	0,2653	0.0058	0.0021	0.0000	0.0398	0.0143	0.0000
720	0.7347	0.2653	0.0029	0.0010	0.0000	0.0183	0.0064	0.0000
730	0.7347	0.2653	0.0014	0.0005	0.0000	0.0085	0.0030	0.0000
740	0.7347	0.2653	0.0007	0.0003	0.0000	0.0040	0.0017	0.0000
750	0.7347	0.2653	0.0003	0.0001	0.0000	0.0017	0.0006	0.0000
760	0.7347	0.2653	0.0002	0.0001	0.0000	0.0008	0.0003	0.0000
770	0.7347	0.2653	0.0001	0.0000	0.0000	0.0003	0.0000	0.0000
780	0.7347	0.2653	0.0000	0.0000	0.0000	0.0000	0.000	0.0000

2.2.2b   Tristimulus Values and Chromaticity Coordinates

By exact analogy with the calculation of the tristimulus values R, G, B, the tristimulus values X, Y, Z of a stimulus $L_e(\lambda)$ are calculated by

$$X = \int_{380}^{780} L_e(\lambda)\bar{x}(\lambda)\,d\lambda \tag{2.2.1}$$

$$Y = \int_{380}^{780} L_e(\lambda)\bar{y}(\lambda)\,d\lambda \tag{2.2.2}$$

$$Z = \int_{380}^{780} L_e(\lambda)\bar{z}(\lambda)\,d\lambda \tag{2.2.3}$$

The chromaticity coordinates x, y are then calculated by

$$x = \frac{X}{X+Y+Z}, y = \frac{Y}{X+Y+Z} \tag{2.2.4}$$

The chromaticity coordinates x, y are plotted as rectangular coordinates to form the CIE 1931 chromaticity diagram, as shown in Figure 2.2.3.

It is important to remember that the CIE chromaticity diagram is not intended to illustrate appearance. The CIE system tells only whether two stimuli match in color, not what they look like. Appearance depends on many factors not taken into account in the chromaticity diagram. Nevertheless, it is often useful to know approximately where colors lie on the diagram. Figure 2.2.4 gives some color names for various parts of the diagram based on observations of self-luminous areas against a dark background.

2.2.2c   Conversion Between Two Systems of Primaries

To transform tristimulus specifications from one system of primaries to another, it is necessary and sufficient to know—in one system—the tristimulus values of the primaries of the other system. For example, consider two systems $\mathbf{R}$, $\mathbf{G}$, $\mathbf{B}$ (in which tristimulus values are represented by R, G, B and chromaticity coordinates by r, g and $\mathbf{R'}$ $\mathbf{G'}$ $\mathbf{B'}$ (in which tristimulus values are represented by R' G' B' and chromaticity coordinates by r' g'). If one system is defined in terms of the other by the match equations

$$\mathbf{R'} \equiv a_{11}\mathbf{R} + a_{21}\mathbf{G} + a_{31}\mathbf{B} \tag{2.2.5}$$

$$\mathbf{G'} \equiv a_{12}\mathbf{R} + a_{22}\mathbf{G} + a_{32}\mathbf{B} \tag{2.2.6}$$

$$\mathbf{B'} \equiv a_{13}\mathbf{R} + a_{23}\mathbf{G} + a_{33}\mathbf{B} \tag{2.2.7}$$

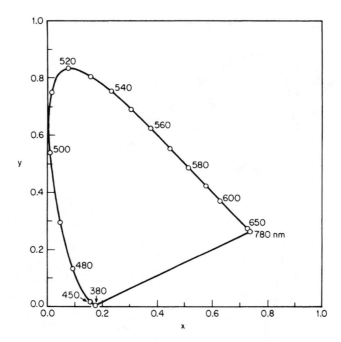

Figure 2.2.3 The CIE 1931 chromaticity diagram showing spectrum locus and wavelengths in nanometers.

Then the following equations can be derived to relate the tristimulus values and chromaticity coordinates of a color stimulus measured in one system to those of the same color stimulus measured in the other system:

$$R = a_{11}R' + a_{12}G' + a_{13}B' \tag{2.2.8}$$

$$G = a_{21}R' + a_{22}G' + a_{23}B' \tag{2.2.9}$$

$$B = a_{31}R' + a_{32}G' + a_{33}B' \tag{2.2.10}$$

$$R' = b_{11}R + b_{12}G + b_{13}B \tag{2.2.11}$$

$$G' = b_{21}R + b_{22}G + b_{23}B \tag{2.2.12}$$

$$B' = b_{31}R + b_{32}G + b_{33}B \tag{2.2.13}$$

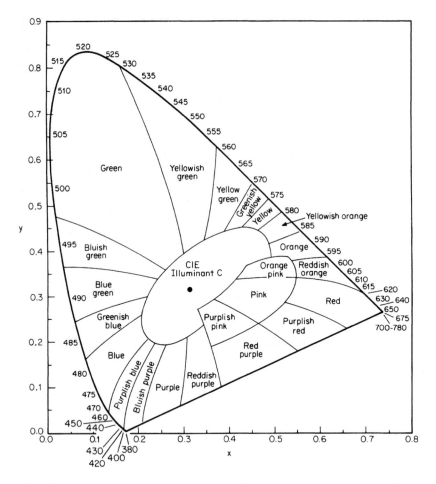

Figure 2.2.4 The CIE 1931 chromaticity diagram divided into various color names derived from observations of self-luminous areas against a dark background. (*After* [3].)

$$r = \frac{\alpha_{11}r' + \alpha_{12}g' + \alpha_{13}}{t}$$

$$(2.2.14)$$

$$g = \frac{\alpha_{21}r' + \alpha_{22}g' + \alpha_{23}}{t}$$

$$(2.2.15)$$

$$t = \alpha_{31}r' + \alpha_{32}g' + \alpha_{33}$$

$$(2.2.16)$$

$$r' = \frac{\beta_{11}r + \beta_{12}g + \beta_{13}}{t'}$$

$$(2.2.17)$$

$$g' = \frac{\beta_{21}r + \beta_{22}g + \beta_{23}}{t'}$$

$$(2.2.18)$$

$$t' = \beta_{31}r + \beta_{32}g + \beta_{33}$$

$$(2.2.19)$$

These equations are known as *projective transformations*, which have the property of retaining straight lines as straight lines. In other words, a straight line in the r, g diagram will transform to a straight line in the r', g' diagram. Another important property is that the center of gravity law continues to apply. The derivation of the CIE **XYZ** color-matching functions and chromaticity diagram from the corresponding data in the **RGB** system is an example of the use of the transformation equations previously given.

2.2.2d   Luminance Contribution of Primaries

Because **X** and **Z** were chosen to be on the alychne, their luminances are zero. Thus, all the luminance of a mixture of **X**, **Y**, and **Z** primaries is contributed by **Y**. This means that the Y tristimulus value is proportional to the luminance of the stimulus.

2.2.2e   Standard Illuminants

The CIE has recommended a number of standard illuminants $E(\lambda)$ for use in evaluating the tristimulus values of reflecting and transmitting objects. Originally, in 1931, it recommended three—known as A, B, and C. These illuminants are specified by tables of relative spectral distribution and were chosen so that they could be reproduced by real physical sources. (CIE terminology distinguishes between *illuminants*, which are tables of numbers, and *sources*, which are physical emitters of light.) The sources are defined as follows:

- **Source A**. A tungsten filament lamp operating at a color temperature of about 2856K. Its chromaticity coordinates are $x = 0.4476$, and $y = 0.4074$. Source A represents incandescent light.

- **Source B**. A source with a composite filter made of two liquid filters of specified chemical composition [4]. The chromaticity coordinates of source B are $x = 0.3484$ and $y = 0.3516$. Source B represents noon sunlight.

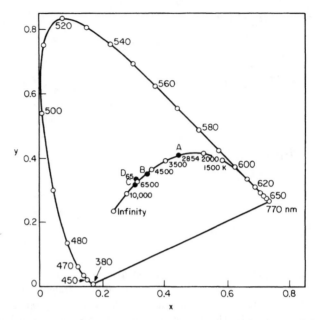

Figure 2.2.5 The relative spectral power distributions of CIE standard illuminants A, B, C, and D$_{65}$.

- **Source C**. This source is also produced by source A with two liquid filters [4]. Its chromaticity coordinates are $x = 0.3101$ and $y = 0.3162$. Source C represents average daylight according to information available in 1931.

In 1971, the CIE introduced a new series of standard illuminants that represented daylight more accurately than illuminants B and C [5]. The improvement is particularly marked in the ultraviolet part of the spectrum, which is important for fluorescent samples. The most important of the D illuminants is D$_{65}$ (sometimes written D6500), which has chromaticity coordinates of $x = 0.3127$ and $y = 0.3290$.

The relative spectral power distributions of illuminants A, B, C, and D$_{65}$ are given in Figure 2.2.5.

2.2.2f   Gamut of Reproducible Colors

In a system that seeks to match or reproduce colors with a set of three primaries, only those colors can be reproduced that lie inside the triangle of primaries. Colors outside the triangle cannot be reproduced because they would require negative amounts of one or two of the primaries.

In a color-reproducing system, it is important to have a triangle of primaries that is sufficiently large to permit a satisfactory gamut of colors to be reproduced. To illustrate the kinds of requirements that must be met, Figure 2.2.6 shows the maximum color gamut for real surface colors and the triangle of typical color television receiver phosphors as standardized by the European Broadcasting Union (EBU). These are shown in the CIE 1976 u', v' chromaticity diagram in

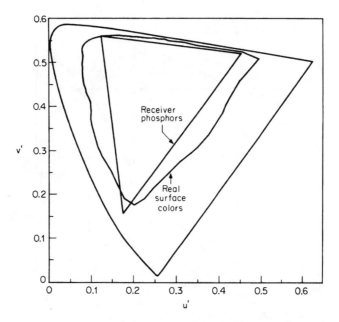

Figure 2.2.6 The color triangle defined by a standard test of color television receiver phosphors compared with the maximum real color gamut on a u′, v′ chromaticity diagram. (*After* [6].)

which the perceptual spacing of colors is more uniform than in the x, y diagram. High-purity blue-green and purple colors cannot be reproduced by these phosphors, whereas the blue phosphor is actually of slightly higher purity than any real surface colors.

2.2.2g   Vector Representation

In the preceding sections the representation of color has been reduced to two dimensions by eliminating consideration of quantity (luminance) and discussing only chromaticity. Because color requires three numbers to specify it fully, a three-dimensional representation can be made, taking the tristimulus values as vectors. For the sake of simplicity, this discussion will be confined to CIE tristimulus values and to a rectangular framework of coordinate axes.

Tristimulus values have been treated as scalar quantities. They may be transformed into vector quantities by multiplying by unit vectors $\mathbf{i}$, $\mathbf{j}$, and $\mathbf{k}$ in the x, y, and z directions. As a result, X, Y, and Z will become vector quantities $\mathbf{i}X$, $\mathbf{j}Y$, and $\mathbf{k}Z$. A color can then be represented by three vectors $\mathbf{i}X$, $\mathbf{j}Y$, and $\mathbf{k}Z$ along the x, y, and z coordinate axes, respectively. Combining these vectors gives a single resultant vector $\mathbf{V}$, represented by the *vector equation*

$$\mathbf{V} = \mathbf{i}X + \mathbf{j}Y + \mathbf{k}Z \tag{2.2.20}$$

The resultant is obtained by the usual vector methods, which are shown in Figure 2.2.7.

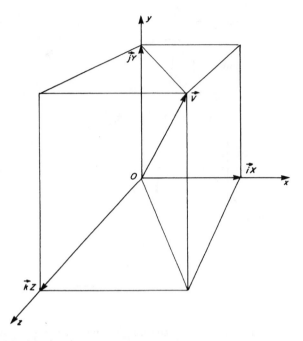

Figure 2.2.7 Combination of vectors.

Some implications of this form of color representations are illustrated in Figure 2.2.8. The diagram shows a vector **V** representing a color, a set of coordinate axes, and the plane $x + y + z = 1$ passing through the points $L(1,0,0)$, $M(0,1,0)$, and $N(0,0,1)$. The vector passes through point Q in this plane, having coordinates x, y, and z. Point P is the projection of point Q into the xy plane, and therefore has coordinates (x, y). From the geometry of the figure it can be seen that

$$\frac{X}{x} = \frac{Y}{y} = \frac{Z}{z} = \frac{X+Y+Z}{x+y+z}$$

$$(2.2.21)$$

Because Q is on the plane $x + y + z = 1$, it follows

$$\frac{X}{x} = \frac{Y}{y} = \frac{Z}{z} = X+Y+Z$$

$$(2.2.22)$$

Therefore

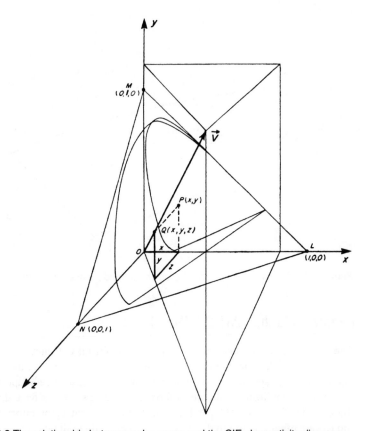

Figure 2.2.8 The relationship between color space and the CIE chromaticity diagram.

$$x = \frac{X}{X+Y+Z}, \quad y = \frac{Y}{X+Y+Z}$$

(2.2.23)

Thus, x and y are the CIE chromaticity coordinates of the color. Therefore, the xy plane in color space represents the CIE chromaticity diagram, when the vectors have magnitudes equal to the CIE tristimulus values and the axes are rectangular. The triangle LMN is a *Maxwell triangle* for CIE primaries. The spectrum locus in the xy plane can be thought of as defining a cylinder with generators parallel to z; this cylinder intersects the plane of the triangle LMN in the spectrum locus for the Maxwell triangle.

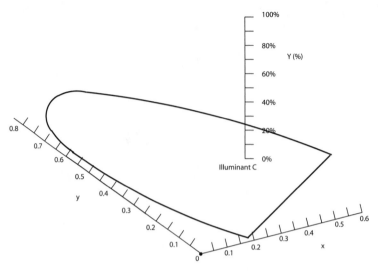

Figure 2.2.9 A drawing of the 1931 CIE color standard illustrating all three dimensions, x, y, and Y.

2.2.3   Refinements to the 1931 CIE Model

The CIE XYZ tristimulus values form the basis for all CIE numerical color descriptors. However, the *XYZ* values are not intuitive for most people. As a result, the CIE continued to develop and refine other approaches to color specifications[2]. Just as geographers use maps to represent geographic coordinates and other information, color scientists have developed two- and three-dimensional diagrams and graphic models to represent color information. Hence, the newer models are often also referred to as *color spaces*. A three-dimensional representation of chromaticity space is shown in Figure 2.2.9. The point at which the luminance axis (*Y*) touches the *x*, *y* plane is dependent upon the chromaticity coordinates of the illuminant or *white point* being referenced.

2.2.3a   Improved Visual Uniformity

The 1931 chromaticity diagram was developed primarily for color specification and was not intended to provide information on color appearance. Consequently, the system does not display perceptual uniformity. That is, colors do not appear to be equally spaced visually. For example, colors that consist of the same visually perceived hue or color family (those associated with a specific wavelength) do not follow straight lines within the diagram, but are curved instead. This non-uniformity is similar to the way a Mercator projection world map distorts what is truly represented on a globe. Another drawback to the 1931 diagram is that black, or the absence of color, does not have a unique position.

---

2. This section was adapted from: "Colorimetry and Television Camera Color Measurement," application note 21 W-7165, Tektronix, Beaverton, Ore., 1992. Used with permission.

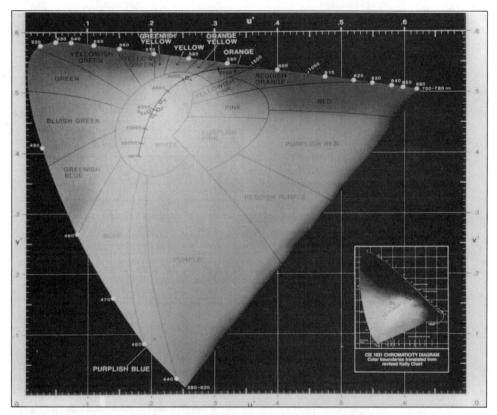

Figure 2.2.10 The 1976 CIE UCS diagram. The *u'*, *v'* chromaticity coordinates for any real color are located within the bounds the horse-shoe-shaped spectrum locus and the line of purples that joins the spectrum ends. (*Courtesy of Photo Research.*)

The search for a chromaticity diagram with greater visual uniformity resulted in the 1976 CIE *u'*, *v'* *uniform chromaticity scales* (UCS) diagram shown in Figure 2.2.10. The *u'* and *v'* chromaticity coordinates for any real color are located within the bounds of the spectrum locus and the line of purples that joins the spectrum ends. As with the *x* and *y* coordinates of the 1931 diagram, the *u'* and *v'* coordinates do not completely describe a color because they contain no information on its inherent lightness. The third dimension of color is again denoted by the tristimulus value *Y*, which represents the luminance factor. The *Y* axis position is perpendicular to the *u'*, *v'* plane, extending up from its surface.

The *u'*, *v'* coordinates of the 1976 UCS diagram can be derived from a simple transformation (defined by the CIE) of the 1931 *x*, *y* coordinates, or, more directly, from a transformation of the *XYZ* tristimulus values:

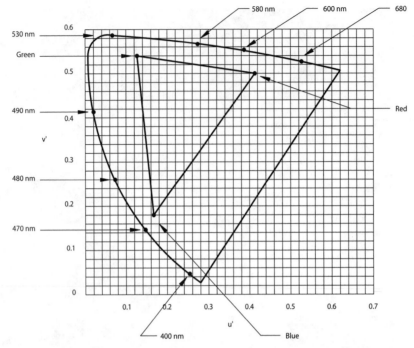

Figure 2.2.11 The triangle representing the range of chromaticities generally achievable using the additive mixture of typical red, green, and blue phosphors on a CRT display.

$$u' = \frac{4x}{-2x+12y+3}, \quad v' = \frac{9y}{-2x+12y+3}$$

$$(2.2.24)$$

These quantities can also be expressed as

$$u' = \frac{4X}{X+15y+3Z}, \quad v' = \frac{9Y}{X+15y+3Z}$$

$$(2.2.25)$$

The 1976 UCS diagram displays marked improvement over the 1931 x, y diagram in overall visual uniformity. It is not without flaws, but it enjoys wide acceptance. For video applications, the CIE 1976 UCS diagram is particularly useful because the region of the space delineated by the chromaticity limits of typical phosphor primaries falls within the most uniform region of the diagram, as illustrated in Figure 2.2.11. The most severe visual nonuniformities occur outside of the phosphor triangles and at the extreme limits of the chromaticity diagram.

2.2.3b   CIELUV

In an ideal color space, the numerical magnitude of a color difference should bear a direct relationship to the difference in color appearance. While the 1976 UCS diagram provides better visual uniformity than earlier approaches, it does not meet the critical goal of full visual uniformity, because uniform differences in chromaticity do not necessarily correspond to equivalent visual differences in Y or luminance. In addition, the diagram lacks a built-in capability to incorporate the important aspect of white reference, which significantly affects color appearance.

The CIE 1976 $L*$, $u*$, $v*$ (CIELUV) color space addresses these concerns. CIELUV is another tristimulus color space that uses three hues to describe a color. It integrates the CIE 1976 u', v' parameters with the 1976 metric lightness function, $L*$.

CIELUV is an *opponent-type color space*. Opponent-color theory is based on the assumption that the eye and brain encode all colors into light-dark, red-green, and yellow-blue signals. In a system of this type, colors are mutually exclusive in that a color cannot be red and green at the same time, or yellow and blue at the same time. A color such as purple can be described as red and blue, however, because these are not opponent colors.

In the CIELUV system, the $u*$ and $v*$ coordinate axes describe the chromatic attributes of color. The $u*$ axis represents the red-green coordinate, while the $v*$ axis represents yellow-blue. Positive values of $u*$ denote red colors, while negative values denote green. Similarly, positive values of $v*$ represent yellows and negative values signify blues. The $L*$ axis denotes variations in lightness or darkness and lies perpendicular to the $u*$, $v*$ plane. The achromatic or neutral colors (black, gray, white) lie on the $L*$ axis (the point where $u*$ and $v*$ intersect, $u* = 0$, $v* = 0$). Figure 2.2.12 shows the basic layout of the axes with respect to one another.

Calculating CIELUV coordinates requires a full chromaticity specification of the reference white point. For calculating CIELUV values in this section, all specifications are in the 1976 u', v', Y format. The following equations are standard u', v', Y to CIELUV transforms defined by the CIE:

$$L_s^* = \sqrt[3]{116\left(\frac{Y_s}{Y_w}\right)} - 16$$

$$(2.2.26)$$

for $(Y_s/Y_w) > 0.008856$

The transform can also be described as

$$L_s^* = 903.29\frac{Y_s}{Y_w}$$

$$(2.2.27)$$

for $(Y_s/Y_w) £ 0.008856$

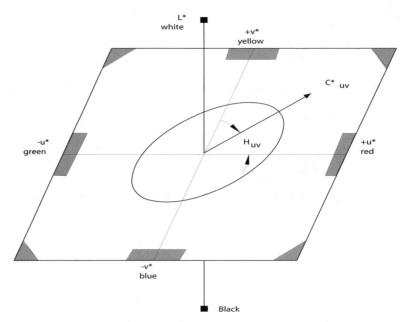

Figure 2.2.12 The CIELUV color space illustrating the relationship between the opponent color axes and the axis representing 1976 CIE metric lightness.

$$u_s^* = 13L^* \left(u_s^* - u_w^* \right), \quad v_s^* = 13L^* \left(v_s^* - v_w^* \right)$$

(2.2.28)

Note that the subscript s (Y_s) refers to color coordinates of the sample color, while coordinates with the subscript w (Y_w) are reference white.

Because each of the opponent coordinates is a function of metric lightness, the CIELUV color space has a unique location for black. As a result, when L^* is equal to zero, the coordinates u^* and v^* are also zero; thus, black lies at a single point on the L^* (neutral) axis.

The total CIELUV color space is represented by an irregularly shaped spheroid, illustrated in Figure 2.2.13. There are two reasons why the space approaches absolute limits at each end of the L^* scale. At the lower end of the color space L^* approaches zero. Thus, by definition, so do u^* and v^* because they are functions of L^*. This is what ensures a unique, single-point definition for black. The top portion of the CIELUV space converges for a different reason. Remember that in the CIE u', v' color standard, as Y increases, the limits on chromaticity become severe and less chromatic variation is possible. This results in a unique white point. The irregularity of the CIE-LUV color solid reflects the fact that certain colors are inherently capable of greater dynamic range than others.

In addition to the basic parameters of color, the CIE has also developed several *psychometric coordinates* designed to equate more with how color is perceived. In the CIELUV system, the two most often used are *psychometric hue angle* and *psychometric chroma*.

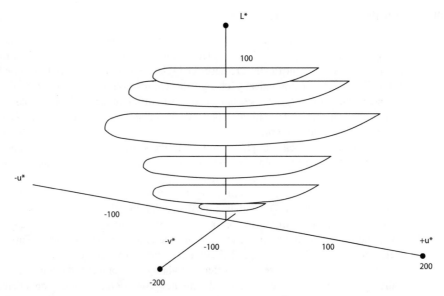

Figure 2.2.13 The CIELUV object-color solid showing constant lightness planes, $L^* = 10.0, 20.0, 40.0, 60.0, 80.0,$ and 90.0.

Psychometric hue angle (h_{uv}) represents an angle in the u^*, v^* plane that is correlated to a color family name or hue. The metric is defined as

$$h_{uv} = \tan^{-1} \frac{u^*}{v^*}$$

$$(2.2.29)$$

Psychometric chroma is a quantity that represents overall vibrancy or colorfulness. It is defined by

$$C_{uv}^* = \sqrt{u^{*2} + v^{*2}}$$

$$(2.2.30)$$

Currently, the CIELUV system and its related quantities are the best visually-uniform color space for additive colors accepted for international use. The deviations from uniformity found in the 1976 CIE u', v' diagram are more obvious in certain regions of color space than in others, particularly towards the limits of the spectrum locus. Fortunately, the gamut of achievable colors on a CRT lie in the most uniform region of the u', v' diagram. Furthermore, because the CIELUV system is based on the CIE u', v' parameters, it provides the same advantages to video-based implementations.

2.2.3c   Colorimetry and Color Targets

Color targets serve as a standardized method for describing specific colors. These targets generally provide reproduction of the resident colors within some well-controlled tolerance. Colors, when expressed scientifically, are described relative to a particular lighting condition. Changing the illuminant can dramatically change the color's composite spectral curve and, hence, the appearance of the color. These changes occur in different areas of the visible spectrum, depending on the character of the illuminant used. As a result, certain colors can be more affected by a shift in illuminant than others.

The colors in a color target are subject to the same laws of physics as any other colors. For example, the Macbeth color checker is widely used as a color reproduction standard in many industries and across applications, most notably as a means of assessing the color balance of photographic films. Like other color targets, however, this chart is standardized for only one specific viewing situation. The widely published color coordinate data for this chart is accurate only when the lighting environment approximates CIE Illuminant C ("average" daylight). The results will be erroneous if the target is used in other lighting environments. Furthermore, it is not enough to simply take the origin (Illuminant C-based) data and transpose it relative to a new illuminant. If the target is used in a different lighting environment, the only way to truly represent the appearance of the color target under the new condition is to obtain data that is representative of each chart color in the new environment.

2.2.4   Color Models

A color model is a specification of a three-dimensional color coordinate system. The model describes a visible subset in the system within which all colors in a particular color gamut lie. The RGB color model, for example, is the *unit cube subset* of the 3D Cartesian coordinate system. The purpose of a model is to permit convenient specification of colors within a given gamut, such as that for CRT monitors. The color gamut is a subset of all visible chromaticities. The model, therefore, can be used to specify all visible colors on a given display.

The primary color models of interest for display technology are:

- *RGB model*, used with color CRT devices

- *YIQ model*, used with conventional video (NTSC) systems

- *CMY model*, used in printing applications

- *HSV model*, used to describe color independent of a given hardware implementation

The RGB, YIQ, and CMY models are hardware-oriented systems. The HSV model, and variations of the model, are not based on a particular hardware system. Instead, they relate to the intuitive notions of hue, saturation, and brightness.

Conversion algorithms permit translation of one color model to another to facilitate comparison of color specifications.

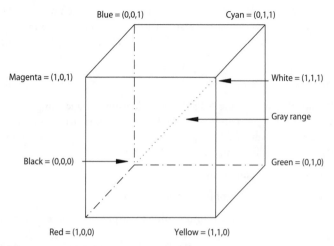

Figure 2.2.14 The RGB color model cube. (*After* [7].)

2.2.4a   RGB Color Model

The RGB (red, green, blue) color model is used to specify color CRT monitors. As illustrated in Figure 2.2.14, it is a unit cube subset of the 3D Cartesian coordinate system. As discussed previously, the RGB color model is additive. Combination of the three primary colors in the proper amounts yields white. As the figure shows, the monochrome (gray scale) vector stretches from black (0, 0, 0 primaries) to white (1, 1, 1 primaries). The color gamut covered by the RGB model is defined by the chromaticities of the CRT phosphors. Therefore, it follows that devices using different phosphors will have different color gamuts.

To convert colors specified in the gamut of one CRT to the gamut of another device, transformations M_1 and M_2 are used from the RGB color space of each monitor to the (X, Y, Z) color space. The form of each transformation is [7]

$$\begin{bmatrix} X \\ Y \\ Z \end{bmatrix} = \begin{bmatrix} X_r & X_g & X_b \\ Y_r & Y_g & Y_b \\ Z_r & Z_g & Z_b \end{bmatrix} \begin{bmatrix} R \\ G \\ B \end{bmatrix} \tag{2.2.31}$$

where X_n = the weights applied to the monitor RGB colors to find X, Y, and Z.

If M is defined as the 3×3 matrix of color-matching coefficients, the preceding equation can be written

$$\begin{bmatrix} X \\ Y \\ Z \end{bmatrix} = M \begin{bmatrix} R \\ G \\ B \end{bmatrix} \tag{2.2.32}$$

Given that M_1 and M_2 are matrices that convert from each of the CRT color gamuts to CIE, then $(M_2 - 1\ M_1)$ converts the RGB specification of monitor 1 to that of monitor 2. As long as the color in question lies in the gamut of both monitors, this matrix product is accurate.

2.2.4b   YIQ Model

The YIQ model describes the NTSC color television broadcasting system used in the U.S. and elsewhere. The NTSC system is optimized for efficient transmission of color information within a limited terrestrial bandwidth. A primary criteria for NTSC is compatibility with monochrome television receivers. The components of YIQ are:

- Y—the luminance component

- I and Q—the encoded chrominance components

The Y signal contains sufficient information to display a black-and-white picture of the encoded color signal.

The YIQ color model uses a 3D Cartesian coordinate system. RGB-to-YIQ mapping is defined by the following [7]

$$\begin{bmatrix} Y \\ I \\ Q \end{bmatrix} = \begin{bmatrix} 0.299 & 0.58 & 0.114 \\ 0.596 & -0.275 & -0.321 \\ 0.212 & -0.528 & 0.311 \end{bmatrix} \begin{bmatrix} R \\ G \\ B \end{bmatrix} \tag{2.2.33}$$

The quantities in the first row of the equation reflect the relative importance of green and red in producing brightness, and the smaller role that blue plays. The equation assumes that the RGB color specification is based on the standard NTSC RGB phosphor set. The CIE coordinates of the set are:

	Red	Green	Blue
x	0.67	0.21	0.14
y	0.33	0.71	0.08

The YIQ model capitalizes on two important properties of the human visual system:

- The eye is more sensitive to changes in luminance than to changes in hue or saturation

- Objects that cover a small part of the field of view produce a limited color sensation

These properties form the basis upon which the NTSC television color system was developed.

2.2.4c   CMY Model

Cyan, magenta, and yellow are complements of red, green, and blue, respectively. When used to subtract color from white light, they are referred to as *subtractive primaries*. The subset of the Cartesian coordinate system for the CMY model is identical to RGB except that white is the origin, rather than black.

In the CMY model, colors are specified by what is removed (subtracted) from white light, rather than what is added to a black screen. The CMY system is commonly used in printing applications, with the white light being that light reflected from paper. For example, when a por-

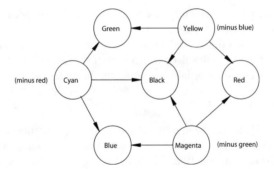

Figure 2.2.15 The CMY color model primaries and their mixtures. (*After* [7].)

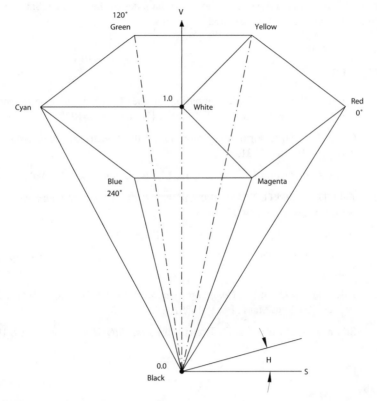

Figure 2.2.16 The single-hexcone HSV color model. The *V* = 1 plane contains the RGB model *R* =1, *G* = 1, and *B* = 1 planes in the regions illustrated. (*After* [8].)

tion of paper is coated with cyan ink, no red light is reflected from the surface. Cyan subtracts red from the reflected white light. The color relationship is illustrated in Figure 2.2.15. As shown, the subtractive primaries may be combined to produce a spectrum of colors. For example, cyan and yellow combine to produce green.

CMYK is a variation of the CMY model used for some color output devices and most color printing presses. *K* refers to the black component of the image. The addition of black to the process is particularly useful for printing applications, where text is almost always printed as black.

HSV Model

The HSV model [8] utilizes the intuitive elements of hue, saturation, and value to describe color. The coordinate system is cylindrical. The subset of the space within which the model is defined is a *hexcone* (six-sided pyramid), as illustrated in Figure 2.2.16. The top of the hexcone corresponds to $V = 1$, which contains the relatively bright colors. Hue (*H*) is measured by the angle around the vertical axis, with red at 0°, green at 120°, and blue at 240°. The complementary colors in the HSV hexcone are 180° opposite one another. The value of *S* is a ratio ranging from 0 on the center line (*V* axis) to 1 on the triangular sides of the hexcone. Saturation is measured relative to the color gamut represented by the model.

2.2.5   References

1. Wright, W. D.: "A Redetermination of the Trichromatic Coefficients of the Spectral Colours," *Trans. Opt. Soc.*, vol. 30, pp. 141–164, 1928–1929.

2. Guild, J.: "The Colorimetric Properties of the Spectrum," *Phil. Trans. Roy. Soc. A.*, vol. 230, pp. 149–187, 1931.

3. Kelly, K. L.: "Color Designations for Lights," *J. Opt. Soc. Am.*, vol. 33, pp. 627–632, 1943.

4. Judd, D. B., and G. Wyszencki: *Color in Business, Science, and Industry,*. 3rd ed., Wiley, New York, N.Y., pp. 44-45, 1975.

5. "Colorimetry," Publication no. 15, CIE, Paris, 1971.

6. Pointer, M. R.: "The Gamut of Real Surface Colours," *Color Res. Appl.*, vol. 5, pp. 145–155, 1980.

7. Foley, James D., et al.: *Computer Graphics: Principles and Practice*, 2nd ed., Addison-Wesley, Reading, Mass., pp. 584–592, 1991.

8. Smith, A. R.: "Color Gamut Transform Pairs," *SIGGRAPH 78*, 12–19, 1978.

2.3

Application of Visual Properties

K. Blair Benson, Jerry C. Whitaker[1]

2.3.1   Introduction

Advanced display systems improve on earlier techniques primarily by better utilizing the resources of human vision. The primary objective of an advanced display is to enhance the visual field occupied by the video image. In many applications this has called for larger, wider pictures that are intended to be viewed more closely than conventional video. In other applications this has called for miniature displays that serve specialized purposes. To satisfy the viewer at this closer inspection, the displayed image must possess proportionately finer detail and sharper outlines.

2.3.2   The Television System

Analog terrestrial broadcast television is the basis of all video display systems. It was the first electronic system to convert an image into electrical signals, encode them for transmission, and display a representative image of the original at a remote location.

The technology of television is based on the conversion of light rays from still or moving scenes and pictures into electronic signals for transmission or storage, and subsequent reconversion into visual images on a screen. A similar function is provided in the production of motion picture film. However, where film records the brightness variations of a complete scene on a single frame in a short exposure no longer than a fraction of a second, the elements of a television picture must be scanned one piece at a time. In the television system, a scene is dissected into a *frame* composed of a mosaic of *picture elements* (*pixels*). A pixel is defined as the smallest area of a television image that can be transmitted within the parameters of the system. This process is accomplished by:

- Analyzing the image with a photoelectric device in a sequence of *horizontal scans* from the top to the bottom of the image to produce an electric signal in which the brightness and color

---

1. From *Standard Handbook of Video and Television Engineering*, 4th. ed., Jerry C. Whitaker (ed.), McGraw-Hill, New York, N.Y., 2003. Used with permission.

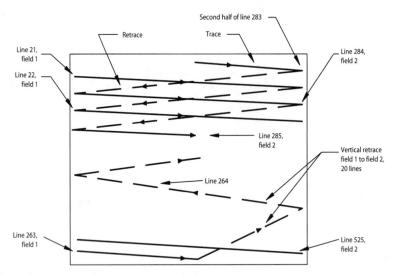

Figure 2.3.1 The interlace scanning pattern (raster) of the television image.

values of the individual picture elements are represented as voltage levels of a video waveform.

- Transmitting the values of the picture elements in sequence as voltage levels of a video signal.

- Reproducing the image of the original scene in a video signal display of parallel scanning lines on a viewing screen.

2.3.2a   Scanning Lines and Fields

The image pattern of electrical charges on a camera pickup element (corresponding to the brightness levels of a scene) are converted to a video signal in a sequential order of picture elements in the scanning process. At the end of each horizontal line sweep, the video signal is *blanked* while the beam returns rapidly to the left side of the scene to start scanning the next line. This process continues until the image has been scanned from top to bottom to complete one *field scan*.

After completion of this first field scan, at the midpoint of the last line, the beam again is blanked as it returns to the top center of the target where the process is repeated to provide a second field scan. The spot size of the beam as it impinges upon the target must be sufficiently fine to leave unscanned areas between lines for the second scan. The pattern of scanning lines covering the area of the target, or the screen of a picture display, is called a *raster*.

2.3.2b   Interlaced Scanning Fields

Because of the half-line offset for the start of the beam return to the top of the raster and for the start of the second field in the NTSC system, the lines of the second field lie in between the lines of the first field. Thus, the lines of the two are *interlaced*. The two interlaced fields constitute a single television *frame*. Figure 2.3.1 shows a frame scan with interlacing of the lines of two fields.

Reproduction of the camera image on a cathode ray tube (CRT) is accomplished by an identical operation, with the scanning beam modulated in density by the video signal applied to an element of the electron gun. This control voltage to the CRT varies the brightness of each picture element on the phosphor screen.

Blanking of the scanning beam during the return trace (*retrace*) is provided for in the video signal by a "blacker-than-black" pulse waveform. In addition, in most receivers and monitors another blanking pulse is generated from the horizontal and vertical scanning circuits and applied to the CRT electron gun to ensure a black screen during scanning retrace.

The interlaced scanning format, standardized for monochrome and compatible color, was chosen primarily for two partially related and equally important reasons:

- To eliminate viewer perception of the intermittent presentation of images, known as *flicker*.

- To reduce the video bandwidth requirements for an acceptable flicker threshold level.

The standards adopted by the FCC for monochrome television in the United States specified a system of 525 lines per frame, transmitted at a frame rate of 30 Hz, with each frame composed of two interlaced fields of horizontal lines. Initially in the development of television transmission standards, the 60 Hz power line waveform was chosen as a convenient reference for vertical scan. Furthermore, in the event of coupling of power line hum into the video signal or scanning/deflection circuits, the visible effects would be stationary and less objectionable than moving *hum bars* or distortion of horizontal-scanning geometry. In much of Europe, a 50 Hz interlaced system was chosen for many of the same reasons. With improvements in television receivers, the power line reference was replaced with a stable crystal oscillator.

The initial 525-line monochrome standard was retained for color in the recommendations of the NTSC for compatible color television. The NTSC system, adopted in 1953 by the FCC, specifies a scanning system of 525 horizontal lines per frame, with each frame consisting of two interlaced fields of 262.5 lines at a field rate of 59.94 Hz. Forty-two of the 525 lines in each frame are blanked as black picture signals and reserved for transmission of the vertical scanning synchronizing signal. This results in 483 visible lines of picture information.

2.3.2c   Synchronizing Video Signals

In monochrome television transmission, two basic synchronizing signals are provided to control the timing of picture-scanning deflection:

- Horizontal sync pulses at the line rate.

- Vertical sync pulses at the field rate in the form of an interval of wide horizontal sync pulses at the field rate. Included in the interval are *equalizing pulses* at twice the line rate to preserve interlace in each frame between the even and odd fields (offset by 1/2 line).

In color transmissions, a third synchronizing signal is added during horizontal scan blanking to provide a frequency and phase reference for signal encoding circuits in cameras and decoding circuits in receivers. These synchronizing and reference signals are combined with the picture video signal to form a *composite video waveform*.

The receiver scanning and color-decoding circuits must follow the frequency and phase of the synchronizing signals to produce a stable and geometrically accurate image of the proper color hue and saturation. Any change in timing of successive vertical scans can impair the interlace of even and odd fields in a frame. Small errors in horizontal scan timing of lines in a field can result in a loss of resolution in vertical line structures. Periodic errors over several lines that may be out of the range of the horizontal-scan automatic frequency control circuit in the receiver will be evident as jagged vertical lines.

2.3.2d   Television Industry Standards

There are three primary analog color transmission standards in use today:

- *NTSC* (National Television Systems Committee)—used in the United States, Canada, Central America, some of South America, and Japan. In addition, NTSC is used in various countries or possessions heavily influenced by the United States.

- *PAL* (Phase Alternate each Line)—used in the United Kingdom, most countries and possessions influenced by England, most European countries, and China. Variation exists in PAL systems.

- *SECAM* (SEquential Color with [Avec] Memory)—used in France, countries and possessions influenced by France, the USSR (generally the former Soviet Bloc nations, including East Germany) and other areas influenced by Russia.

The three standards are incompatible for a variety of reasons.

2.3.2e   Composite Video

The term *composite* is used to denote a video signal that contains:

- Picture luminance and chrominance information

- Timing information for synchronization of scanning and color signal processing circuits

The negative-going portion of the NTSC composite waveform, shown in Figure 2.3.2, is used to transmit information for synchronization of scanning circuits. The positive-going portion of the amplitude range is used to transmit luminance information representing brightness and—for color pictures—chrominance.

At the completion of each line scan in a receiver or monitor, a horizontal synchronizing (*H-sync*) pulse in the composite video signal triggers the scanning circuits to return the beam rapidly to the left of the screen for the start of the next line scan. During the return time, a horizontal blanking signal at a level lower than that corresponding to the blackest portion of the scene is added to avoid the visibility of the retrace lines. In a similar manner, after completion of each field, a vertical blanking signal blanks out the retrace portion of the scanning beam as it returns to the top of the picture to start the scan of the next field. The small-level difference between video reference black and the blanking level is called *setup*. Setup is used in the NTSC system as a guard band to ensure separation of the synchronizing and video-information functions, and to ensure adequate blanking of the scanning retrace lines on receivers.

The waveforms of Figure 2.3.3 show the various reference levels of video and sync in the composite signal. The unit of measurement for video level was specified initially by the Institute of Radio Engineers (IRE). These *IRE units* are still used to quantify video signal levels. Primary IRE values are given in Table 2.3.1.

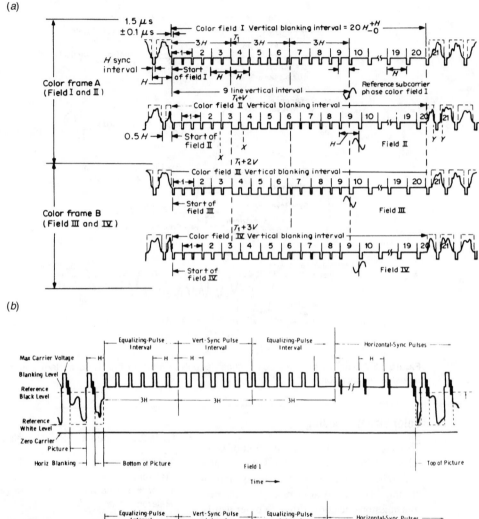

Figure 2.3.2 The NTSC color television waveform: (*a*) principle components, (*b*) detail of picture elements. (*Source: Electronic Industries Association.*)

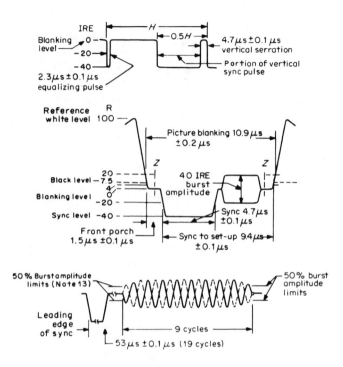

Figure 2.3.3 Detail of sync and color subcarrier pulse widths for the NTSC system. (*Source: Electronic Industries Association.*)

Color Signal Encoding

To facilitate an orderly introduction of color television broadcasting in the United States and other countries with existing monochrome services, it was essential that new transmissions be compatible. In other words, color pictures would provide acceptable quality on unmodified monochrome receivers. In addition, because of the limited availability of RF spectrum, another related requirement was fitting approximately 2 MHz bandwidth of color information into the 4.2 MHz video bandwidth of the then existing 6 MHz broadcasting channels with little or no modification of existing transmitters. This is accomplished using the band-sharing color system developed by the NTSC, and by taking advantage of the fundamental characteristics of the eye regarding color sensitivity and resolution.

Table 2.3.1 Video and Sync Levels in IRE Units

Signal Level	IRE Level
Reference white	100
Color burst sine wave peak	+20 to –20
Reference black	7.5
Blanking	0
Sync level	–40

The video-signal spectrum generated by scanning an image consists of energy concentrated near harmonics of the 15,734 Hz line scanning frequency. Additional lower amplitude sideband components exist at multiples of 59.94 Hz (the field scan frequency) from each line scan harmonic. Substantially no energy exists halfway between the line scan harmonics, that is, at odd harmonics of one-half the line frequency. Thus, these blank spaces in the spectrum are available for the transmission of a signal for carrying color information and its sideband. In addition, a signal modulated with color information injected at this frequency is of relatively low visibility in the reproduced image because the odd harmonics are of opposite phase on successive scanning lines and in successive frames, requiring four fields to repeat. Furthermore, the visibility of the color video signal is reduced further by the use of a subcarrier frequency near the cutoff of the video bandpass.

In the NTSC system, color is conveyed using two elements:

- A luminance signal

- A chrominance signal

The luminance signal is derived from components of the three primary colors, red, green, and blue in the proportions for *reference white, E_y*, as follows

$$E_y = 0.3E_R + 0.59E_G + 0.11E_B \qquad (2.3.1)$$

Where:
E_R = the red chrominance component
E_G = green chrominance component
E_B = blue chrominance component

These transmitted values equal unity for white and thus result in the reproduction of colors on monochrome receivers at the proper luminance level (the *constant-luminance* principle).

The color signal consists of two chrominance components, I and Q, transmitted as amplitude modulated sidebands of two 3.579545 MHz subcarriers in quadrature (differing in phase by 90°). The subcarriers are suppressed, leaving only the sidebands in the color signal. Suppression of the carriers permits demodulation of the components as two separate color signals in a receiver by reinsertion of a carrier of the phase corresponding to the desired color signal. This system for recovery of the color signals is called *synchronous demodulation*.

I and Q signals are composed of red, green, and blue primary color components produced by color cameras and other signal generators. The phase relationship among the I and Q signals, the derived primary and complementary colors, and the color synchronizing burst can be shown graphically on a *vectorscope* display. The horizontal and vertical sweep signals on a vectorscope are produced from $R - Y$ and $B - Y$ subcarrier sine waves in quadrature, producing a circular display. The chrominance signal controls the intensity of the display. A vectorscope display of an EIA color bar signal is shown in Figure 2.3.4.

Color Signal Decoding

Each of the two chrominance signal carriers can be recovered individually by means of synchronous detection. A reference subcarrier of the same phase as the desired chroma signal is

Figure 2.3.4 Vectorscope representation of vector and chroma amplitude relationships in the NTSC system for a color bars signal. (*Courtesy of Tektronix.*)

applied as a gate to a balanced demodulator. Only the modulation of the signal in the same phase as the reference will be present in the output. A lowpass filter can be added to remove second harmonic components of the chroma signal generated in the process.

2.3.2f   Deficiencies of Conventional Video Signals

The composite transmission of luminance and chrominance in a single channel is achieved in the NTSC system by choosing the chrominance subcarrier to be an odd multiple of one-half the line-scanning frequency. This causes the component frequencies of chrominance to be interleaved with those of luminance. The intent of this arrangement is to make it possible to easily separate the two sets of components at the receiver, thus avoiding interference prior to the recovery of the primary color signals for display.

In practice, this process has been fraught with difficulty. The result has been a substantial limitation on the horizontal resolution available in consumer receivers. Signal intermodulation arising in the bands occupied by the chrominance subcarrier signal produces degradations in the image known as *cross color* and *cross luminance*. Cross color causes a display of false colors to be superimposed on repetitive patterns in the luminance image. Cross luminance causes a crawling dot pattern that is primarily visible around colored edges.

These effects have been sufficiently prominent that manufacturers of NTSC receivers have tended to limit the luminance bandwidth to less than 3 MHz (below the 3.58 MHz subcarrier frequency). This is far short of the 4.2 MHz maximum potential of the broadcast signal. The end result is that the horizontal resolution in such receivers is confined to about 250 lines. The filtering typically employed to remove chrominance from luminance is a simple notch filter tuned to the subcarrier frequency.

Comb Filtering

It is clear that the quality of conventional receivers would improve if the signal mixture between luminance and chrominance could be substantially reduced, if not eliminated. This became possible with the development and widespread use of the *comb filter*. A common version of the comb filter consists of a glass fiber connected between two transducers—a video-to-acoustic transducer and an acoustic-to-video transducer. The composite signal fed to the input transducer produces an acoustic analog version of the signal, which reaches the far end of the fiber after an acoustic delay equal to one line-scan interval (63.555 µs for NTSC and 64 µs for PAL). The delayed electrical output version of the signal is then removed.

When the delayed output signal is added to the input, the sum represents luminance nearly devoid of chrominance content. This process is illustrated in Figure 2.3.5. Conversely, when the delayed output is subtracted from the input, the sum represents chrominance similarly devoid of luminance. When these signals are used to recover the primary-color information, cross-color and cross-luminance effects are largely removed. Newer versions of comb filters use charge-coupled devices that perform the same function without acoustical treatment.

Similar improvements in video quality can be realized through improved encoding techniques. Comb filters of advanced design have been produced that adapt their characteristics to changes in the image content. By these means, luminance/chrominance component separation is greatly increased. Filters using two or more line and/or field delays have been used. The greater the number of delays, the sharper the cutoff of the filter passband. Figure 2.3.6 shows a simplified diagram of a 2-*H* comb filter, and its luminance passband.

(*a*)

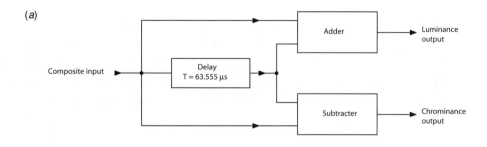

(*b*)

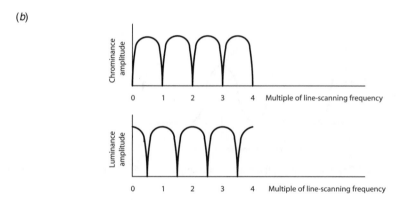

Figure 2.3.5 Comb filtering: (*a*) circuit introducing a one scan-line delay, (*b*) the luminance and chrominance passband.

2.3.3 Video Colorimetry

A video display can be regarded as a series of small visual colorimeters. In each picture element, a colorimetric match is made to an element of the original scene. The primaries are the red, green, and blue phosphors. The mixing occurs inside the eye of the observer because the eye cannot resolve the individual phosphor dots; they are too closely spaced. The outputs (R, G, B) of the three phosphors may be regarded as tristimulus values. Coefficients in the related equations depend on the chromaticity coordinates of the phosphors and on the luminous outputs of each phosphor for unit electrical input. Usually the gains of each of the three channels are set so that equal electrical inputs to the three produce a standard displayed white such as CIE illuminant D_{65}.

Therefore, a video camera must produce, for each picture element, three electrical signals representative of the three tristimulus values (R, G, B) of the required display. To accomplish this the system must have three optical channels with spectral sensitivities equal to the color-matching functions $\bar{r}(\lambda)$, $\bar{g}(\lambda)$, and $\bar{b}(\lambda)$ corresponding to the three primaries of the display.

Thus, the information to be conveyed by the electronic circuits comprising the camera, transmission system, and receiver/display is the amount of each of the three primaries (phosphors) required to match the input color. This information is based on the following items:

- An agreement concerning the chromaticities of the three primaries to be used.

(a)

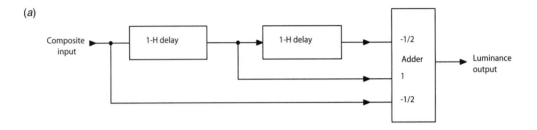

(b)

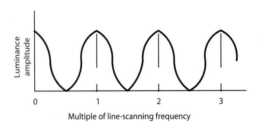

Figure 2.3.6 Comb filtering: (a) circuit introducing a two scan-line delay; (b) the luminance passband.

- The representation of the amounts of these three primaries by electrical signals suitably related to them.

- The specification (typically) that the electrical signals shall be equal at some specified chromaticity.

The electrical signal voltages are then representative of the tristimulus values of the original scene. They obey all the laws to which tristimulus values conform, including being transformable to represent the amounts of primaries of other chromaticities than those for which the signals were originally composed. Such transformations can be arranged by forming three sets of linear combinations of the original signals.

Unfortunately the simple objective of producing an exact colorimetric match between each picture element in the display and the corresponding element of the original scene is difficult to fulfill, and, in any case, may not achieve the ultimate objective of equality of appearance between the display and the original scene. There are several reasons for this, including:

- It may be difficult to achieve the luminance of the original scene because of the limitation of the maximum luminance that can be generated by the reproducing system.

- The adaptation of the eye may be different for the reproduction than it is for the original scene because the surrounding conditions are different.

- Ambient light complicates viewing the reproduced picture and changes its effective contrast ratio.

- The angle subtended by the reproduced picture may be different from that of the original scene.

Although an oversimplification, it is often considered that adequate reproduction is achieved when the chromaticity is accurately reproduced, while the luminance is reproduced *proportionally* to the luminance of the original scene. Even though the adequacy of this approach is somewhat questionable, it provides a starting point for system designers and enables the establishment of targets for system performance.

A block diagram of a simplified color television system is shown in Figure 2.3.7. Light reflected (or transmitted) by an object is split into three elements so that a portion strikes each camera tube, producing outputs proportional to the tristimulus values R, G, and B. Gain controls are provided so that the three signals can be made equal when the camera is viewing a standard white object. For transmission, the signals are encoded into three different waveforms and then decoded back to R, G, and B at the receiver. The primary purpose of encoding is to enable the signal to be transmitted within a limited bandwidth and to maintain compatibility with monochrome reception systems. The decoded R, G, and B signals are applied to the picture tube to excite the three phosphors. Gain controls are provided so equal inputs of R, G, and B will produce a standard white on the display.

2.3.3a Gamma

So far in this discussion, a linear relationship has been assumed between corresponding electrical and optical quantities in both the camera and the receiver. In practice, however, the *transfer function* is often not linear. For example, over the useful operating range of a typical color receiver, the light output of each phosphor follows a power-law relationship to the video voltage applied to the grid or cathode of the CRT. The light output (L) is proportional to the video-driving voltage (E_v) raised to the power γ

$$L = KE_v{}^{\gamma}$$

$$(2.3.2)$$

Where γ is typically about 2.5 for a color CRT. This produces black compression and white expansion. Compensation for these three nonlinear transfer functions is accomplished by three electronic *gamma correctors* in the color camera video processing amplifiers. Thus, the three signals that are encoded, transmitted, and decoded are not in fact R, G, and B, but rather R', G', and B', given by

$$R' = R^{1/\gamma}, \quad G' = G^{1/\gamma}, \quad B' = B^{1/\gamma}$$

$$(2.3.3)$$

If the rest of the system is linear, application of these signals to the color picture tube causes light outputs that are linearly related to the R, G, and B tristimulus inputs to the color camera, and so the correct reproduction is achieved.

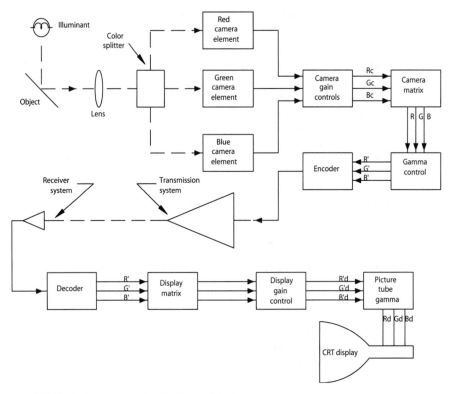

Figure 2.3.7 Block diagram of a simplified color television system.

2.3.3b   Display White

The NTSC signal specifications were designed so that equal signals ($R = G = B$) would produce a display white of the chromaticity of illuminant C. For many years, most home receivers (but not studio monitors) were set so that equal signals produced a much bluer white. The correlated color temperature was about 9300K and the chromaticity was usually slightly on the green side of the Planckian locus. The goal of this practice was to achieve satisfactorily high brightness and avoid excessive red/green current ratios with available phosphors. With modern phosphors, high brightness and red/green current equality can be achieved for a white at the chromaticity of D_{65} so that both monitors and receivers are now usually balanced to D_{65}. Because D_{65} is close to illuminant C, the color rendition is generally better than with the bluer balance of older display systems.

2.3.3c   Scene White

When the original scene is illuminated by daylight (of which D_{65} is representative), it is a clearly reasonable aim to reproduce the chromaticities of each object exactly in the final display. However, many video images are taken in a studio with incandescent illumination of about 3000K. In viewing the original scene, the eye adapts to a great extent so that most objects have similar appearance in both daylight and incandescent light. In particular, whites appear white under both types of illumination. However, if the chromaticities were to be reproduced exactly, studio whites would appear much yellower than the outdoor whites. This is because the viewer's adaptation is controlled more by the ambient viewing illuminant than by the scene illuminant and, therefore, does not correct fully for the change of scene illuminant. Because of this property, exact reproduction of chromaticities is not necessarily a good objective.

The ideal objective is for the reproduction to have exactly the same appearance as the original scene, but not enough is yet known about the chromatic adaptation of the human eye to define what this means in terms of chromaticity. A simpler criterion is to aim to reproduce objects with the same chromaticity they would have if the original scene were illuminated by D_{65}. This can be achieved by placing an optical filter (colored glass) in front of the camera with the spectral transmittance of the filter being equal to the ratio of the spectral power distributions of D_{65} and the actual studio illumination. As far as the camera is concerned, this has exactly the same effect as putting the same filter over every light source.

This solution has disadvantages because a different correction filter (in effect, a different set of camera sensitivities) is required for every scene illuminant. For example, every phase of daylight requires a special filter. In addition, insertion of a filter increases light scattering and can slightly degrade the contrast, resolution, and S/N.

An alternative method involves the adjustment of gain controls in the camera so that a white object produces equal signals in the three channels irrespective of the actual chromaticity of the illuminant. For colors other than white this solution does not produce exactly the same effect as a correction filter, but in practice it is satisfactory. Typically, the camera operator focuses a white reference of the scene inside a cursor on the monitor, and a microprocessor in a the camera performs white balancing automatically.

2.3.3d   Phosphor Chromaticities

The phosphor chromaticities specified by the NTSC in 1953 were based on phosphors in common use for color television displays at that time. Since then, different phosphors have been introduced, mainly to increase the brightness of displays. These modern phosphors, especially the green ones, have different chromaticities so that the gamut of reproducible chromaticities has been reduced. However, because of the increased brightness, the overall effect on color rendition has been beneficial. Figure 2.3.8 shows two sets of modern phosphors plotted on the CIE chromaticity diagram.

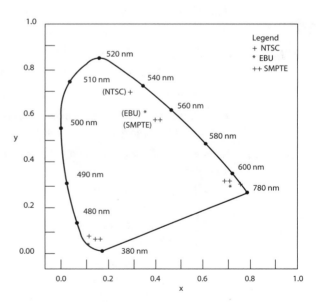

Figure 2.3.8 CIE 1931 chromaticity diagram showing three sets of phosphors used in color television displays.

2.3.4   Bibliography

Baldwin, M., Jr.: "The Subjective Sharpness of Simulated Television Images," *Proceedings of the IRE*, vol. 28, July 1940.

Belton, J.: "The Development of the CinemaScope by Twentieth Century Fox," *SMPTE Journal*, vol. 97, SMPTE, White Plains, N.Y., September 1988.

Benson, K. B., and D. G. Fink: *HDTV: Advanced Television for the 1990s*, McGraw-Hill, New York, N.Y., 1990.

Bingley, F. J.: "Colorimetry in Color Television—Pt. I," *Proc. IRE*, vol. 41, pp. 838–851, 1953.

Bingley, F. J.: "Colorimetry in Color Television—Pts. II and III," *Proc. IRE*, vol. 42, pp. 48–57, 1954.

Bingley, F. J.: "The Application of Projective Geometry to the Theory of Color Mixture," *Proc. IRE*, vol. 36, pp. 709–723, 1948.

DeMarsh, L. E.: "Colorimetric Standards in US Color Television," *J. SMPTE*, Society of Motion Picture and Television Engineers, White Plains, N.Y., vol. 83, pp. 1–5, 1974.

Epstein, D. W.: "Colorimetric Analysis of RCA Color Television System," *RCA Review*, vol. 14, pp. 227–258, 1953.

Fink, D. G.: "Perspectives on Television: The Role Played by the Two NTSCs in Preparing Television Service for the American Public," *Proceedings of the IEEE*, vol. 64, IEEE, New York, N.Y., September 1976.

Fink, D. G.: *Color Television Standards*, McGraw-Hill, New York, N.Y., 1955.

Fink, D. G.: *Color Television Standards*, McGraw-Hill, New York, N.Y., 1986.

Herman, S.: "The Design of Television Color Rendition," *J. SMPTE*, SMPTE, White Plains, N.Y., vol. 84, pp. 267–273, 1975.

Hubel, David H.: *Eye, Brain and Vision*, Scientific American Library, New York, N.Y., 1988.

Hunt, R. W. G.: *The Reproduction of Colour*, 3d ed., Fountain Press, England, 1975.

Judd, D. B.: "The 1931 C.I.E. Standard Observer and Coordinate System for Colorimetry," *Journal of the Optical Society of America*, vol. 23, 1933.

Kelly, K. L.: "Color Designation of Lights," *Journal of the Optical Society of America*, vol. 33, 1943.

Kelly, R. D., A. V. Bedbord and M. Trainer: "Scanning Sequence and Repetition of Television Images," *Proceedings of the IRE*, vol. 24, April 1936.

Neal, C. B.: "Television Colorimetry for Receiver Engineers," *IEEE Trans. BTR*, vol. 19, pp. 149–162, 1973.

Pearson, M. (ed.): *Proc. ISCC Conf. on Optimum Reproduction of Color*, Williamsburg, Va., 1971, Graphic Arts Research Center, Rochester, N.Y., 1971.

Pointer, R. M.: "The Gamut of Real Surface Colors, *Color Res. App.*, vol. 5, 1945.

Pritchard, D. H.: "U.S. Color Television Fundamentals—A Review," *IEEE Trans. CE*, vol. 23, pp. 467–478, 1977.

Sproson, W. N.: *Colour Science in Television and Display Systems*, Adam Hilger, Bristol, England, 1983.

Uba, T., K. Omae, R. Ashiya, and K. Saita: "16:9 Aspect Ratio 38V-High Resolution Trinitron for HDTV," *IEEE Transactions on Consumer Electronics*, IEEE, New York, N.Y., February 1988.

Wentworth, J. W.: *Color Television Engineering*, McGraw-Hill, New York, N.Y., 1955.

Wintringham, W. T.: "Color Television and Colorimetry," *Proc. IRE*, vol. 39, pp. 1135–1172, 1951.

Essential Video System Characteristics

Laurence J. Thorpe, Donald G. Fink[1]

2.4.1 Introduction

The central objective of any video service is to offer the viewer a sense of presence in the scene, and of participation in the events portrayed. To meet this objective, the video image should convey as much of the spatial and temporal content of the scene as is economically and technically feasible. Experience in the motion picture industry has demonstrated that a larger, wider picture, viewed closely, contributes greatly to the viewer's sense of presence and participation.

The deployment of HDTV services for consumer applications is directed toward this same end. From the visual point of view, the term "high-definition" is, to some extent, a misnomer as the primary visual objective of the system is to provide an image that occupies a larger part of the visual field. Higher definition is secondary; it need be no higher than is adequate for the closer scrutiny of the image.

2.4.2 Visual Acuity

Visual acuity is the ability of the eye to distinguish between small objects (and hence, to resolve the details of an image) and is expressed in reciprocal minutes of the angle subtended at the eye by two objects that can be separately identified. When objects and background are displayed in black and white (as in monochrome television), at 100 percent contrast, the range of visual acuity extends from 0.2 to about 2.5 reciprocal minutes (5 to 0.4 minutes of arc, respectively). An acuity of 1 reciprocal minute is usually taken as the basis of television system design. At this value, stationary white points on two scanning lines separated by an intervening line (the remainder of the scanning pattern being dark) can be resolved at a distance of about 20 times the picture height. Adjacent scanning lines, properly interlaced, cannot ordinarily be distinguished at distances greater than six or seven times the picture height.

Visual acuity varies markedly as a function of the following:

1. From *Standard Handbook of Video and Television Engineering*, 4th. ed., Jerry C. Whitaker (ed.), McGraw-Hill, New York, N.Y., 2003. Used with permission.

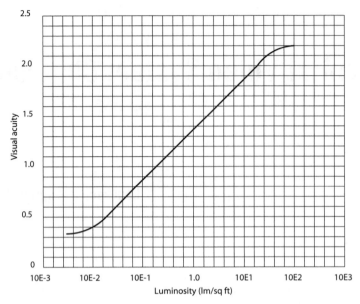

Figure 2.4.1 Visual acuity (the ability to resolve details of an image) as a function of the luminosity to which the eye is adapted. (*After R. J. Lythgoe.*)

- Luminance of the background
- Contrast of the image
- Luminance of the area surrounding the image
- Luminosity to which the eye is adapted

The acuity is approximately proportional to the logarithm of the background luminance, increasing from about 1 reciprocal minute at 1 ft·L (3.4 cd/m^2) background brightness and 100 percent contrast, to about 2 reciprocal minutes at 100 ft·L (340 cd/m^2). Figure 2.4.1 shows experimental data on visual acuity as a function of luminance, when the contrast is nearly 100 percent, using incandescent lamps as the illumination. These data apply to details viewed in the center of the field of view (by cone vision near the fovea of the retina). The acuity of rod vision, outside the foveal region, is poorer by a factor of about five times.

Visual acuity falls off rapidly as the contrast of the image decreases. Under typical conditions (1 ft·L background luminance) acuity increases from about 0.2 reciprocal minute at 10 percent contrast to about 1.0 reciprocal minute at 100 percent contrast, the acuity being roughly proportional to the percent contrast. Figure 2.4.2 shows experimental measurements of this effect.

When colored images are viewed, visual acuity depends markedly on the color. Acuity is higher for objects illuminated by monochromatic light than those illuminated by a source of the same color having an extended spectrum. This improvement results from the lack of chromatic aberration in the eye. In a colored image reproduced by primary colors, acuity is highest for the green primary and lowest for the blue primary. This is partly explained by the relative luminances of the primaries. The acuity for blue and red images is approximately two-thirds that for a

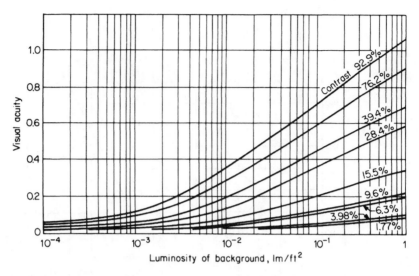

Figure 2.4.2 Visual acuity of the human eye as a function of luminosity and contrast (experimental data). (*After J. P. Conner and R. E. Ganoung.*)

white image of the same luminance. For the green primary, acuity is about 90 percent that of a white image of the same luminance. When the effect of the relative contributions to luminance of the standard FCC/NTSC color primaries (approximately, green:red:blue = 6:3:1) is taken into account, acuities for the primary images are in the approximate ratio green:red:blue = 8:3:1.

Visual acuity is also affected by *glare*, that is, regions in or near the field of view whose luminance is substantially greater than the object viewed. When viewing a video image, acuity fails rapidly if the area surrounding the image is brighter than the background luminance of the image. Acuity also decreases slightly if the image is viewed in darkness, that is, if the surround luminance is substantially lower than the average image luminance. Acuity is not adversely affected if the surround luminance has a value in the range from 0.1 to 1.0 times the average image luminance.

2.4.2a   Contrast Sensitivity

The ability of the eye to distinguish between the luminances of adjacent areas is known as *contrast sensitivity*, expressed as the ratio of the luminance to which the eye is adapted (usually the same as the background luminance) to the least perceptible luminance difference between the background luminance of the scene and the object luminance.

Two forms of contrast vision must be distinguished. In rod vision, which occurs at a background luminance of less than about 0.01 ft·L, the contrast sensitivity ranges from 2 to 5 (the average luminance is two to five times the least perceptible luminance difference), increasing slowly as the background luminance increases from 0.0001 to 0.01 ft·L. In cone vision, which takes place above approximately 0.01 ft·L, the contrast sensitivity increases proportionately to background luminance from a value of 30 at 0.01 ft·L to 150 at 10 ft·L. Contrast sensitivity does

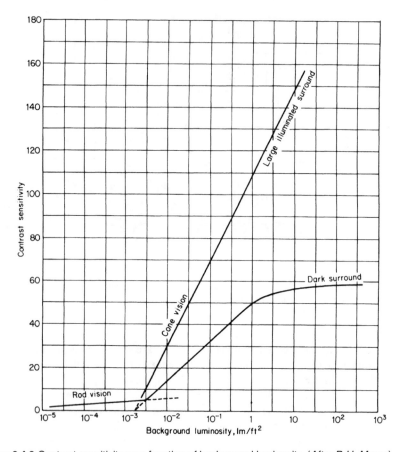

Figure 2.4.3 Contrast sensitivity as a function of background luminosity. (*After P. H. Moon.*)

not vary markedly with the color of the light. It is somewhat higher for blue light than for red at low background luminance. The opposite case applies at high background luminance levels. Contrast sensitivity, like visual acuity, is reduced if the surround is brighter than the background. Figures 2.4.3 and 2.4.4 illustrate experimental data on contrast sensitivity.

2.4.2b   Flicker

The perceptibility of flicker varies so widely with viewing conditions that it is difficult to describe quantitatively. In one respect, however, flicker phenomena can be readily compared. When the viewing situation is constant (no change in the image or surround other than a proportional change in the luminance of all their parts, and no change in the conditions of observation), the luminance at which flicker just becomes perceptible varies logarithmically with the luminance (the *Ferry-Porter law*). Numerically, a positive increment in the flicker frequency of 12.6 Hz raises the luminance flicker threshold 10 times. In video (NTSC) scanning, the applicable

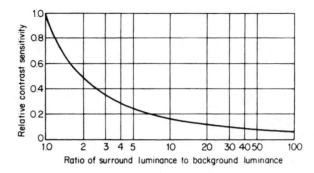

Figure 2.4.4 The effects of surround luminance on contrast sensitivity of the human eye. (*After P. H. Moon.*)

Table 2.4.1 Relative Flicker Threshold for Various Luminances
(The luminances tabulated have only relative significance; they are based on a value of 180 ft·L for a flicker frequency of 60 Hz, which is typical performance under NTSC.)

Flicker Frequency (Hz)	System	Frames/s	Flicker Threshold Luminance (ft·L)
48	Motion pictures	24	20
50	Television scanning	25	29
60	Television scanning	30	180

flicker frequency is the field frequency (the rate at which the area of the image is successively illuminated). Table 2.4.1 lists the flicker-threshold luminance for various flicker frequencies, based on 180 ft·L for 60 fields per second. The following factors affect the flicker threshold:

• The luminance of the flickering area

• Color of the area

• Solid angle subtended by the area at the eye

• Absolute size of the area

• Luminance of the surround

• Variation of luminance with time and position within the flickering area

• Adaptation and training of the observer

2.4.3 Foveal and Peripheral Vision

As stated previously, there are two areas of the retina of the eye that must be satisfied by video images: the *fovea* and the areas peripheral to it. Foveal vision extends over approximately one degree of the visual field, whereas the total field to the periphery of vision extends about 160°

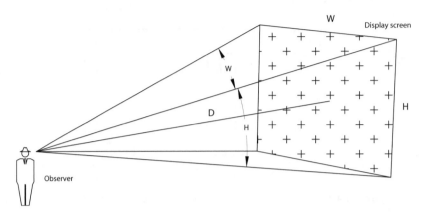

Figure 2.4.5 Geometry of the field of view occupied by a television image.

horizontally and 80° vertically. Motions of the eye and head are necessary to assure that the fovea is positioned on that part of the retinal image where the detailed structure of the scene is to be discerned.

The portion of the visual field outside the foveal region provides the remaining visual information. Thus, a large part of visual reality is conveyed by this extra-foveal region. The vital perceptions of extra-foveal vision, notably motion and flicker, have received only secondary attention in the development of video engineering. Attention has first been paid to satisfying the needs of foveal vision. This is true because designing a system capable of resolving fine detail presents the major technical challenge, requiring a transmission channel that offers essentially no discrimination in the amplitude or time delay among the signals carried over a wide band of frequencies.

The properties of peripheral vision impose a number of constraints. Peripheral vision has great sensitivity to even modest changes in brightness and position. Thus, the bright portions of a wide image viewed closely are more apt to flicker at the left and right edges than the narrower image of conventional systems.

Figure 2.4.5 illustrates the geometry of the field occupied by the video image. The viewing distance D determines the angle h subtended by the picture height. This angle is usually measured as the ratio of the viewing distance to the picture height (D/H). The smaller the ratio, the more fully the image fills the field of view.

The useful limit to the viewing range is that distance at which the eye can just perceive the finest details in the image. Closer viewing serves no purpose in resolving detail, while more distant viewing prevents the eye from resolving all the detailed content of the image. The preferred viewing distance, expressed in picture heights, is the *optimal viewing ratio*, commonly referred to as the *optimal viewing distance*. It defines the distance at which a viewer with normal vision would prefer to see the image, when pictorial clarity is the criterion.

The optimal viewing ratio is not a fixed value; it varies with subject matter, viewing conditions, and visual acuity of the viewer. It does serve, however, as a convenient basis for comparing the performance of conventional and advanced display systems.

Computation of the optimal viewing ratio depends upon the degree of detail offered in the vertical dimension of the image, without reference to its pictorial content. The discernible detail is limited by the number of scanning lines presented to the eye, and by the ability of those lines to present the image details separately.

Ideally, each detail of a given scene would be reproduced by one pixel. That is, each scanning line would be available for one picture element along any vertical line in the image. In practice, however, some of the details in the scene inevitably fall between scanning lines. Two or more lines, therefore, are required for such picture elements. Some vertical resolution is lost in the process. Measurements of this effect show that only about 70 percent of the vertical detail is presented by the scanning lines. This ratio is known as the *Kell factor*; it applies irrespective of the manner of scanning, whether the lines follow each other sequentially (progressive scan) or alternately (interlaced scan).

When interlaced scanning is used, the 70 percent figure applies only when the image is fully stationary and the line of sight from the viewer does not move vertically. In practice these conditions are seldom met, so an additional loss of vertical resolution, the *interlace factor*, occurs under typical conditions.

This additional loss depends upon many aspects of the subject matter and viewer attention. Under favorable conditions, the additional loss reduces the effective value of vertical resolution to not more than 50 percent; that is, no more than half the scanning lines display the vertical detail of an interlaced image. Under unfavorable conditions, a larger loss can occur. The effective loss also increases with image brightness because the scanning beam becomes larger.

Because interlacing imposes this additional detail loss, it was decided by some system designers early in the development of advanced systems to abandon interlaced scanning for image display. Progressive-scanned displays can be derived from interlaced transmissions by digital image storage techniques. Such scanning conversion improves the vertical resolution by about 40 percent.

2.4.3a   Horizontal Detail and Picture Width

Because the fovea is approximately circular in shape, its vertical and horizontal resolutions are nearly the same. This would indicate that the horizontal resolution of a display should be equal to its vertical resolution. Such equality is the usual basis of video system design, but it is not a firm requirement. Considerable variation in the shape of the picture element produces only a minor degradation in the sharpness of the image, provided that its area is unchanged. This being the case, image sharpness depends primarily on the *product* of the resolutions, that is, the total number of picture elements in the image.

The ability to extend the horizontal dimensions of the picture elements has been applied in wide-screen motion pictures. For example, the CinemaScope system used anamorphic optical projection to enlarge the image in the horizontal direction. Because the emulsion of the film had equal vertical and horizontal resolution, the enlargement lowered the horizontal resolution of the image.

When wide-screen motion pictures are displayed at the conventional aspect ratio, the full width of the film cannot be shown. This requires that the viewed area be moved laterally when scanned to keep the center of interest within the area displayed by the picture monitor. The interrelation of the aspect ratios for various services is shown in Figure 2.4.6.

The picture width-to-height ratio chosen for HDTV (widescreen) service is 1.777. This is a compromise with some of the wider aspect ratios of the film industry, imposed by constraints on

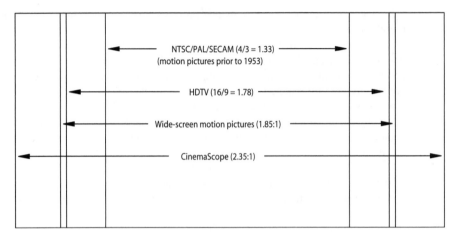

Figure 2.4.6 Comparison of the aspect ratios of television and motion pictures.

the bandwidth of the HDTV channel. Other factors being equal, the video baseband increases in direct proportion to the picture width.

The 16:9 aspect ratio of HDTV offers an optimal viewing distance of 3.3 times the picture height, and a viewing angle of 39°, which is about 20 percent of the total horizontal visual field. While this is a small percentage, it covers that portion of the field within which the majority of visual information is conveyed.

2.4.3b   Perception of Depth

Perception of depth depends on the angular separation of the images received by the eyes of the viewer. Successful binocular video systems have been produced, but their cost and inconvenience have restricted large scale adoption. A considerable degree of depth perception is inferred in the flat image of video from the following:

•   The perspective appearance of the subject matter.

•   Camera techniques through the choice of the focal length of lenses and by changes in depth of focus.

Continuous adjustment of focal length by the zoom lens provides the viewer with an experience in depth perception wholly beyond the scope of natural vision. No special steps have been taken in the design of advanced visual services to offer depth clues not previously available. However, the wide field of view offered by HDTV and similar displays significantly improves depth perception, compared with that of conventional systems.

2.4.4   Contrast and Tonal Range

The range of brightness accommodated by video displays is limited compared to natural vision. This limitation of video has not been overcome in advanced display systems. Moreover, the display brightness of wide-screen images may be restricted by the need to spread the available light over a large area.

Within the upper and lower limits of display brightness, the quality of the image depends on the relationship between changes in brightness in the scene and corresponding changes in the image. It is an accepted rule of image reproduction that the brightness be directly proportional to the source; the curve relating input and output brightness should be a straight line. Because the output versus input curves of cameras and displays are, in many cases, not straight lines, intermediate adjustment (*gamma correction*) is required. Gamma correction in wide-screen displays requires particular attention to avoid excessive noise and other artifacts evident at close scrutiny.

2.4.4a Chrominance Properties

Two aspects of color science are of particular interest with regard to advanced display systems. The first is the total range of colors perceived by the camera and offered by the display. The second is the ability of the eye to distinguish details presented in different colors. The *color gamut* defines the realism of the display, while attention to color acuity is essential to avoid presenting more chromatic detail than the eye can resolve.

In the earliest work on color television, equal bandwidth was devoted to each primary color, although it had long been known by color scientists that the color acuity of the normal eye is greatest for green light, less for red, and much less for blue. The NTSC system made use of this fact by devoting much less bandwidth to the red-minus-luminance channel, and still less to blue-minus-luminance. All color television services employ this device in one form or another. Properly applied, it offers economy in the use of bandwidth, without loss of resolution or color quality in the displayed image.

2.4.4b Temporal Factors in Vision

The response of the eye over time is essentially continuous. The cones of the retina convey nerve impulses to the visual cortex of the brain in pulses recurring about 1000 times per second. Fortunately, video display systems can take advantage of a temporal property of the eye, persistence of vision. The special properties of this phenomenon must be carefully observed to assure that motion of the scene is free from the effects of flicker in all parts of the image. A common conflict encountered in the design of advanced imaging systems is the need to reduce the rate at which some of the spatial aspects of the image are reproduced, thus conserving bandwidth while maintaining temporal integrity, particularly the position and shape of objects in motion. To resolve such conflicts, it has become standard practice to view the video signal and its processing simultaneously in three dimensions: width, height, and time.

Basic to video and motion pictures is the presentation of a rapid succession of slightly different frames. The human visual system retains the image during the dark interval. Under specific conditions, the image appears to be continuously present. Any difference in the position of an object from one frame to the next is interpreted as motion of that object. For this process to represent visual reality, two conditions must be met:

- The rate of repetition of the images must be sufficiently high so that motion is depicted smoothly, with no sudden jumps from frame to frame.

- The repetition rate must be sufficiently high so that the persistence of vision extends over the interval between flashes. Here an idiosyncrasy of natural vision appears: the brighter the flash, the shorter the persistence of vision.

Continuity of Motion

Early in the development of motion pictures it was found that movement could be depicted smoothly at any frame rate faster than about 15 frames/s. To accurately portray rapid motion, a worldwide standard of 24 frames/s was selected for the motion picture industry. This frame rate does not solve all the problems of reproducing fast action. Restrictive production techniques (such as limited camera angles and restricted rates of panning) must be observed when rapid motion is encountered.

The acuity of the eye when viewing objects in motion is impaired because the temporal response of the fovea is slower than the surrounding regions of the retina. Thus, a loss of sharpness in the edges of moving objects is an inevitable aspect of natural vision. This property represents an important component of video system design.

Much greater losses of sharpness and detail, under the general term *smear*, occur whenever the image moves across the sensitive surface of the camera. Each element in the surface then receives light not from one detail, but from a succession of them. The signal generated by the camera is the sum of the passing light, not that of a single detail, and smear results. As in photography, this effect can be reduced by using a short exposure, provided there is sufficient light relative to the sensitivity of the camera. Electronic shutters can be used to limit the exposure to 1/1000 s or less when sufficient light is available.

Another source of smear occurs if the camera response carries over from one frame scan to the next. The retained signal elements from the previous scan are then added to the current scan, and any changes in their relative position causes a misalignment and consequent loss of detail. Such *carry-over smear* occurs when the exposure occupies the full scan time. A similar carry-over smear can occur in the display, when the light given off by one line persists long enough to be appreciably present during the successive scan of that line. Such carry-over also helps reduce flicker in the display; there is room for compromise between flicker reduction and loss of detail in moving objects.

Flicker Effects

The process of interlaced scanning is designed to eliminate flicker in the displayed image. Perception of flicker is primarily dependent upon two conditions:

- The brightness level of an image

- The relative area of an image in a picture

The 30 Hz transmission rate for a full 525 line conventional video frame is comparable to the highly successful 24-frame-per-second rate of motion-picture film. However, at the higher brightness levels produced on video screens, if all 483 lines (525 less blanking) of an image were to be presented sequentially as single frames, viewers would observe a disturbing flicker in picture areas of high brightness. For comparison, motion-picture theaters, on average, produce a screen brightness of 10 to 25 ft·L, whereas a direct-view CRT may have a highlight brightness of 50 to 80 ft·L.

Through the use of interlaced scanning, single field images with one-half the vertical resolution capability of the 525 line system are provided at the high flicker-perception threshold rate of 60 Hz. Higher resolution of the full 490 lines (525 less vertical blanking) of vertical detail is provided at the lower flicker-perception threshold rate of 30 Hz. The result is a relatively flickerless picture display at a screen brightness of well over 50 to 75 ft·L, more than double that of motion-picture film projection. Both 60 Hz fields and 30 Hz frames have the same horizontal resolution capability.

The second advantage of interlaced scanning, compared to progressive scanning, is a reduction in video bandwidth for an equivalent flicker threshold level. Progressive scanning of 525 lines would have to be completed in 1/60 s to achieve an equivalent level of flicker perception. This would require a line scan to be completed in half the time of an interlaced scan. The bandwidth would then double for an equivalent number of pixels per line.

Interlace scanning, however, is the source of several degradations of image quality. While the total area of the image flashes at the rate of the field scan, twice that of the frame scan, the individual lines repeat at the slower frame rate. This gives rise to several degradations associated with the effect known as *interline flicker*. This causes small areas of the image, particularly when they are aligned horizontally, to display a shimmering or blinking that is visible at the usual viewing distance. A related effect is unsteadiness in extended horizontal edges of objects, as the edge is portrayed by a particular line in one field and by another line in the next. These effects become more pronounced as the vertical resolution provided by the camera is increased.

Interlacing also produces aberrations in the vertical and diagonal outlines of moving objects. This occurs because vertically-adjacent picture elements appear at different times in successive fields. An element on one line appears 1/60 s (actually 1/59.95 s) later than the vertically-adjacent element on the preceding field. If the objects in the scene are stationary, no adverse effects arise from this time difference. If an object is in rapid motion, however, the time delay causes the elements of the second field to be displaced to the right of, instead of vertically or diagonally adjacent to, those of the first field. Close inspection of such moving images shows that their vertical and diagonal edges are not sharp, but actually a series of step-wide serrations, usually coarser than the basic resolution of the image. Because the eye loses some acuity as it follows objects in motion, these serrations are often overlooked. They represent, however, an important impairment compared with motion picture images. All of the picture elements in a film frame are exposed and displayed simultaneously so the impairments resulting from interlacing do not occur.

As previously noted, the defects of interlacing have been an important target in advanced display system development. To avoid them, scanning at the camera must be progressive, using only one set of adjacent lines per frame. At the receiver, the display scan must match the camera.

2.4.5   Bibliography

Belton, J.: "The Development of the CinemaScope by Twentieth Century Fox," *SMPTE Journal*, vol. 97, SMPTE, White Plains, N.Y., September 1988.

Benson, K. B., and D. G. Fink: *HDTV: Advanced Television for the 1990s*, McGraw-Hill, New York, N.Y., 1990.

Fink, D. G., et. al.: "The Future of High Definition Television," *SMPTE Journal*, vol. 89, SMPTE, White Plains, N.Y., February/March 1980.

Fujio, T., J. Ishida, T. Komoto and T. Nishizawa: "High Definition Television Systems—Signal Standards and Transmission," *SMPTE Journal*, vol. 89, SMPTE, White Plains, N.Y., August 1980.

Miller, Howard: "Options in Advanced Television Broadcasting in North America," *Proceedings of the ITS*, International Television Symposium, Montreux, Switzerland, 1991.

Morizono, M.: "Technological Trends in High-Resolution Displays Including HDTV," *SID International Symposium Digest*, paper 3.1, May 1990.

Pitts, K. and N. Hurst: "How Much Do People Prefer Widescreen (16 × 9) to Standard NTSC (4 × 3)?," *IEEE Transactions on Consumer Electronics*, IEEE, New York, N.Y., August 1989.

Sproson, W. N.: *Colour Science in Television and Display Systems*, Adam Hilger, Bristol, England, 1983.

Uba, T., K. Omae, R. Ashiya, and K. Saita: "16:9 Aspect Ratio 38V-High Resolution Trinitron for HDTV," *IEEE Transactions on Consumer Electronics*, IEEE, New York, N.Y., February 1988.

van Raalte, John A.: "CRT Technologies for HDTV Applications," *1991 HDTV World Conference Proceedings*, National Association of Broadcasters, Washington, D.C., April 1991.

2.5

The Principles of Video Compression

Peter D. Symes, Jerry C. Whitaker[1]

2.5.1 Introduction

Compression is the science of reducing the amount of data used to convey information. There are a wide range of techniques available to accomplish this task—some quite simple, some very complex. Compression relies on the fact that information, by its very nature, is not random but exhibits order and patterning. If we can extract this order and patterning, we can often represent and transmit the information using less data than would be needed for the original. We can then reconstruct the original, or a close approximation to it, at the receiving point.

There are several families of compression techniques, fundamentally different in their approach to the problem—so different, in fact, that often they can be used sequentially to good advantage. Sophisticated compression systems use one or more techniques from each family to achieve the greatest possible reduction in data. The JPEG and MPEG standards are the most common for imaging applications.

One important categorization of compression systems is the degree of symmetry or asymmetry. For applications such as video conferencing, for example, there are similar numbers of transmitters and receivers, while for broadcast applications, there are far more receivers than transmitters. Table 2.5.1 compares various bit-rate-reduction applications.

2.5.2 Information and Data

One of the more important concerns in the technical world today is the handling of information. This information may be of many types—written text, the spoken word, music, still pictures, and moving pictures are just a few examples. Whatever the type of information, we can represent it by electrical signals or data, and transmit it or store it.

The more complex the information, the more data are needed to represent it. Plain text can be represented by eight bits per character, or about 20 kilobits for a page. CD-quality music requires

---

1. From *Standard Handbook of Video and Television Engineering*, 4th. ed., Jerry C. Whitaker (ed.), McGraw-Hill, New York, N.Y., 2003. Used with permission.

nearly 1500 kilobits for each second, and full motion 525-line component video needs over 200 megabits for each second.

When we need to transmit a given amount of data, we use a certain data bandwidth (measured in bits per second) for the time necessary to transmit all the bits. In the real world, there is always some restriction on the available bandwidth. Sometimes it is a practical restriction, like the use of a modem, that sets the requirements for compression. We could send high quality uncompressed audio data over a dial-up modem, but each second's worth of audio would take about 50 seconds to transmit. In other words, we would have to receive the data gradually, store it away, then play the resulting file at the correct rate to hear the sound. However, we may want to transmit sound so that we can listen to it as it arrives in "real time," in which case we could choose to reduce the quality that we are trying to send.

Another example of this type of requirement is the digital television system developed for North America. When the early work began in 1977, few thought it would be possible to transmit high-definition television (characterized by the 1125-line system) over a single 6 MHz channel as used for today's 525-line television. When digital techniques were considered, the problem looked even worse, as the high-definition signal represented over 1 gigabits/s of data. In the end, two teams of engineers combined to produce the answer: the transmission team designed a system to reliably deliver nearly 20 Mbits/s over a 6 MHz channel, and the compression engineers achieved data reduction of about sixty to one, while maintaining excellent picture quality.

Sometimes the motive for compression is purely financial. Bandwidth costs money, and so compression reduces the cost of transmission. A satellite link with an effective data bandwidth of 40 Mbits/s will earn more money if it can carry ten television channels instead of just one.

Transmission and storage of information may seem to be very different problems, but when considering the quantity of data, the economics are very much the same. Just as there is a cost for each bit transmitted, there is a cost for each bit stored in memory, on disk, or on tape. If we can use less data, both transmission and storage will be cheaper. There are some differences in how compressed data is best structured, but otherwise everything said about compression applies equally to transmission or storage. Table 2.5.1 lists the typical storage space requirements for various signal formats.

2.5.2a Signal Conditioning

It is extremely important to ensure that the signal or data stream we want to compress represents the information we want to transmit, and *only* the information we want to transmit. If there is additional information of any nature this will take bits to transmit, and result in fewer bits being available for the information we do need. Surplus information is irrelevant because the intended recipient(s) can make no use of it.

Surplus information can take many forms. It can be information in the original signal or data stream that exceeds the capabilities of the receiving device. For example, there is no point in transmitting more resolution than the receiving device can process and display.

Another form of surplus information is *artifacts*—features or elements of the input that are not truly part of the information. Noise is the most obvious example. Noise is by nature random or nearly so and this makes it essentially incompressible.

Many other types of artifact exist, ranging from filter ringing to film scratches. Some may seem trivial, but in the field of compression they can be very important. Compression relies on order and self-consistency in a signal, and artifacts damage or destroy this order.

Table 2.5.1 Storage Space Requirements for Audio and Video Signals (*After*[1].)

Media Signals	Specifications	Data Rate
Audio voice	1 ch; 8-bit @ 8 kHz	64 kbits/s
MPEG audio Layer II	1 ch; 16-bit @ 48 kHz	128 kbits/s
MPEG audio Layer III	1 ch; 16-bit @ 48 kHz	64 kbits/s
AC-3	6 ch; 16-bit @ 48 kHz	384 kbits/s
CD	2 ch; 16-bit @ 44.1 kHz	1.4 Mbits/s
AES/EBU	2 ch; 24-bit @ 48 kHz	3.07 Mbits/s
MPEG-1 (Video)	352 × 288, 30 f/s, 8-bit	1.5 Mbits/s
MPEG-2 (MP@ML)	720 × 576, 30 f/s, 8-bit	Max. 15 Mbits/s
MPEG-2 (4:2:2 P@ML)	720 × 608, 30 f/s, 8-bit	Max. 50 Mbits/s
CCIR-601	720 × 480, 30 f/s, 8-bit	216 Mbits/s
HDTV	1920 × 1080, 30 f/s, 8-bit	995 Mbits/s

This statement is just one aspect of a more general rule. Compression systems are designed for particular tasks, and make use of assumptions about the nature of the data being compressed.

Video is generally an environment where images are sampled according to the limitations of the Nyquist theorem. The theorem can be expressed as follows:

To ensure that a signal can be recovered from a series of samples, the sampling frequency must be more than double the highest frequency in the signal.

The frequency transforms widely used in video compression, particularly the *discrete cosine transform* (DCT), depend for their action on the fact that the information is band-limited, and that sampling was performed according to Nyquist. If these rules are broken, the transform does not perform as expected, and the compression system may fail.

Correctly used, signal conditioning can provide remarkable increases in efficiency at minimal cost. Equipment available today incorporates many different types of filters targeted at different types of artifacts. The benefits of appropriate conditioning are twofold: because the artifacts are unwanted, there is a clear advantage in avoiding the use of bits to transmit them; and, because the artifacts do not "belong," they generally break the rules or assumptions of the compression system. For this reason, artifacts do not compress well and use a disproportionately high number of bits to transmit.

2.5.2b   Lossless Compression

Some compression techniques are truly lossless. In other words, when we have compressed some data, we can reverse the process (de-compress) and get the same data we started with, exactly and precisely. Lossless compression works by removing *redundant* information—information that if removed can be re-created from the remaining data.

In one sense, this is the ultimate ideal of compression—there is no cost to using it other than the cost of the compression and decompression processes. Unfortunately, lossless compression suffers from two significant disadvantages. First, lossless compression typically offers only relatively small compression ratios, so used alone it often does not meet certain economic requirements. Second, the compression ratio is dependent on the input data. Used alone, lossless compression cannot guarantee a constant output data rate, which might be required for a given transmission channel.

For some applications, lossless is the only type of compression that can be used. If we wish to transmit a binary file such as a computer program, receiving an approximation to the original is of no use whatsoever. The program will only run if replicated exactly. Fortunately, many computer files have a high degree of order and patterning, so lossless compression techniques do yield very useful results. Compression programs like the well-known PKZIP use a combination of lossless techniques, and are particularly effective on graphics image files.

One advantage of lossless compression is that it can be applied to any data stream. Most video compression schemes use lossy techniques to achieve large degrees of compression. Lossless techniques are then applied to the resulting data stream to reduce the data rate even further.

The simplest form of lossless compression is *run length encoding*. Many data streams contain long runs of a specific value. For example, a graphics file will typically contain values for each pixel, ordered line by line. A red object will likely be represented by several long runs of the value "red." With appropriate coding, it can require much less data to say "Red 23 times" instead of "R R R R R R R R R R R R R R R R R R R R R R R." This is run length encoding—we code the length of the "run" of a certain value.

A more sophisticated type of lossless compression is *entropy encoding*, also know as *variable length encoding*. This is usually the last step in compression. Where run length encoding relies on adjacent values being the same, entropy encoding looks at the overall frequency of specific values, wherever they are located. The implementation of entropy encoding is quite complex, but the principle is easy to understand if we define our terms carefully.

If the data is represented by bytes (groups of 8 bits), each byte can have any *value* between 0 and 255. When we come to transmit this data, we will choose a *symbol* to represent each possible value, and then transmit a sequence of *symbols* until all the *values* have been sent.

The simplest way to code the data is to send the bytes themselves; in other words, each symbol is the same as the value it represents. This is so obvious that we tend not to think of anything else, and make no distinction between the value and the symbol. But, it does not have to be like this. So long as both the transmitting and receiving points know and apply the same set of rules, we can use any set of 256 unique symbols to represent the 256 possible values. This is not very helpful until we add another factor: provided we have a set of rules that lets us know when one symbol ends and another begins, *not all the symbols need to be the same length.*

This is the essence of entropy encoding. It compresses the data by using short symbols to represent values that occur frequently, and longer symbols to represent the values that occur less frequently. Unless the input data is very close to random in its distribution of values, this technique means that total data used for the symbols will be less than the total of the original values.

Optimization of entropy encoding is complex, but the basic idea is neither difficult nor new. Let us compare two ways of coding the alphabet. Anyone who has spent much time with computers has come across the ASCII encoding system. In ASCII we represent any letter by a particular set of eight bits, or one byte. It does not matter what *value* (letter) we choose, the *symbol* is eight bits long. A much earlier method of encoding the alphabet was invented by Samuel Morse in 1838. The Morse code is a simple but useful example of an entropy coding system.

The Morse code uses combinations of dots and dashes as symbols to represent letters. A dash is three times as long as a dot; each dot or dash within a symbol is separated by one dot length; each symbol is separated by one dash length. Morse designed his code to transmit English language text, and to maximize the speed of transmission he looked at the frequency with which the various letters of the alphabet appear in typical text. The letter "e" is the most common letter in English, so Morse assigned the shortest possible symbol—one dot (actually three dots in length because of the inter-symbol space). The next most common letter "t" received the next shortest symbol, one dash. At the other extreme, an infrequent letter like "z" receives three dashes and a dot, a symbol four times as long as that used for "e."

The Morse code also illustrates one of the problems of entropy encoding. For transmission to be successful, obviously, both the transmitter and the receiver must use the same mapping of symbols to values. The relationship between values and symbols is known as the *code table*. But the most efficient mapping can only be determined when the frequency of values is known. The code table for Morse is derived from the frequency of letters in English text. If we wanted to send Polish language text, where the letters "c" "y" and 'z' are used much more frequently than in English, this code table would not be efficient because those letters have long symbols. If we wanted to transmit a few lines of Polish in the middle of many pages of English, this is probably not a major concern. However, if we had to transmit a very large volume of Polish, we might want to change the code table. We can do this if we have established rules at the transmitter and receiver to allow for different code tables, but we would need to evaluate this decision carefully. If we decide to change code tables, we have to transmit the new table, and that means transmitting a considerable amount of additional data.

There is no perfect solution to this problem, and the best compromise depends on the application. Some coding schemes use only one code table; this may not always be the most efficient table, but the coding system does not carry any of the overhead necessary to change tables. Other systems have a small number of fixed code tables, known to both the transmitter and the receiver. The encoder can examine the data to be transmitted, and select the most efficient code table. The only overhead is the provision of a message that tells the receiver "use table number 3." Still other schemes allow for the transmission of custom code tables whenever the characteristics of the data change substantially.

In 1952, Huffman demonstrated a method for deriving an efficient set of (variable length) symbols for a data source with known statistics. This method is widely used today, and entropy encoding is frequently referred to as *Huffman encoding*.

2.5.2c   Lossy Compression

As mentioned in the previous section, lossless compression would be an ideal answer, except that it rarely provides large degrees of compression, and it cannot be used alone to guarantee a fixed bit-rate. Lossless compression is an important part of our tool kit, but on its own it does not provide a solution to many practical problems. If we want to put digital audio over a modem link, or high-definition television through a 6 MHz channel, we have to accept that the compression process will result in some loss—what we get out will not be exactly the same as what we put in. This is the field of *lossy compression*. Ideally, lossy compression—like signal conditioning—removes irrelevant information. Some information is truly irrelevant in that the intended recipient cannot perceive that it is missing. In most cases, we look also for information that is close to irrelevant, where the quality loss is small compared to the data savings.

The objective of lossy compression is simple. We want to get maximum benefit (compression ratio, or bit rate reduction) at minimum cost (the loss in quality). However, the realization of this objective is not simple. A very large number of parameters have to be chosen for any given implementation, and many of these parameters must be varied according to the dynamic characteristics of the data. Even with a simple measure of the resulting quality (or lack of it), optimizing such a large number of variables a complex task.

Unfortunately, there is as yet no simple measure that can be applied to the quality of a compression system, particularly for video. The only measure that really matters is the subjective effect as perceived by a viewer, and this is a very complex function. If compression loss were confined to one characteristic of a picture, it would be relatively simple to derive a way of measuring this characteristic objectively. We could then perform careful subjective tests on a representative sample of viewers and arrive at a calibrated relationship between this characteristic and the subjective quality. Unfortunately, compression loss results in not one but many changes to the characteristics of the picture, each of these has a complex (non-linear) relationship to subjective quality, and they interact with each other.

Compression loss in images involves two main components: things that should be in the picture but are lost; and things that are added to the picture (artifacts) that should not be there. Areas where we would expect loss include spatial resolution (luminance and color—probably to different degrees), and shadow and highlight detail. Where extreme compression is needed for applications such as video conferencing or the Internet, temporal resolution would also be sacrificed. Artifacts that might be added include blocking, "mosquito noise" on edges, quantization noise, stepping of gray scales, patterning, ringing, and so on. As a further complication, in the presence of loss—a loss in resolution, for example—the addition of small amounts of artifact, such as noise or ringing, may actually improve the subjective quality of the picture.

Even within "standardized" compression schemes such as MPEG, the tuning of a system to yield the maximum quality-per-bit is a black art involving models of the human psycho-visual system that are closely guarded commercial secrets. Evaluation of the effects of parameter changes is a very complex subject.

2.5.2d   Quantization

For lossy compression, information in some form has to be discarded, and *quantization* is the tool most frequently applied. Quantization determines the precision with which we represent values. As mentioned previously, we need to represent information with sufficient precision, but it is important not to transmit unnecessary information. Properly applied, quantization can ensure that the information transmitted or stored has just enough precision for the intended application. We also use quantization to reduce information beyond this point, knowing that impairments will be introduced. Part of the job of a compression system is to ensure that the data is arranged so that the information discarded by quantization has the minimum possible subjective effect on the delivered image.

The simplest form of quantizer is the *scalar quantizer*, and the most common implementation is the *uniform scalar quantizer*. It operates on a one-dimensional variable, such as intensity, and corresponds to a staircase function with equal spacing of the steps. A continuum of input values is divided into a number of ranges of equal size. Figure 2.5.1 shows a range of input values from 0 to 255, divided by a

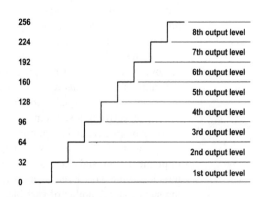

Figure 2.5.1 A uniform scalar quantizer.

Table 2.5.2 The Choice of Reconstruction Values and its Effect on Mean Square Error

Input Range	Output Code	Reconstruction Value	MSE	Reconstruction Value	MSE
0-31	000	00000000 (0)	325.5	00010000 (16)	85.5
32-63	001	00100000 (32)		00110000 (48)	
64-95	010	01000000 (64)		01010000 (80)	
96-127	011	01100000 (96)		01110000 (112)	
128-159	100	10000000 (128)		10010000 (144)	
160-191	101	10100000 (160)		10110000 (176)	
192-223	110	11000000 (192)		11010000 (208)	
224-255	111	11100000 (224)		11110000 (240)	

uniform scalar quantizer into eight equal regions. There are 256 possible input values, but only eight possible output symbols.

As the individual input values are processed by the quantizer, they are compared with *decision values* representing the steps of the staircase. Any input value between the first two decision levels will be assigned the first output symbol, and so on.

The other parameter that has to be considered is the *reconstruction value*. When we apply the quantizer shown in the figure, there are only eight possible output values, so we could represent each of these by a three bit code, or a value between zero and seven, as shown in Table 2.5.2. When we want to use the quantized data, it is of no use using intensity values zero to seven: these are all almost black. We need to get back to an approximation of the original range of input values, so for each of the quantized levels, we need to choose a single value in the range 0 to 255. We could just add five zeros to each three-bit symbol, but this would make all of the output values less than the input values; all the errors would be in the same direction.

It is reasonably intuitive that if the input values are evenly distributed, a reconstruction value at the center of each step is likely to give smaller overall errors, and this is illustrated in Table 2.5.2 where the overall error is expressed as a mean square error (MSE).

The most obvious implementation of quantization is to reduce directly the precision of the intensity values of the image. The necessary precision depends upon the application. Too few bits will produce a "paint by numbers" effect because the boundaries between adjacent values will be clearly visible. Assuming all the quantization steps are equally visible, the eye can resolve intensity differences of about one percent. To avoid visible boundaries, we should keep the quantization step smaller than this, preferably by a factor of about two to one. If we are trying to achieve a contrast ratio in a displayed image of 100:1 (about the best that can be achieved in a movie theater) the 256 values offered by 8-bit quantization should be adequate. For applications with a lower contrast ratio, such as the printed page, a coarser quantization such as 6 bits may be sufficient.

2.5.2e   Data Manipulation

In an earlier section, we described the two most common methods of lossless compression—run length encoding and variable length encoding. We have also discussed reducing the amount of information through the use of quantization. These techniques can all be used directly on image data, but far greater efficiency can be obtained if the form of the data is manipulated so as to take maximum advantage of these techniques. Two manipulation techniques are commonly used in compression systems: *predictive coding* and the *transform*.

Predictive coding relies on the fact that continuous-tone images have a high degree of correlation between the values of nearby pixels. In other words, gradual shading is more common than abrupt changes in intensity. Various prediction schemes can be used; the simplest is just to use the preceding pixel according to the scanning system used. The technique is to predict or "guess" each pixel value according to a known rule, then transmit just the error between the predicted and actual values. At the receiving end, the decoder uses the same rule to make the same prediction, and uses the transmitted error value to obtain the correct value.

The range of possible errors is actually twice as large as the range of intensity values, so on the face of it this may not be a good technique to use in compression. The intensity values in a typical image are scattered fairly randomly across all or most permissible values. However, if we choose a good predictor, most of the error values will be small—low values will be much more common than high values. This distribution is well suited to variable length encoding. In practice, a typical continuous-tone image may be transmitted losslessly with about half the number of bits per pixel. If the error signal is quantized, then variable length encoded, an image can be transmitted to any desired quality level with fewer bits than would be needed for the same quality using direct quantization.

Another powerful means of data manipulation is the transform, and the transform most commonly used is the *discrete cosine transform* (DCT). DCT takes a block of (typically) 8 × 8 pixel intensity values and transforms these into an 8 × 8 block of DCT coefficients, representing a range of horizontal, vertical, and diagonal frequencies within the original block. Another way of describing the process is to say that the DCT transform can represent any 8 × 8 block of pixel values by an appropriate mix of the 64 basis patterns shown in Figure 2.5.2. The "appropriate mix" is the 8 × 8 array of DCT coefficients. This process is reversible given sufficient arithmetic precision, but this requires 11-bit DCT coefficients to represent 8-bit intensity values. Again, the first step has increased the amount of data.

The benefit of DCT comes from a remarkable correspondence with the sensitivity of the human psycho-visual system. When a continuous tone image is transformed by DCT, most of the energy is represented by a few high-value coefficients. Many of the coefficients will be close to zero, and if these coefficients are forced to zero, the effect on the reconstructed image to a human observer is negligible. Furthermore, it is found that the higher frequency coefficients may be very coarsely quantized with only small effects on the quality of the reconstructed image.

2.5.3　Applying the Basic Principles

As outlined in some detail already, a compression system reduces the volume of data by exploiting spatial and temporal redundancies and by eliminating the data that cannot be displayed suitably by the associated display or imaging devices. The main objective of compression is to retain as little data as possible, just sufficient to reproduce the original images without causing unacceptable distortion of the images [2]. In general, a compression system consists of the following components:

- *Digitization, sampling, and segmentation*: Steps that convert analog signals on a specified grid of picture elements into digital representations and then divide the video input—first into frames, then into blocks.

- *Redundancy reduction*: The decorrelation of data into fewer useful data bits using certain invertible transformation techniques.

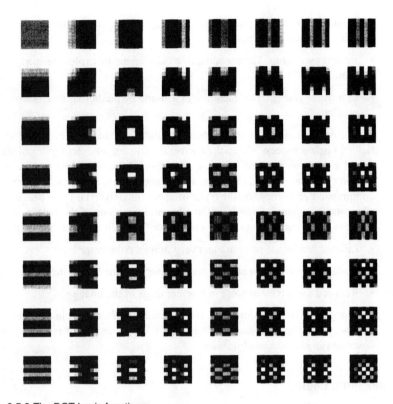

Figure 2.5.2 The DCT basis functions.

- *Entropy reduction*: The representation of digital data using fewer bits by dropping less significant information. This component causes distortion; it is the main contributor in *lossy* compression.

- *Entropy coding*: The assignment of code words (bit strings) of shorter length to more likely image symbols and code words of longer length to less likely symbols. This minimizes the average number of bits needed to code an image.

Key terms important to the understanding of this topic include the following:

- *Motion compensation*: The coding of video segments with consideration to their displacements in successive frames.

- *Spatial correlation*: The correlation of elements within a still image or a video frame for the purpose of bit-rate reduction.

- *Spectral correlation*: The correlation of different color components of image elements for the purpose of bit-rate reduction.

- *Temporal correlation*: The correlation between successive frames of a video file for the purpose of bit-rate reduction.

- *Quantization compression*: The dropping of the less significant bits of image values to achieve higher compression.

- *Intraframe coding*: The encoding of a video frame by exploiting spatial redundancy within the frame.

- *Interframe coding*: The encoding of a frame by predicting its elements from elements of the previous frame.

The removal of spatial and temporal redundancies that exist in natural video imagery is essentially a lossless process. Given the correct techniques, an exact replica of the image can be reproduced at the viewing end of the system. Such lossless techniques are important for medical imaging applications and other demanding uses. These methods, however, may realize only low compression efficiency (on the order of approximately 2:1). For video, a much higher compression ratio is required. Exploiting the inherent limitations of the *human visual system* (HVS) can result in compression ratios of 50:1 or higher. These limitations include the following:

- Limited luminance response and very limited color response

- Reduced sensitivity to noise in high frequencies, such as at the edges of objects

- Reduced sensitivity to noise in brighter areas of the image

The goal of compression, then, is to discard all information in the image that is not absolutely necessary from the standpoint of what the HVS is capable of resolving. Such a system can be described as *psychovisually lossless*.

2.5.3a   Transform Coding

In technical literature, countless versions of different coding techniques can be found [3]. Despite the large number of techniques available, one that comes up regularly (in a variety of flavors) during discussions about transmission standards is *transform coding* (TC).

Transform coding is a universal bit-rate-reduction method that is well suited for both large and small bit rates. Furthermore, because of several possibilities that TC offers for exploiting the visual inadequacies of the human eye, the subjective impression given by the resulting picture is frequently better than with other methods. If the intended bit rate turns out to be insufficient, the effect is seen as a lack of sharpness, which is less disturbing (subjectively) than coding errors such as frayed edges or noise with a structure. Only at very low bit rates does TC produce a particularly noticeable artifact: the *blocking effect*.

Because all pictures do not have the same statistical characteristics, the optimum transform is not constant, but depends on the momentary picture content that has to be coded. It is possible, for example, to recalculate the optimum transform matrix for every new frame to be transmitted, as is performed in the *Karhunen-Loeve transform* (KLT). Although the KLT is efficient in terms of ultimate performance, it is not typically used in practice because investigating each new picture to find the best transform matrix is usually too demanding. Furthermore, the matrix must be indicated to the receiver for each frame, because it must be used in decoding of the relevant inverse transform. A practical compromise is the *discrete cosine transform* (DCT). This transform matrix is constant and is suitable for a variety of images; it is sometimes referred to as "quick KLT."

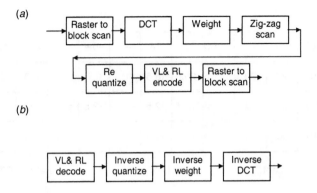

Figure 2.5.3 Block diagram of a sequential DCT codec: (*a*) encoder, (*b*) decoder. (*From* [3]. *Used with permission.*)

The DCT is a near relative of the *discrete Fourier transform* (DFT), which is widely used in signal analysis. Similar to DFT techniques, DCT offers a reliable algorithm for quick execution of matrix multiplication.

The main advantage of DCT is that it *decorrelates* the pixels efficiently; put another way, it efficiently converts statistically dependent pixel values into independent coefficients. In so doing, DCT packs the signal energy of the image block onto a small number of coefficients. Another significant advantage of DCT is that it makes available a number of fast implementations. A block diagram of a DCT-based coder is shown in Figure 2.5.3.

In addition to DCT, other transforms are practical for data compression, such as the *Slant transform* and the *Hadamard transform* [4].

Planar Transform

The similarities of neighboring pixels in a video image are not only line- or column-oriented, but also area-oriented [3]. To make use of these *neighborhood relationships*, it is desirable to transform not only in lines and columns, but also in areas. This can be achieved by a *planar transform*. In practice, *separable transforms* are used almost exclusively. A separable planar transform is nothing more than the repeated application of a simple transform. It is almost always applied to square picture segments of size $N \times N$, and it progresses in two steps, as illustrated in Figure 2.5.4. First, all lines of the picture segments are transformed in succession, then all rows of the segments calculated in the first step are transformed.

In textbooks, the planar transform frequently is called a *2D transform*. The transform is, in principle, possible for any segment forms—not just square ones [5]. Consequently, for a segment of the size $N \times N$, $2N$ transforms are used. The coefficients now are no longer arranged as vectors, but as a matrix. The coefficients of the i lines and the j columns are called c_{ij} ($i, j = 1 \dots N$). Each of these coefficients no longer represents a basic vector, but a *basic picture*. In this way, each $N \times N$ picture segment is composed of $N \times N$ different basic pictures, in which each coefficient gives the weighting of a particular basic picture. Figure 2.5.5 shows the basic pictures of the coefficients c_{11} and c_{23} for a planar 4×4 DCT. Because c_{11} represents the dc part, it is called the *dc coefficient*; the others are appropriately called the *ac coefficients*.

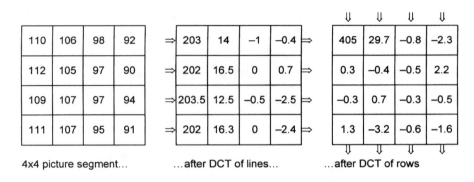

Figure 2.5.4 A simplified search of a best-matched block. (*From* [3]. *Used with permission.*)

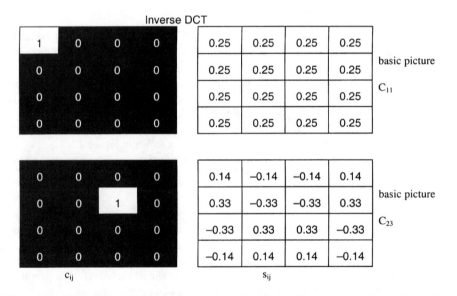

Figure 2.5.5 The mechanics of motion-compensated prediction. Shown are the pictures for a planar 4 × 4 DCT. Element C_{11} is located at row 1, column 1; element C_{23} is located at row 2, column 3. Note that picture C_{11} values are constant, referred to as dc coefficients. The changing values shown in picture C_{23} are known as ac coefficients. (*From* [3]. *Used with permission.*)

The planar transform of television pictures in the interlaced format is somewhat problematic. In moving regions of the picture, depending on the speed of motion, the similarities of vertically neighboring pixels of a frame are lost because changes have occurred between samplings of the two halves of the picture. Consequently, interlaced scanning may cause the performance of the system (or *output concentration*) to be greatly weakened, compared with progressive scanning. Well-tuned algorithms, therefore, try to detect stronger movements and switch to a transform in one picture half (i.e., field) for these picture regions [6]. However, the coding in one-half of the

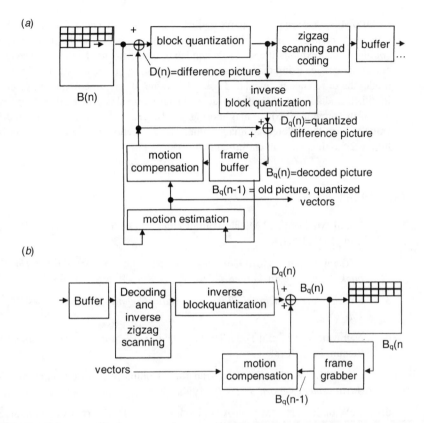

Figure 2.5.6 Overall block diagram of a DPCM system: (*a*) encoder, (*b*) decoder. (*From* [3]. *Used with permission.*)

picture is less efficient because the correlation of vertically neighboring pixels is weaker than in the full picture of a static scene. Simply stated, if the picture sequences are interlaced, the picture quality may be influenced by the motion content of the scene to be coded.

Interframe Transform Coding

With common algorithms, compression factors of approximately 8 can be achieved while maintaining good picture quality [3]. To achieve higher factors, the similarities between successive frames must be exploited. The nearest approach to this goal is the extension of the DCT in the time dimension. A drawback of such *cubic* transforms is the increase in calculation effort, but the greatest disadvantage is the higher memory requirement: for an $8 \times 8 \times 8$ DCT, at least seven frame memories would be needed. Much simpler is the *hybrid DCT*, which also efficiently codes pictures with moving objects. This method comprises, almost exclusively, a motion-compensated *difference pulse-code-modulation* (DPCM) technique; instead of each picture being transferred individually, the motion-compensated difference of two successive frames is coded.

DPCM is, in essence, predictive coding of sample differences. DPCM can be applied for both *interframe coding*, which exploits the temporal redundancy of the input image, and *intraframe coding*, which exploits the spatial redundancy of the image. In the intraframe mode, the difference is calculated using the values of two neighboring pixels of the same frame. In the interframe mode, the difference is calculated using the value of the same pixel on two consecutive frames. In either mode of operation, the value of the target pixel is predicted using the reconstructed values of the previously coded neighboring pixels. This value is then subtracted from the original value to form the differential image value. The differential image is then quantized and encoded. Figure 2.5.6 illustrates an end-to-end DPCM system.

2.5.4   References

1.   Robin, Michael, and Michel Poulin: *Digital Television Fundamentals*, McGraw-Hill, New York, N.Y., 1998.

2.   Lakhani, Gopal: "Video Compression Techniques and Standards," *The Electronics Handbook*, Jerry C. Whitaker (ed.), CRC Press, Boca Raton, Fla., pp. 1273–1282, 1996.

3.   Solari, Steve. J.: *Digital Video and Audio Compression*, McGraw-Hill, New York, N.Y., 1997.

4.   Netravali, A. N., and B. G. Haskell: *Digital Pictures, Representation, and Compression*, Plenum Press, 1988.

5.   Gilge, M.: "Region-Oriented Transform Coding in Picture Communication," *VDI-Verlag, Advancement Report, Series 10*, 1990.

6.   DeWith, P. H. N.: "Motion-Adaptive Intraframe Transform Coding of Video Signals," *Philips J. Res.*, vol. 44, pp. 345–364, 1989.

2.5.5   Bibliography

Symes, Peter D.: *Video Compression*, McGraw-Hill, N.Y., 1998.

Symes, Peter D.: "Video Compression Systems," *NAB Engineering Handbook*, 9th Ed., Jerry C. Whitaker (ed.), National Association of Broadcasters, Washington, D.C., pp. 907–922, 1999.

Fundamental Audio Principles

Sound would be of little interest if we could not hear. It is through the production and perception of sounds that it is possible to communicate and monitor events in our surroundings. Some sounds are functional, others are created for aesthetic pleasure, and still others yield only annoyance. Obviously a comprehensive examination of sound must embrace not only the physical properties of the phenomenon but also the consequences of interaction with listeners.

This section deals with sound in its various forms, beginning with a description of what it is and how it is generated, how it propagates in various environments, and finally what happens when sound impinges on the ears and is transformed into a perception. Part of this examination is a discussion of the factors that influence the opinions about sound and spatial qualities that so readily form when listening to music, whether live or reproduced.

Audio engineering, in virtually all its facets, benefits from an understanding of these basic principles. A foundation of technical knowledge is a useful instrument, and fortunately most of the important ideas can be understood without recourse to complex mathematics. It is the intuitive interpretation of the principles that is stressed in this section; more detailed information can be found in the reference material.

As with video compression, audio compression relies on the fundamental capabilities and limitations of human hearing.

In This Section:

3.1

The Physical Nature of Sound

Floyd E. Toole, E. A. G. Shaw, G. A. Daigle, M. R. Stinson[1]

3.1.1 Introduction

Sound is a physical disturbance in the medium through which it is propagated. Although the most common medium is air, sound can travel in any solid, liquid, or gas. In air, sound consists of localized variations in pressure above and below normal atmospheric pressure (*compressions* and *rarefactions*).

Air pressure rises and falls routinely, as environmental weather systems come and go, or with changes in altitude. These fluctuation cycles are very slow, and no perceptible sound results, although it is sometimes evident that the ears are responding in a different way to these *infrasonic* events. At fluctuation frequencies in the range from about 20 cycles per second up to about 20,000 cycles per second the physical phenomenon of sound can be perceived as having pitch or tonal character. This generally is regarded as the *audible* or *audio-frequency range*, and it is the frequencies in this range that are the concern of this chapter. Frequencies above 20,000 cycles per second are classified as *ultrasonic*.

3.1.2 Sound Waves

The essence of sound waves is illustrated in Figure 3.1.1, which shows a tube with a piston in one end. Initially, the air within and outside the tube is all at the prevailing atmospheric pressure. When the piston moves quickly inward, it compresses the air in contact with its surface. This energetic compression is rapidly passed on to the adjoining layer of air, and so on, repeatedly. As it delivers its energy to its neighbor, each layer of air returns to its original uncompressed state. A longitudinal sound pulse is moving outward through the air in the tube, causing only a passing disturbance on the way. It is a pulse because there is only an isolated action, and it is longitudinal because the air movement occurs along the axis of sound propagation. The rate at which the pulse propagates is the speed of sound. The pressure rise in the compressed air is proportional to the velocity with which the piston moves, and the perceived loudness of the resulting sound pulse

1. From *Standard Handbook of Audio and Radio Engineering*, 2nd. ed., Jerry C. Whitaker and K. Blair Benson (eds.), McGraw-Hill, New York, N.Y., 2002. Used with permission.

is related to the incremental amplitude of the pressure wave above the ambient atmospheric pressure.

Percussive or impulsive sounds such as these are common, but most sounds do not cease after a single impulsive event. Sound waves that are repetitive at a regular rate are called *periodic.* Many musical sounds are periodic, and they embrace a very wide range of repetitive patterns. The simplest of periodic sounds is a pure tone, similar to the sound of a tuning fork or a whistle. An example is presented when the end of the tube is driven by a loudspeaker reproducing a recording of such a sound (Figure 3.1.2). The pattern of displace-

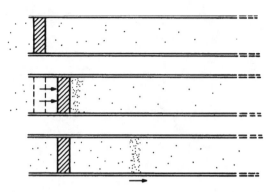

Figure 3.1.1 Generation of a longitudinal sound wave by the rapid movement of a piston in the end of a tube, showing the propagation of the wave pulse at the speed of sound down the length of the tube.

ment versus time for the loudspeaker diaphragm, shown in Figure 3.1.2*b*, is called a *sine wave* or *sinusoid.*

If the first diaphragm movement is inward, the first event in the tube is a pressure compression, as seen previously. When the diaphragm changes direction, the adjacent layer of air undergoes a *pressure rarefaction.* These cyclic compressions and rarefactions are repeated, so that the sound wave propagating down the tube has a regularly repeated, periodic form. If the air pressure at all points along the tube were measured at a specific instant, the result would be the graph of air pressure versus distance shown in Figure 3.1.2*c.* This reveals a smoothly sinusoidal waveform with a repetition distance along the tube symbolized by λ, the *wavelength* of the periodic sound wave.

If a pressure-measuring device were placed at some point in the tube to record the instantaneous changes in pressure at that point as a function of time, the result would be as shown in Figure 3.1.2*d.* Clearly, the curve has the same shape as the previous one except that the horizontal axis is time instead of distance. The periodic nature of the waveform is here defined by the time period *T,* known simply as the *period* of the sound wave. The inverse of the period, $1/T$, is the *frequency* of the sound wave, describing the number of repetition cycles per second passing a fixed point in space. An ear placed in the path of a sound wave corresponding to the musical tone middle C would be exposed to a frequency of 261.6 cycles per second or, using standard scientific terminology, a frequency of 261.6 hertz (Hz). The perceived loudness of the tone would depend on the magnitude of the pressure deviations above and below the ambient air pressure.

The parameters discussed so far are all related by the *speed of sound.* Given the speed of sound and the duration of one period, the wavelength can be calculated as

$$\lambda = cT \qquad\qquad (3.1.1)$$

where:
λ = wavelength
c = speed of sound
T = period

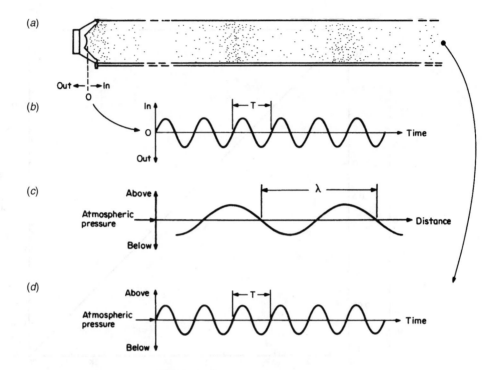

Figure 3.1.2 Characteristics of sound waves: (*a*) A periodic sound wave, a sinusoid in this example, is generated by a loudspeaker placed at the end of a tube. (*b*) Waveform showing the movement of the loudspeaker diaphragm as a function of time: displacement versus time. (*c*) Waveform showing the instantaneous distribution of pressure along a section of the tube: pressure versus distance. (*d*) Waveform showing the pressure variation as a function of time at some point along the tube: pressure versus time.

By knowing that the frequency $f = 1/T$, the following useful equation and its variations can be derived:

$$\lambda = \frac{c}{f} \quad f = \frac{c}{\lambda} \quad c = f\lambda \tag{3.1.2}$$

The speed of sound in air at a room temperature of 22°C (72°F) is 345 m/s (1131 ft/s). At any other ambient temperature, the speed of sound in air is given by the following approximate relationships [1, 2]:

$$c(m/s) = 331.29 + 0.607t(°C) \tag{3.1.3}$$

or

$$c(m/s) = 1051.5 + 1.106t(°F) \tag{3.1.4}$$

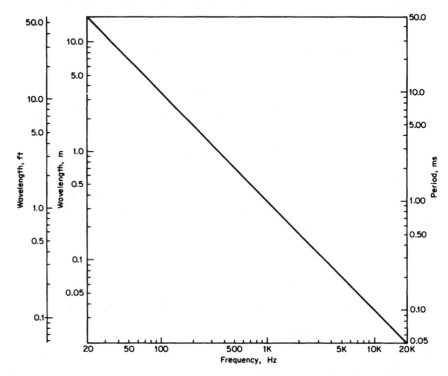

Figure 3.1.3 Relationships between wavelength, period, and frequency for sound waves in air.

where t = ambient temperature.

The relationships between the frequency of a sound wave and its wavelength are essential to understanding many of the fundamental properties of sound and hearing. The graph of Figure 3.1.3 is a useful quick reference illustrating the large ranges of distance and time embraced by audible sounds. For example, the tone middle C with a frequency of 261.6 Hz has a wavelength of 1.3 m (4.3 ft) in air at 20°C. In contrast, an organ pedal note at Cl, 32.7 Hz, has a wavelength of 10.5 m (34.5 ft), and the third-harmonic overtone of C8, at 12,558 Hz, has a wavelength of 27.5 mm (1.1 in). The corresponding periods are, respectively, 3.8 ms, 30.6 ms, and 0.08 ms. The contrasts in these dimensions are remarkable, and they result in some interesting and trouble-some effects in the realms of perception and audio engineering. For the discussions that follow it is often more helpful to think in terms of wavelengths rather than in frequencies.

3.1.2a   Complex Sounds

The simple sine waves used for illustration reveal their periodicity very clearly. Normal sounds, however, are much more complex, being combinations of several such pure tones of different frequencies and perhaps additional transient sound components that punctuate the more sustained elements. For example, speech is a mixture of approximately periodic vowel sounds and staccato

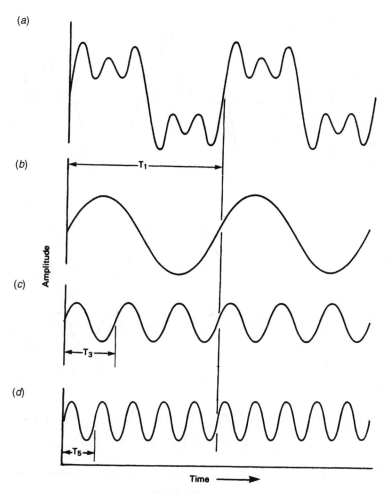

Figure 3.1.4 A complex waveform constructed from the sum of three harmonically related sinusoidal components, all of which start at the origin of the time scale with a positive-going zero crossing. Extending the series of odd-harmonic components to include those above the fifth would result in the complex waveform progressively assuming the form of a square wave. (*a*) Complex waveform, the sum of *b, c,* and *d.* (*b*) Fundamental frequency. (*c*) Third harmonic. (*d*) Fifth harmonic.

consonant sounds. Complex sounds can also be periodic; the repeated wave pattern is just more intricate, as is shown in Figure 1.1.4a. The period identified as T_1 applies to the *fundamental frequency* of the sound wave, the component that normally is related to the characteristic pitch of the sound. Higher-frequency components of the complex wave are also periodic, but because they are typically lower in amplitude, that aspect tends to be disguised in the summation of several such components of different frequency. If, however, the sound wave were analyzed, or broken down into its constituent parts, a different picture emerges: Figure 1.1.4b, c, and d. In this example, the analysis shows that the components are all *harmonics,* or whole-number multiples,

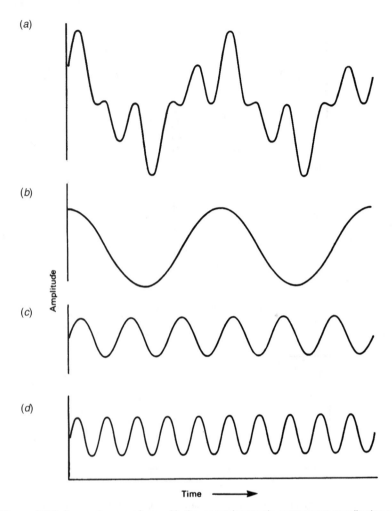

Figure 3.1.5 A complex waveform with the same harmonic-component amplitudes as in Figure 3.1.4, but with the starting time of the fundamental advanced by one-fourth period: a phase shift of 90°.

of the fundamental frequency; the higher-frequency components all have multiples of entire cycles within the period of the fundamental.

To generalize, it can be stated that all *complex periodic waveforms* are combinations of several harmonically related sine waves. The shape of a complex waveform depends upon the relative amplitudes of the various harmonics and the position in time of each individual component with respect to the others. If one of the harmonic components in Figure 3.1.4 is shifted slightly in time, the shape of the waveform is changed, although the frequency composition remains the same (Figure 3.1.5). Obviously a record of the time locations of the various harmonic components is required to completely describe the complex waveform. This information is noted as the *phase* of the individual components.

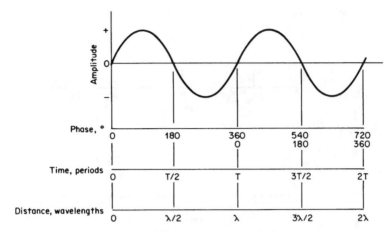

Figure 3.1.6 The relationship between the period *T* and wavelength λ of a sinusoidal waveform and the phase expressed in degrees. Although it is normal to consider each repetitive cycle as an independent 360°, it is sometimes necessary to sum successive cycles starting from a reference point in one of them.

3.1.2b   Phase

Phase is a notation in which the time of one period of a sine wave is divided into 360°. It is a relative quantity, and although it can be defined with respect to any reference point in a cycle, it is convenient to start (0°) with the upward, or positive-going, zero crossing and to end (360°) at precisely the same point at the beginning of the next cycle (Figure 3.1.6). *Phase shift* expresses in degrees the fraction of a period or wavelength by which a single-frequency component is shifted in the time domain. For example, a phase shift of 90° corresponds to a shift of one-fourth period. For different frequencies this translates into different time shifts. Looking at it from the other point of view, if a complex waveform is time-delayed, the various harmonic components will experience different phase shifts, depending on their frequencies.

A special case of phase shift is a *polarity reversal*, an inversion of the waveform, where all frequency components undergo a 180° phase shift. This occurs when, for example, the connections to a loudspeaker are reversed.

3.1.2c   Spectra

Translating time-domain information into the frequency domain yields an *amplitude-frequency spectrum* or, as it is commonly called, simply a *spectrum*. Figure 3.1.7a shows the spectrum of the waveform in Figures 3.1.4 and 3.1.5, in which the height of each line represents the amplitude of that particular component and the position of the line along the frequency axis identifies its frequency. This kind of display is a *line spectrum* because there are sound components at only certain specific frequencies. The phase information is shown in Figure 1.1.7b, where the difference between the two waveforms is revealed in the different *phase-frequency spectra*.

The equivalence of the information presented in the two domains—the waveform in the time domain and the amplitude- and phase-frequency spectra in the frequency domain—is a matter of considerable importance. The proofs have been thoroughly worked out by the French mathemati-

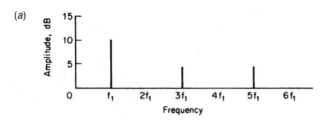

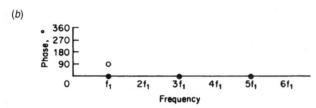

Figure 3.1.7 The amplitude-frequency spectra (*a*) and the phase-frequency spectra (*b*) of the complex waveforms shown in Figures 3.1.4 and 3.1.5. The amplitude spectra are identical for both waveforms, but the phase-frequency spectra show the 90° phase shift of the fundamental component in the waveform of Figure 3.1.5. Note that frequency is expressed as a multiple of the fundamental frequency f_1. The numerals are the harmonic numbers. Only the fundamental f_1 and the third and fifth harmonics (f_3 and f_5) are present.

cian Fourier, and the well-known relationships bear his name. The breaking down of waveforms into their constituent sinusoidal parts is known as *Fourier analysis*. The construction of complex waveshapes from summations of sine waves is called *Fourier synthesis*. *Fourier transformations* permit the conversion of time-domain information into frequency-domain information, and vice versa. These interchangeable descriptions of waveforms form the basis for powerful methods of measurement and, at the present stage, provide a convenient means of understanding audio phenomena. In the examples that follow, the relationships between time-domain and frequency-domain descriptions of waveforms will be noted.

Figure 3.1.8 illustrates the sound waveform that emerges from the larynx, the buzzing sound that is the basis for vocalized speech sounds. This sound is modified in various ways in its passage down the vocal tract before it emerges from the mouth as speech. The waveform is a series of periodic pulses, corresponding to the pulses of air that are expelled, under lung pressure, from the vibrating vocal cords. The spectrum of this waveform consists of a harmonic series of components, with a fundamental frequency, for this male talker, of 100 Hz. The gently rounded contours of the waveform suggest the absence of strong high-frequency components, and the amplitude-frequency spectrum confirms it. The *spectrum envelope*, the overall shape delineating the amplitudes of the components of the line spectrum, shows a progressive decline in amplitude as a function of frequency. The amplitudes are described in *decibels*, abbreviated dB. This is the common unit for describing sound-level differences. The rate of this decline is about –12 dB per octave (an *octave* is a 2:1 ratio of frequencies).

Increasing the pitch of the voice brings the pulses closer together in time and raises the fundamental frequency. The harmonic-spectrum lines displayed in the frequency domain are then spaced farther apart but still within the overall form of the spectrum envelope, which is defined by the shape of the pulse itself. Reducing the pitch of the voice has the opposite effect, increasing the spacing between pulses and reducing the spacing between the spectral lines under the envelope. Continuing this process to the limiting condition, if it were possible to emit just a single pulse, would be equivalent to an infinitely long period, and the spacing between the spectral lines would vanish. The discontinuous, or *aperiodic,* pulse waveform therefore yields a *continuous* spectrum having the form of the spectrum envelope.

Isolated pulses of sound occur in speech as any of the variations of consonant sounds, and in music as percussive sounds and as transient events punctuating more continuous melodic lines. All these aperiodic sounds exhibit continuous spectra with shapes that are dictated by the waveforms. The leisurely undulations of a bass drum waveform contain predominantly low-frequency energy, just as the more rapid pressure changes in a snare drum waveform require the presence of higher frequencies with their more rapid rates of change. A technical waveform of considerable use in measurements consists of a very brief impulse which has the important feature of containing equal amplitudes of all frequencies within the audio-frequency bandwidth. This is moving toward a limiting condition in which an infinitely short event in the time domain is associated with an infinitely wide amplitude-frequency spectrum.

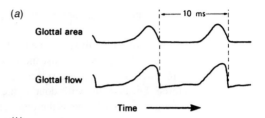

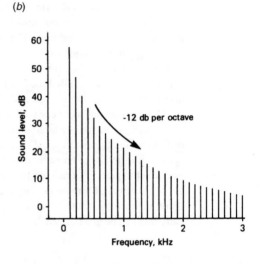

Figure 3.1.8 Characteristics of speech. (*a*) Waveforms showing the varying area between vibrating vocal cords and the corresponding airflow during vocalized speech as a function of time. (*b*) The corresponding amplitude-frequency spectrum, showing the 100-Hz fundamental frequency for this male speaker. (*From* [3]. *Used with permission.*)

3.1.3 Dimensions of Sound

The descriptions of sound in the preceding section involved only pressure variation, and while this is the dimension that is most commonly referred to, it is not the only one. Accompanying the pressure changes are temporary movements of the air "particles" as the sound wave passes (in this context a particle is a volume of air that is large enough to contain many molecules while its dimensions are small compared with the wavelength). Other measures of the magnitude of the sound event are the displacement amplitude of the air particles away from their rest positions and the velocity amplitude of the particles during the movement cycle. In the physics of sound, the *particle displacement* and the *particle velocity* are useful concepts, but the difficulty of their measurement limits their practical application. They can, however, help in understanding other concepts.

In a normally propagating sound wave, energy is required to move the air particles; they must be pushed or pulled against the elasticity of the air, causing the incremental rises and falls in pressure. Doubling the displacement doubles the pressure change, and this requires double the force. Because the work done is the product of force times distance and both are doubled, the energy in a sound wave is therefore proportional to the square of the particle displacement amplitude or, in more practical terms, to the square of the sound pressure amplitude.

Sound energy spreads outward from the source in the three dimensions of space, in addition to those of amplitude and time. The energy of such a sound field is usually described in terms of the energy flow through an imaginary surface. The sound energy transmitted per unit of time is called *sound power*. The sound power passing through a unit area of a surface perpendicular to a specified direction is called the *sound intensity*. Because intensity is a measure of energy flow, it also is proportional to the square of the sound pressure amplitude.

The ear responds to a very wide range of sound pressure amplitudes. From the smallest sound that is audible to sounds large enough to cause discomfort, there is a ratio of approximately 1 million in sound pressure amplitude, or 1 trillion (10^{12}) in sound intensity or power. Dealing routinely with such large numbers is impractical, so a logarithmic scale is used. This is based on the *bel,* which represents a ratio of 10:1 in sound intensity or sound power (the power can be acoustical or electrical). More commonly the decibel, one-tenth of a bel, is used. A difference of 10 dB therefore corresponds to a factor-of-10 difference in sound intensity or sound power. Mathematically this can be generalized as

$$\text{Level difference} = \log\frac{P_1}{P_2} \text{ bels} \tag{3.1.5}$$

or

$$\text{Level difference} = 10 \log\frac{P_1}{P_2} \text{ decibels} \tag{3.1.6}$$

where P_1 and P_2 are two levels of power.

For ratios of sound pressures (analogous to voltage or current ratios in electrical systems) the squared relationship with power is accommodated by multiplying the logarithm of the ratio of pressures by 2, as follows:

$$\text{Level difference} = 10 \log\frac{p_1^2}{p_2^2} = 20 \log\frac{p_1}{p_2} \text{ dB} \tag{3.1.7}$$

where *P1* and *P2* are sound pressures.

The relationship between decibels and a selection of power and pressure ratios is given in Table 3.3.1. The footnote to the table describes a simple process for interpolating between these values, an exercise that helps to develop a feel for the meaning of the quantities.

Table 3.1.1 Various Power and Amplitude Ratios and their Decibel Equivalents*

Sound or Electrical Power Ratio	Decibels	Sound Pressure, Voltage, or Current Ratio	Decibels
1	0	1	0
2	3.0	2	6.0
3	4.8	3	9.5
4	6.0	4	12.0
5	7.0	5	14.0
6	7.8	6	15.6
7	8.5	7	16.9
8	9.0	8	18.1
9	9.5	9	19.1
10	10.0	10	20.0
100	20.0	100	40.0
1,000	30.0	1,000	60.0
10,000	40.0	10,000	80.0
100,000	50.0	100,000	100.0
1,000,000	60.0	1,000,000	120.0

Notes:

* Other values can be calculated precisely by using Eqs. (3.1.6) and (3.1.7) or estimated by using this table and the following rules:

1) Power ratios that are multiples of 10 are converted into their decibel equivalents by multiplying the appropriate exponent by 10. For example, a power ratio of 1000 is 10^3, and this translates into $3 \times 10 = 30$ dB. Since power is proportional to the square of amplitude, the exponent of 10 must be doubled to arrive at the decibel equivalent of an amplitude ratio.

2) Intermediate values can be estimated by combining values in this table by means of the rule that the multiplication of power or amplitude ratios is equivalent to adding level differences in decibels. For example, increasing a sound level by 27 dB requires increasing the power by a ratio of 500 (20 dB is a ratio of 100, and 7 dB is a ratio of 5; the product of the ratios is 500). The corresponding increase in sound pressure or electrical signal amplitude is a factor of just over 20 (20 dB is a ratio of 10, and 7 dB falls between 6.0 and 9.5 and is therefore a ratio of something in excess of 2); the calculated value is 22.4.

3) Reversing the process, if the output from a power amplifier is increased from 40 to 800 W, a ratio of 20, the sound pressure level would be expected to increase by 13 dB (a power ratio of 10 is 10 dB, a ratio of 2 is 3 dB, and the sum is 13 dB). The corresponding voltage increase measured at the output of the amplifier would be a factor of between 4 and 5 (by calculation, 4.5).

The representation of the relative magnitudes of sound pressures and powers in decibels is important, but there is no indication of the absolute magnitude of either quantity being compared. This limitation is easily overcome by the use of a universally accepted reference level with which others are compared. For convenience the standard reference level is close to the smallest sound that is audible to a person with normal hearing. This defines a scale of *sound pressure level* (SPL), in which 0 dB represents a sound level close to the hearing-threshold level for middle and high frequencies (the most sensitive range). The SPL of a sound therefore describes, in decibels, the relationship between the level of that sounds and the reference level. Table 3.1.2 gives examples of SPLs of some common sounds with the corresponding intensities and an indication of listener reactions. From this table it is clear that the musically useful range of SPLs extend from the level of background noises in quiet surroundings to levels at which listeners begin to experience auditory discomfort and nonauditory sensations of feeling or pain in the ears themselves.

Table 3.1.2 Typical Sound Pressure Levels and Intensities for Various Sound Sources*

Sound Source	Sound Pressure Level, dB	Intensity W/m^2	Listener Reaction
	160		Immediate damage
	150	10^3	
Jet engine at 10 m	140		
	130		Painful feeling
Jet aircraft takeoff at 500 m	120		Discomfort
Amplified rock music	110	1	
Chain saw at 1 m	100		Very loud
Power mower at 1.5 m	90		
75-piece orchestra at 7 m	80	10^{-3}	Loud
City traffic at 15 m	70		
Normal speech at 1 m	60		
Suburban residence	50	10^{-6}	Comfortable listening level
Library	40		
Empty auditorium	30		
Recording studio	20	10^{-9}	Quiet
Breathing	10		
Silence	0**	10^{-12}	Inaudible

Notes:

* The relationships illustrated in this table are necessarily approximate because the conditions of measurement are not defined. Typical levels should, however, be within about 10 dB of the stated values.

** 0 dB sound pressure level (SPL) represents a reference sound pressure of 0.0002 μbar, or 0.00002 N/m^2.

While some sound sources, such as chain saws and power mowers, produce a relatively constant sound output, others, like a 75-piece orchestra, are variable. The sound from such an orchestra might have a *peak factor* of 20 to 30 dB; the momentary, or peak, levels can be this amount higher than the long-term average SPL indicated [4].

The sound power produced by sources gives another perspective on the quantities being described. In spite of some impressively large sounds, a full symphony orchestra produces only about 1 acoustic watt when working through a typical musical passage. On crescendos with percussion, though, the levels can be of the order of 100 W. A bass drum alone can produce about 25 W of acoustic power of peaks. All these levels are dependent on the instruments and how they are played. Maximum sound output from cymbals might be 10 W; from a trombone, 6 W; and from a piano, 0.4 W [5]. By comparison, average speech generates about 25 μW, and a present-day jet liner at takeoff between 50 and 100 kW. Small gasoline engines produce from 0.001 to 1.0 acoustic watt, and electric home appliances less than 0.01 W [6].

3.1.4   References

1.   Beranek, Leo L: *Acoustics,* McGraw-Hill, New York, N.Y., 1954.

2.   Wong, G. S. K.: "Speed of Sound in Standard Air," *J. Acoust. Soc. Am.*, vol. 79, pp. 1359–1366, 1986.

3.   Pickett, J. M.: *The Sounds of Speech Communications*, University Park Press, Baltimore, MD, 1980.

4.   Ehara, Shiro: "Instantaneous Pressure Distributions of Orchestra Sounds," *J. Acoust. Soc. Japan*, vol. 22, pp. 276–289, 1966.

5.   Stephens, R. W. B., and A. E. Bate: *Acoustics and Vibrational Physics,* 2nd ed., E. Arnold (ed.), London, 1966.

6.   Shaw, E. A. G.: "Noise Pollution—What Can be Done?" *Phys. Today,* vol. 28, no. 1, pp. 46–58, 1975.

Chapter 3.2
Sound Propagation

Floyd E. Toole, E. A. G. Shaw, G. A. Daigle, M. R. Stinson[1]

3.2.1 Introduction

Sound propagating away from a source diminishes in strength at a rate determined by a variety of circumstances. It also encounters situations that can cause changes in amplitude and direction. Simple reflection is the most obvious process for directional change, but with sound there are also some less obvious mechanisms.

3.2.2 Inverse-Square and Other Laws

At increasing distances from a source of sound the level is expected to decrease. The rate at which it decreases is dictated by the directional properties of the source and the environment into which it radiates. In the case of a source of sound that is small compared with the wavelength of the sound being radiated, a condition that includes many common situations, the sound spreads outward as a sphere of ever-increasing radius. The sound energy from the source is distributed uniformly over the surface of the sphere, meaning that the intensity is the sound power output divided by the surface area at any radial distance from the source. Because the area of a sphere is $4\pi r^2$, the relationship between the sound intensities at two different distances is

$$\frac{I_1}{I_2} = \frac{r_2^2}{r_1^2} \tag{3.2.1}$$

where I_1 = intensity at radius r_1, I_2 = intensity at radius r_2, and

---

1. From *Standard Handbook of Audio and Radio Engineering*, 2nd. ed., Jerry C. Whitaker and K. Blair Benson (eds.), McGraw-Hill, New York, N.Y., 2002. Used with permission.

$$\text{Level difference} = 10 \log \frac{r_2^2}{r_1^2} = 20 \log \frac{r_2}{r_1} \text{ dB} \tag{3.2.2}$$

This translates into a change in sound level of 6 dB for each doubling or halving of distance, a convenient mnemonic.

In practice, however, this relationship must be used with caution because of the constraints of real environments. For example, over long distances outdoors the absorption of sound by the ground and the air can modify the predictions of simple theory [1]. Indoors, reflected sounds can sustain sound levels to greater distances than predicted, although the estimate is correct over moderate distances for the *direct sound* (the part of the sound that travels directly from source to receiver without reflection). Large sound sources present special problems because the sound waves need a certain distance to form into an orderly wave-front combining the inputs from various parts of the source. In this case, measurements in what is called the *near field* may not be representative of the integrated output from the source, and extrapolations to greater distances will contain errors. In fact, the *far field* of a source is sometimes defined as being distances at which the inverse-square law holds true. In general, the far field is where the distance from the source is at least 2 to 3 times the distance between the most widely separated parts of the sound source that are radiating energy at the same frequency.

If the sound source is not small compared with the wavelength of the radiated sound, the sound will not expand outward with a spherical wavefront and the rate at which the sound level reduces with distance will not obey the inverse-square law. For example, a sound source in the form of a line, such as a long column of loudspeakers or a long line of traffic on a highway, generates sound waves that expand outward with a cylindrical wavefront. In the idealized case, such sounds attenuate at the rate of 3 dB for each doubling of distance.

3.2.3 Sound Reflection and Absorption

A sound source suspended in midair radiates into a *free field* because there is no impediment to the progress of the sound waves as they radiate in any direction. The closest indoor equivalent of this is an *anechoic room*, in which all the room boundaries are acoustically treated to be highly absorbing, thus preventing sounds from being reflected back into the room. It is common to speak of such situations as sound propagation in *full space*, or 4π *steradians* (sr; the units by which solid angles are measured).

In normal environments sound waves run into obstacles, such as walls, and the direction of their propagation is changed. Figure 3.2.1 shows the *reflection* of sound from various surfaces. In this diagram the pressure crests of the sound waves are represented by the curved lines, spaced one wavelength apart. The radial lines show the direction of sound propagation and are known as *sound rays*. For reflecting surfaces that are large compared with the sound wavelength, the normal *law of reflection* applies: the angle that the incident sound ray makes with the reflecting surface equals the angle made by the reflected sound ray.

This law also holds if the reflecting surface has irregularities that are small compared with the wavelength, as shown in Figure 3.2.1c, where it is seen that the irregularities have negligible effect. If, however, the surface features have dimensions similar to the wavelength of the incident sound, the reflections are *scattered* in all directions. At wavelengths that are small compared with

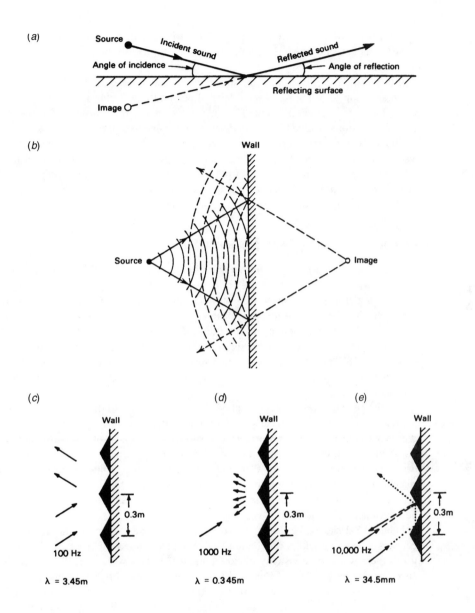

Figure 3.2.1 (*a*) The relationship between the incident sound, the reflected sound, and a flat reflecting surface, illustrating the law of reflection. (*b*) A more elaborate version of (*a*), showing the progression of wavefronts (the curved lines) in addition to the sound rays (arrowed lines). (*c*) The reflection of sound having a frequency of 100 Hz (wavelength 3.45 m) from a surface with irregularities that are small compared with the wavelength. (*d*) When the wavelength of the sound is similar to the dimensions of the irregularities, the sound is scattered in all directions. (*e*) When the wavelength of the sound is small compared with the dimensions of the irregularities, the law of reflection applies to the detailed interactions with the surface features.

the dimensions of the surface irregularities, the sound is also sent off in many directions but, in this case, as determined by the rule of reflections applied to the geometry of the irregularities themselves.

If there is perfect reflection of the sound, the reflected sound can be visualized as having originated at an image of the real source located behind the reflector and emitting the same sound power. In practice, however, some of the incident sound energy is *absorbed* by the reflecting surface; this fraction is called the *sound absorption coefficient* of the surface material. A coefficient of 0.0 indicates a perfect reflector, and a coefficient of 1.0 a perfect absorber; intermediate values indicate the portion of the incident sound energy that is dissipated in the surface and is not reflected. In general, the sound absorption coefficient for a material is dependent on the frequency and the angle of incidence of the sound. For simplicity, published values are normally given for octave bands of frequencies and for random angles of incidence.

3.2.3a   Interference: The Sum of Multiple Sound Sources

The principle of *superposition* states that multiple sound waves (or electrical signals) appearing at the same point will add linearly. Consider two sound waves of identical frequency and amplitude arriving at a point in space from different directions. If the waveforms are exactly in step with each other, i.e., there is no phase difference, they will add perfectly and the result will be an identical waveform with double the amplitude of each incoming sound (6-dB-higher SPL). Such *in-phase* signals produce *constructive interference*. If the waveforms are shifted by one-half wavelength (180° phase difference) with respect to each other, they are *out of phase*; if the pressure fluctuations are precisely equal and opposite, *destructive interference* occurs, and perfect cancellation results.

In practice, interference occurs routinely as a consequence of direct and reflected sounds adding at a microphone or a listener's ear. The amplitude of the reflected sound is reduced because of energy lost to absorption at the reflecting surface and because of inverse-square-law reduction related to the additional distance traveled. This means that constructive interference yields sound levels that are increased by less than 6 dB and that destructive interference results in imperfect cancellations that leave a residual sound level. Whether the interference is constructive or destructive depends on the relationship between the extra distance traveled by the reflection and the wavelength of the sound.

Figure 3.2.2 shows the direct and reflected sound paths for an omnidirectional source and receivers interacting with a reflecting plane. Note that there is an acoustically mirrored source, just as there would be a visually mirrored one if the plane were optically reflecting. If the distance traveled by the direct sound and that traveled by the reflected sound are different by an amount that is small and is also small compared with a wavelength of the sound under consideration (receiver R_1), the interference at the receiver will be constructive. If the plane is perfectly reflecting, the sound at the receiver will be the sum of two essentially identical sounds and the SPL will be about 6 dB higher than the direct sound alone. Constructive interference will also occur when the difference between the distances is an even multiple of half wavelengths. Destructive interference will occur for odd multiples of half wavelengths.

As the path length difference increases, or if there is absorption at the reflective surface, the difference in the sound levels of the direct and reflected sounds increases. For receivers R_2 and R_3 in Figure 3.2.2, the situation will differ from that just described only in that, because of the

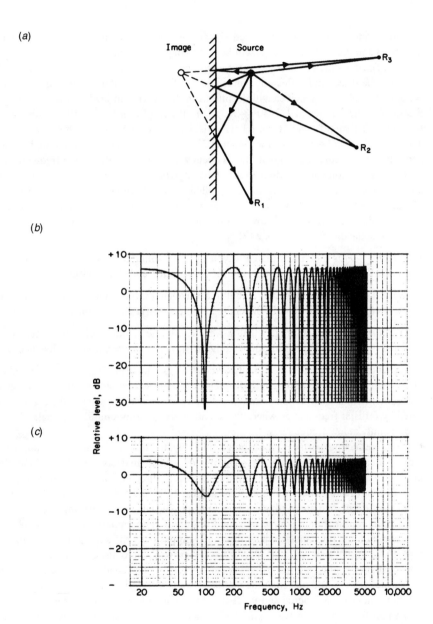

Figure 3.2.2 (*a*) Differing direct and reflected path lengths as a function of receiver location. (*b*) The interference pattern resulting when two sounds, each at the same sound level (0 dB) are summed with a time delay of just over 5 ms (a path length difference of approximately 1.7 m). (*c*) The reflection signal has been attenuated by 6 dB (it is now at a relative level of –6 dB, while the direct sounds remains at 0 dB); the maximum sound level is reduced, and perfect nulls are no longer possible. The familiar comb-filtering pattern remains.

additional attenuation of the reflected signal, the constructive peaks will be significantly less than 6 dB and the destructive dips will be less than perfect cancellations.

For a fixed geometrical arrangement of source, reflector, and receiver, at sufficiently low frequencies the direct and reflected sounds add. As the wavelength is reduced (frequency rising), the sound level at the receiver will decline from the maximum level in the approach to the first destructive interference at $\lambda/2 = r_2 - r_1$, where the level drops to a null. Continuing upward in frequency, the sound level at the receiver rises to the original level when $\lambda = r_2 - r_1$, falls to another null at $3\lambda/2 = r_2 - r_1$, rises again at $2\lambda = r_2 - r_1$, and so on, alternating between maxima and minima at regular intervals in the frequency domain. The plot of the frequency response of such a transmission path is called an *interference pattern*. It has the visual appearance of a comb, and the phenomenon has also come to be called *comb filtering* (see Figure 3.2.2*b*).

Taking a more general view and considering the effects averaged over a range of frequencies, it is possible to generalize as follows for the influence of a single reflecting surface on the sound level due to the direct sound alone [2].

- When $r_2 - r_1$ is much less than a wavelength, the sound level at the receiver will be elevated by 6 dB or less, depending on the surface absorption and distances involved.

- When $r_2 - r_1$ is approximately equal to a wavelength, the sound level at the receiver will be elevated between 3 and 6 dB, depending on the specific circumstances.

- When $r_2 - r_1$ is much greater than a wavelength, the sound level at the receiver will be elevated by between 0 and 3 dB, depending on the surface absorption and distances involved.

A special case occurs when the sound source, such as a loudspeaker, is mounted in the reflecting plane itself. There is no path length difference, and the source radiates into a hemisphere of free space, more commonly called a *half space*, or 2π sr. The sound level at the receiver is then elevated by 6 dB at frequencies where the sound source is truly omnidirectional, which—in practice—is only at low frequencies.

Other reflecting surfaces contribute additively to the elevation of the sound level at the receiver in amounts that can be arrived at by independent analysis of each. Consider the situation in which a simple point monopole (omnidirectional) source of sound is progressively constrained by reflecting planes intersecting at right angles. In practice this could be the boundaries of a room that are immediately adjacent to a loudspeaker which, at very low frequencies, is effectively an omnidirectional source of sound. Figure 3.2.3 summarizes the relationships between four common circumstances, where the sound output from the source radiates into solid angles that reduce in stages by a factor of 2. These correspond to a loudspeaker radiating into free space (4π sr), placed against a large reflecting surface (2π sr), placed at the intersection of two reflecting surfaces (π sr), and placed at the intersection of three reflecting surfaces ($\pi/2$ sr). In all cases the dimensions of the source and its distance from any of the reflecting surfaces are assumed to be a small fraction of a wavelength. The source is also assumed to produce a constant volume velocity of sound output; i.e., the volumetric rate of air movement is constant throughout.

By using the principles outlined here and combining the outputs from the appropriate number of image sources that are acoustically mirrored in the reflective surfaces, it is found that the sound pressure at a given radius increases in inverse proportion to the reduction in solid angle; sound pressure increases by a factor of 2, or 6 dB, for each halving of the solid angle.

The corresponding sound intensity (the sound power passing through a unit surface area of a sphere of the given radius) is proportional to pressure squared. Sound intensity therefore

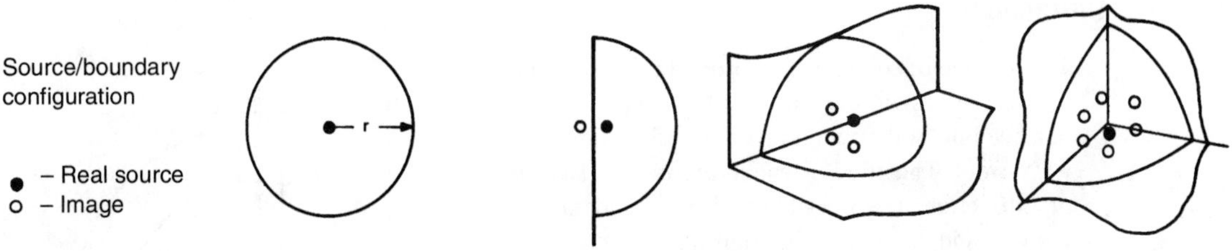

Source/boundary configuration				
● – Real source o – Image				
Solid angle seen by source (steradians)	4π	2π	π	$\pi/2$
Sound pressure at radius r	P_r (0 dB)	$2\,P_r$ (+6 dB)	$4\,P_r$ (+12 dB)	$8\,P_r$ (+18 dB)
Sound intensity at radius r	I_r (0 dB)	$4\,I_r$ (+6 dB)	$16\,I_r$ (+12 dB)	$64\,I_r$ (+18 dB)
Surface area at radius r	A_r	$A_r/2$	$A_r/4$	$A_r/8$
Sound power radiated into the solid angle $P = IA$	$= I_r\,A_r$ $= P_r$ (0 dB)	$= 4\,I_r\,A_r/2$ $= 2\,P_r$ (+3 dB)	$= 16\,I_r\,A_r/4$ $= 4\,P_r$ (+6 dB)	$= 64\,I_r\,A_r/8$ $= 8\,P_r$ (+9 dB)

Figure 3.2.3 Behavior of a point monopole sound source in full space (4π) and in close proximity to reflecting surfaces that constrain the sound radiation to progressively smaller solid angles. (*After* [3].)

increases by a factor of 4 for each halving of the solid angle. This also is 6 dB for each reduction in angle because the quantity is power rather than pressure.

Finally, multiplying the sound intensity by the surface area at the given radius yields the total sound power radiated into the solid angle. Because the surface area at each transition is reduced by a factor of 2, the total sound power radiated into the solid angle increases by a factor of 2, or 3 dB, for each halving of the solid angle.

By applying the reverse logic, reducing the solid angle by half increases the rate of energy flow into the solid angle by a factor of 2. At a given radius, this energy flows through half of the surface area that it previously did, so that the sound intensity is increased by a factor of 4; i.e., pressure squared is increased by a factor of 4. This means that sound pressure at that same radius is increased by a factor of 2.

The simplicity of this argument applies when the surfaces shown in Figure 3.2.3 are the only ones present; this can only happen outdoors. In rooms there are the other boundaries to consider, and the predictions discussed here will be modified by the reflections, absorption, and standing-wave patterns therein.

3.2.3b   Diffraction

The leakage of sound energy around the edges of an opening or around the corners of an obstacle results in a bending of the sound rays and a distortion of the wave-front. The effect is called *diffraction*. Because of diffraction it is possible to hear sounds around corners and behind walls—anywhere there might have been an "acoustical shadow." In fact, acoustical shadows exist, but to an extent that is dependent on the relationship between the wavelength and the dimensions of the objects in the path of the sound waves.

When the openings or obstructions are small compared with the wavelength of the sound, the waves tend to spread in all directions and the shadowing effect is small. At higher frequencies, when the openings or obstructions are large compared with the wavelengths, the sound waves tend to continue in their original direction of travel and there is significant shadowing. Figure 3.2.4 illustrates the effect.

The principle is maintained if the openings are considered to be the diaphragms of loudspeakers. If one wishes to maintain wide dispersion at all frequencies, the radiating areas of the driver units must progressively reduce at higher frequencies. Conversely, large radiating areas can be used to restrict the dispersion, though the dimensions required may become impractically large at low frequencies. As a consequence, most loudspeakers are approximately omnidirectional at low frequencies.

Sounds radiated by musical instruments obey the same laws. Low-frequency sounds from most instruments and the human voice radiate in all directions. Higher-frequency components can exhibit quite strong directional biases that are dependent on the size and orientation of the major sound-radiating elements. Figure 3.2.5a shows the frequency-dependent directivities of a trumpet, a relatively simple source. Compare this with the complexity of the

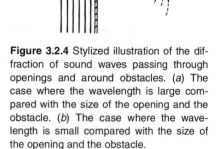

Figure 3.2.4 Stylized illustration of the diffraction of sound waves passing through openings and around obstacles. (*a*) The case where the wavelength is large compared with the size of the opening and the obstacle. (*b*) The case where the wavelength is small compared with the size of the opening and the obstacle.

directional characteristics of a cello (Figure 3.2.5b). It is clear that no single direction is representative of the total sound output from complex sound sources—a particular difficulty when it comes to choosing microphone locations for sound recordings. Listeners at a live performance hear a combination of all the directional components as spatially integrated by the stage enclosure and the hall itself.

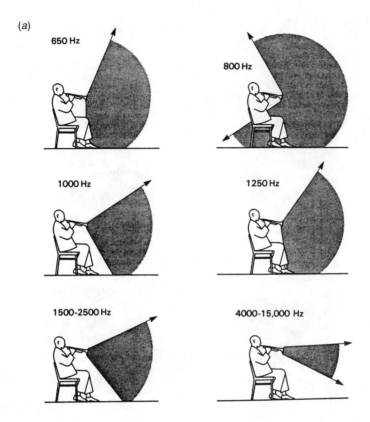

Figure 3.2.5 A simplified display of the main sound radiation directions at selected frequencies for: (*a*) a trumpet, (*b*, next page) a cello. (*From* [4]. *Used with permission.*)

3.2.3c   Refraction

Sound travels faster in warm air than in cold and faster downwind than upwind. These factors can cause sound rays to be bent, or *refracted*, when propagating over long distances in vertical gradients of wind or temperature. Figure 3.2.6 shows the downward refraction of sound when the propagation is downwind or in a *temperature inversion*, as occurs at night when the temperature near the ground is cooler than the air higher up. Upward refraction occurs when the propagation is upwind or in a *temperature lapse*, a typical daytime condition when the air temperature falls with increasing altitude. Thus, the ability to hear sounds over long distances is a function of local climatic conditions; the success of outdoor sound events can be significantly affected by the time of day and the direction of prevailing winds.

Figure 3.2.5b

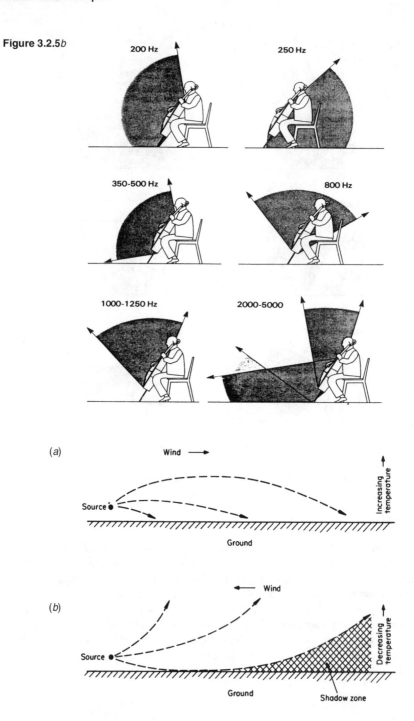

Figure 3.2.6 The refraction of sound by wind and by temperature gradients: (*a*) downwind or in a temperature inversion, (*b*) upwind or in a temperature lapse. (*From* [1]. *Used with permission.*)

3.2.4   References

1. Piercy, J. E., and T. F. W. Embleton: "Sound Propagation in the Open Air," in *Handbook* of *Noise Control*, 2d ed., C. M. Harris (ed.), McGraw-Hill, New York, N.Y., 1979.

2. Waterhouse, R. V., and C. M. Harris: "Sound in Enclosed Spaces," in *Handbook of Noise Control*, 2d ed., C. M. Harris (ed.), McGraw-Hill, New York, N.Y., 1979.

3. Olson, Harry F.: *Acoustical Engineering*, Van Nostrand, New York, N.Y., 1957.

4. Meyer, J.: *Acoustics and the Performance of Music*, Verlag das Musikinstrument, Frankfurt am Main, 1987.

3.3

The Physical Nature of Hearing

Floyd E. Toole, E. A. G. Shaw, G. A. Daigle, M. R. Stinson[1]

3.3.1 Introduction

The process of hearing begins with acoustical modifications to the sound waves as they interact with the head and the external ear, the visible portion of the system. These acoustical changes are followed by others in the ear canal and by a conversion of the sound pressure fluctuations into mechanical displacements by the eardrum. Transmitted through the mechanical coupling system of the middle ear to the inner ear, the displacement patterns are partially analyzed and then encoded in the form of neural signals. The signals from the two ears are cross-compared at several stages on the way to the auditory centers of the brain, where finally there is a transformation of the streams of data into perceptions of sound and acoustical space.

By these elaborate means we are able to render intelligible acoustical signals that, in technical terms, can be almost beyond description. In addition to the basic information, the hearing process keeps us constantly aware of spatial dimensions, where sounds are coming from, and the general size, shape, and decor of the space around us—a remarkable process indeed.

3.3.2 Anatomy of the Ear

Figure 1.4.1*a* shows a cross section of the ear in a very simplified form in which the outer, middle, and inner ear are clearly identified. The head and the outer ear interact with the sound waves, providing acoustical amplification that is dependent on both direction and frequency, in much the same way as an antenna. At frequencies above about 2 kHz there are reflections and resonances in the complex folds of the *pinna* [1]. Consequently, sounds of some frequencies reach the *tympanic membrane* (eardrum) with greater amplitude than sounds of other frequencies. The amount of the sound pressure gain or loss depends on both the frequency and the angle of incidence of the incoming sound. Thus, the external ear is an important first step in the perceptual process, encoding sounds arriving from different directions with distinctive spectral characters. For example, the primary resonance of the external ear, at about 2.6 kHz, is most sensitive to

---

1. From *Standard Handbook of Audio and Radio Engineering*, 2nd. ed., Jerry C. Whitaker and K. Blair Benson (eds.), McGraw-Hill, New York, N.Y., 2002. Used with permission.

sounds arriving from near 45° azimuth. This can be demonstrated by listening to a source of broadband sound while looking directly at it and then slowly rotating the head until one ear is pointing toward it. As the head is rotated through 45°, the sound should take on a "brighter" character as sounds in the upper midrange are accentuated. People with hearing problems use this feature of the ear to improve the intelligibility of speech when they unconsciously tilt the head, directing the ear toward the speaker. Continuing the rotation reveals a rapid dulling of the sound as the source moves behind the head. This is caused by acoustical shadowing due to diffraction by the pinna, a feature that helps to distinguish between front and back in sound localization.

At the eardrum the sound pressure fluctuations are transformed into movement that is coupled by means of the middle-ear bones (the *ossicular chain*) to the *oval window*, the input to the inner ear (*cochlea*). The middle ear increases the efficiency of sound energy transfer by providing a partial impedance match between sound in air, on the one hand, and wave motion in the liquid-filled inner ear, on the other. The inner ear performs the elaborate function of analyzing the sound into its constituent frequencies and converting the result into neural signals that pass up the auditory (eighth) nerve to the auditory cortex of the brain. From there sound is transformed into the many and varied perceptions that we take for granted. In the following discussions we shall be dealing with some of these functions in more detail.

(a)

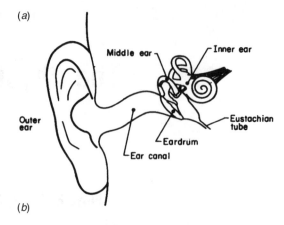

(b)

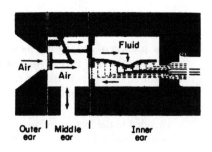

Figure 3.3.1 The human ear: (*a*) cross-sectional view showing the major anatomical elements, (*b*) a simplified functional representation.

3.3.3   Psychoacoustics and the Dimensions of Hearing

The physical dimensions of sound have parallels in the perceptual processes. The relationships are usually nonlinear, more complex than at first appearance, and somewhat variable among individuals as well as with time and experience. Nevertheless, they are the very essence of hearing.

The study of these relationships falls under the general umbrella of *psycho-acoustics*. A more specialized study, known as *psychophysics* or *psychometrics*, is concerned with quantification of the magnitudes of the sensation in relation to the magnitude of the corresponding physical stimulus.

3.3.3a   Loudness

Loudness is the term used to describe the magnitude of an auditory sensation. It is primarily dependent upon the physical magnitude (sound pressure) of the sound producing the sensation, but many other factors are influential.

Sounds come in an infinite variety of frequencies, timbres, intensities, temporal patterns, and durations; each of these, as well as the characteristics of the individual listener and the context within which the sound is heard, has an influence on loudness. Consequently, it is impossible for a single graph or equation to accurately express the relationship between the physical quality and quantity of sound and the subjective impression of loudness. Our present knowledge of the phenomenon is incomplete, but there are some important experimentally determined relationships between loudness and certain measurable quantities of sound. Although it is common to present and discuss these relationships as matters of fact, it must always be remembered that they have been arrived at through the process of averaging the results of many experiments with many listeners. These are not precise engineering data; they are merely indicators of trends.

Loudness as a Function of Frequency and Amplitude

The relationship between loudness and the frequency and SPL of the simplest of sounds, the pure tone, was first established by Fletcher and Munson, in 1933 [2]. There have been several subsequent redeterminations of loudness relationships by experimenters incorporating various refinements in their techniques. The data of Robinson and Dadson [3], for example, provide the basis for the International Organization for Standardization (ISO) recommendation R226 [4]. The presentation of loudness data is usually in the form of *equal-loudness contours*, as shown in Figure 3.3.2. Each curve shows the SPLs at which tones of various frequencies are judged to sound equal in loudness to a 1-kHz reference tone; the SPL of the reference tone identifies the curve in units called *phons*. According to this method, the *loudness level* of a sound, in phons, is the SPL level of a 1-kHz pure tone that is judged to be equally loud.

The equal-loudness contours of Figure 3.3.2 show that the ears are less sensitive to low frequencies than to middle and high frequencies and that this effect increases as sound level is reduced. In other words, as the overall sound level of a broadband signal such as music is reduced, the bass frequencies will fade faster than middle or high frequencies. In the curves, this appears as a crowding together of the contours at low frequencies, indicating that, at the lower sound levels, a small change in SPL of low-frequency sounds produces the same change in loudness as a larger change in SPL at middle and high frequencies. This may be recognized as the basis for the loudness compensation controls built into many audio amplifiers, the purpose of which is to boost progressively the bass frequencies as the overall sound level is reduced. The design and use of such compensation have often been erroneous because of a confusion between the shape of the loudness contours themselves and the *differences* between curves at various phon levels [5]. Sounds reproduced at close to realistic levels should need no compensation, since the ears will respond to the sound just as they would to the "live" version of the program. By the same token, control-room monitoring at *very* high sound levels can result in program equalization that is note appropriate to reproduction at normal domestic sound levels (combined

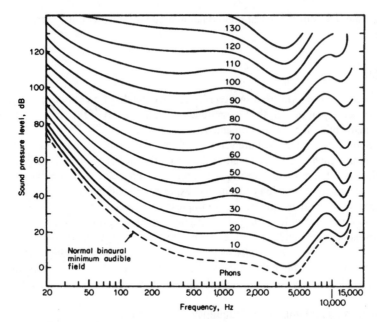

Figure 3.3.2 Contours of equal loudness showing the sound pressure level required for pure tones at different frequencies to sound as loud as a reference tone of 1000 Hz. (*From ISO Recommendation R226.*)

with this are the effects of temporary and permanent changes in hearing performance caused by exposure to loud sounds).

It is difficult to take the interpretations of equal-loudness contours much beyond generalizations since, as mentioned earlier, they are composites of data from many individuals. There is also the fact that they deal with pure tones and the measurements were done either through headphones (Fletcher and Munson [2]) or in an anechoic chamber (Robinson and Dadson [3]). The relationship between these laboratory tests and the common application for these data, the audition of music in normal rooms, is one that is only poorly established.

The lowest equal-loudness contour defines the lower limit of perception: the *hearing-threshold level*. It is significant that the ears have their maximum sensitivity at frequencies that are important to the intelligibility of speech. This optimization of the hearing process can be seen in various other aspects of auditory performance as well.

The rate of growth of loudness as a function of the SPL is a matter of separate interest. Units of *sones* are used to describe the magnitude of the subjective sensation. One sone is defined as the loudness of a tone at the 40-phon loudness level. A sound of loudness 2 sones would be twice as loud, and a sound of 0.5 sone would be half as loud. The *loudness function* relating the subjective sensation to the physical magnitude has been studied extensively [6], and while there are consistencies in general behavior, there remain very large differences in the performance of individuals and in the effect of the temporal and spectral structure of the sound. A common approximation relates a change of 10 dB in SPL to a doubling or halving of loudness. Individual variations on this may be a factor of 2 or more, indicating that one is not dealing with precise data. For example, the growth of loudness at low frequencies, as shown in the curves of Figure

3.3.2, indicates a clear departure from the general rule. Nevertheless, it is worth noting that significant changes in loudness require large differences in SPL and sound power; a doubling of loudness that requires a 10-dB increase in sound level translates into a factor of 3.16 in sound pressure (or voltage) and a factor of 10 in power.

Loudness as a Function of Bandwidth

The studies of loudness that used pure tones leave doubts about how they relate to normal sounds that are complexes of several frequencies or continuous bands of sound extending over a range of frequencies. If the bandwidth of a sound is increased progressively while maintaining a constant overall measured sound level, it is found that loudness remains constant from narrow bandwidths up to a value called the *critical bandwidth*. At larger bandwidths, the loudness increases as a function of bandwidth because of a process known as *loudness summation*. For example, the broadband sound of an orchestra playing a chord will be louder than the simple sound of a flute playing a single note even when the sounds have been adjusted to the same SPL.

The critical bandwidth varies with the center frequency of the complex sound being judged. At frequencies below about 200 Hz it is fairly constant at about 90 Hz; at higher frequencies the critical bandwidth increases progressively to close to 4000 Hz at 15 kHz. The sound of the orchestra therefore occupies many critical bandwidths while the sound of the flute is predominantly within one band.

Loudness as a Function of Duration

Brief sounds can appear to be less loud than sounds with the same maximum sound level but longer duration. Experiments show that there is a progressive growth of loudness as signal duration is increased up to about 200 ms; above that, the relationship levels out. The implication is that the hearing system integrates sound energy over a time interval of about 200 ms. In reality, the integration is likely to be of neural energy rather than acoustical energy, which makes the process rather complicated, since it must embrace all the nonlinearities of the perceptual mechanism.

The practical consequence of this is that numerous temporal factors, such as duration, intermittency, repetition rate, and so on, all influence the loudness of sounds that are separate from SPL.

Measuring the Loudness of Complex Sounds

Given the numerous variables and uncertainties in ascertaining the loudness of simple sounds, it should come as no surprise that measuring the loudness of the wideband, complex, and ever-changing sounds of real life is a problem that has resisted simple interpretation. Motivated by the need to evaluate the annoyance value of sounds as well as the more neutral quantity of loudness, various methods have been developed for arriving at single-number ratings of complex sounds. Some methods make use of spectral analysis of the sound, adjusted by correction factors and weighting, to compute a single-number loudness rating.

Simplifying the loudness compensation permits the process to be accomplished with relatively straightforward electronics providing a direct-reading output in real time, a feature that makes the device practical for recording and broadcasting applications. Such devices generally provide better indications of the loudness of typical music and speech program material than the very common and even simpler *volume-unit* (VU) meters or *sound-level meters* [7].

The VU meter responds to the full audio-frequency range, with a flat frequency response but with some control of its dynamic (time) response. A properly constructed VU meter should exhibit a response time of close to 300 ms, with an overswing of not more than 1.5 percent, and a return time similar to the response time. The dial calibrations and reference levels are also standardized. Such devices are therefore useful for measuring the magnitudes of steady-state signals and for giving a rough indication of the loudness of complex and time-varying signals, but they fail completely to take into account the frequency dependence of loudness.

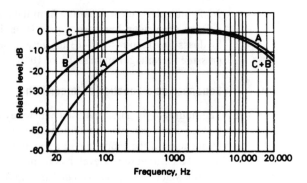

Figure 3.3.3 The standard frequency-weighting networks used in sound-level meters.

The sound-level meters used for acoustical measurements are adjustable in both amplitude and time response. Various *frequency-weighting* curves, *A-weighting* being the most popular, acknowledge the frequency-dependent aspects of loudness, and "fast" and "slow" time responses deal differently with temporal considerations. Although these instruments are carefully standardized and find extensive use in acoustics, noise control, and hearing conservation, they are of limited use as program-level indicators. Figure 3.3.3 shows the common frequency-weighting options found in sound-level meters. A-weighting has become the almost universal choice for measurements associated with loudness, annoyance, and the assessment of hearing-damage risk.

Peak program meters (PPM) are also standardized [8], and they find extensive use in the recording and broadcast industries. However, they are used mainly as a means of avoiding overloading recorders and signal-processing equipment. Consequently, the PPM has a very rapid response (an integration time of about 10 ms in the normal mode), so that brief signal peaks are registered, and a slow return (around 3 s), so that the peak levels can·be easily seen. These devices therefore are not useful indicators of loudness of fluctuating signals.

3.3.3b   Masking

Listening to a sound in the presence of another sound, which for the sake of simplicity we shall call noise, results in the desired sound being, to some extent, less audible. This effect is called *masking*. If the noise is sufficiently loud, the signal can be completely masked, rendering it inaudible; at lower noise levels the signal will be partially masked, and only its apparent loudness may be reduced. If the desired sound is complex, it is possible for masking to affect only portions of the total sound. All this is dependent on the specific nature of both the signal and the masking sound.

In audio it is possible for the low-level sounds of music, for example, to be masked by background noise in a sound system. That same noise can mask distortion products, so the effects need not be entirely undesirable. In addition to the unwanted noises that have been implied so far, there can be masking of musical sounds by other musical sounds. Thus we encounter the interesting situation of the perceived sound of a single musical instrument modified by the sounds of other instruments when it is joined in an ensemble.

In addition to the partial and complete masking that occurs when two sounds occur simultaneously, there are instances of *temporal masking*, when the audibility of a sound is modified by a sound that precedes it in time (*forward masking*) or, strange as it may seem, by a sound that follows it (*backward masking*).

Simultaneous Masking

At the lowest level of audibility, the threshold, the presence of noise can cause a *threshold shift* wherein the amplitude of the signal must be increased to restore audibility. At higher sound levels the masked sound may remain audible but, owing to partial masking, its loudness can be reduced.

In *simultaneous masking* the signal and the masking sound coexist in the time domain. It is often assumed that they must also share the same frequency band. While this seems to be most effective, it is not absolutely necessary. The effect of a masking sound can extend to frequencies that are both higher and lower than those in the masking itself. At low sound levels a

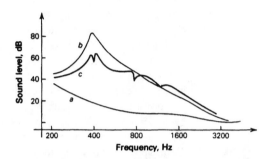

Figure 3.3.4 Detection threshold for pure tones of various frequencies: (*a*) in insolation, (*b*) in the presence of a narrow band (365 to 455 Hz) of masking noise centered on 400 Hz at a sound level of 80 dB, (*c*) in the presence of a making tone of 400 Hz at 80 dB. (*From* [40]. *Used with permission.*)

masking sound tends to influence signals with frequencies close to its own, but at higher sound levels the masking effect spreads to include frequencies well outside the spectrum of the masker. The dominant effect is an *upward spread* of masking that can extend several octaves above the frequency of the masking sound. There is also a *downward spread* of masking, but the effect is considerably less. In other words, a low-frequency masking sound can reduce the audibility of higher-frequency signals, but a high-frequency masking sound has relatively little effect on signals of lower frequency. Figure 3.3.4 shows that a simple masking sound elevates the hearing threshold over a wide frequency range but that the elevation is greater for frequencies above the masking sound.

In the context of audio, this means that we have built-in noise and distortion suppression. Background noises of all kinds are less audible while the music is playing but stand out clearly during the quiet intervals. Distortions generated in the recording and reproduction processes are present only during the musical sound and are therefore at least partially masked by the music itself. This is especially true for harmonic distortions, in which the objectionable distortion products are at frequencies higher than the masking sound—the sound that causes them to exist. Intermodulation-distortion products, on the other hand, are at frequencies both above and below the frequencies of the signals that produce the distortion. In this case, the upper distortion products will be subject to greater masking by the signal than the lower distortion products.

Studies of distortion have consistently noted that all forms of distortion are less audible with music than with simple signals such as single tones or combinations of tones; the more effective masking of the spectrally complex music signal is clearly a factor in this. Also noted is that intermodulation distortion is more objectionable than its harmonic equivalent. A simple explanation for this may be that not only are the difference-frequency components of intermodulation distortion unmusical, but they are not well masked by the signals that produce them.

Temporal Masking

The masking that occurs between signals not occurring simultaneously is known as *temporal masking*. It can operate both ways, from an earlier to a later sound (forward masking) or from a later to an earlier sound (backward masking). The apparent impossibility of backward masking (going backward in time) has a physiological explanation. It takes time for sounds to be processed in the peripheral auditory system and for the neural information to travel to the brain. If the later sound is substantially more intense than the earlier sound, information about it can take precedence over information about the earlier sound. The effect can extend up to 100 to 200 ms, but because such occurrences are rare in normal hearing, the most noteworthy auditory experiences are related to forward masking.

Forward masking results from effects of a sound that remain after the physical stimulus has been removed. The masking increases with the sound level of the masker and diminishes rapidly with time, although effects can sometimes be seen for up to 500 ms [9]. Threshold shifts of 10 to 20 dB appear to be typical for moderate sound levels, but at high levels these may reach 40 to 50 dB. Combined with these substantial effects is a broadening of the frequency range of the masking; at masker sound levels above about 80 dB maximum masking no longer occurs at the frequency of the masking sound but at higher frequencies.

There are complex interactions among the numerous variables in the masking process, and it is difficult to translate the experimental findings into factors specifically related to audio engineering. The effects are not subtle, however, and it is clear that in many ways they influence what we hear.

3.3.3c   Acoustic Reflex

One of the less-known features of hearing is the *acoustic reflex*, an involuntary activation of the middle-ear muscles in response to sound. These tiny muscles alter the transmission of sound energy through the middle ear, changing the quantity and quality of the sound that reaches the inner ear. As the muscles tighten, there may be a slight reduction in the overall sound level reaching the inner ear, but mainly there is a change in spectral balance as the low frequencies are rolled off. Below approximately 1 kHz the attenuation is typically 5 to 10dB, but it can be as much as 30 dB.

The reflex is activated by sounds above 80- to 85-dB SPL, which led to the early notion that it was a protective mechanism; however, the most hazardous sounds are at frequencies that are little affected by the reflex, and, furthermore, the reflex is too slow to block the passage of loud transients. The reflex activates rather slowly, in 10 to 20 ms for loud sounds and up to 150 ms for sounds near the activation threshold; then, after an interval, it slowly relaxes. Obviously there have to be other reasons for its existence. Although there is still some speculation as to its purpose, the fact that it is automatically activated when we talk and when we chew suggests that part of the reason is simply to reduce the auditory effects of our own voice and eating sounds.

Some people can activate the reflex voluntarily, and they report a reduction in the loudness of low frequencies during the period of activation. The behavior of the reflex also appears to depend on the state of the listener's attention to the sound itself This built-in tone control clearly is a complication in sound quality assessments since the spectral balance appears to be a function of sound level, the pattern of sound-level fluctuations in time, and the listener's attitude or attention to the sound.

3.3.3d   Pitch

Pitch is the subjective attribute of frequency, and while the basic correspondence between the two domains is obvious—low pitch to low frequencies and high pitch to high frequencies—the detailed relationships are anything but simple.

Fortunately waveforms that are periodic, however complex they may be, tend to be judged as having the same pitch as sine waves of the same repetition frequency. In other words, when a satisfactory pitch match has been made, the fundamental frequency of a complex periodic sound and a comparison sinusoid will normally be found to have the same frequency.

The exceptions to this simple rule derive from those situations where there is no physical energy at the frequency corresponding to the perceived pitch. Examples of pitch being associated with a *missing fundamental* are easily demonstrated by using groups of equally spaced tones, such as 100, 150, and 200 Hz, and observing that the perceived pitch corresponds to the difference frequency, 50 Hz. Common experience with sound reproducers, such as small radios, that have limited low-frequency bandwidth, illustrates the strength of the phenomenon, as do experiences with musical instruments, such as some low-frequency organ sounds, that may have little energy at the perceived fundamental frequency.

Scientifically, pitch has been studied on a continuous scale, in units of *mels*. It has been found that there is a highly nonlinear relationship between subjectively judged ratios of pitch and the corresponding ratios of frequency, with the subjective pitch interval increasing in size with increasing frequency. All this, though, is of little interest to traditional musicians, who have organized the frequency domain into intervals having special tonal relationships. The *octave* is particularly notable because of the subjective similarity of sounds spaced an octave apart and the fact that these sounds commonly differ in frequency by factors of 2. The musical fifth is a similarly well-defined relationship, being a ratio of 3:2 in repetition frequencies. These and the other intervals used in musical terminology gain meaning as one moves away from sine waves, with their one-frequency purity, into the sounds of musical instruments with their rich collection of overtones, many of which are harmonically related. With either pure tones [10] or some instrumental sounds in which not all the overtones are exactly harmonically related, the subjective octave may differ slightly from the physical octave; in the piano this leads to what is called *stretched tuning* [11].

The incompatibility of the mel scale of pitch and the hierarchy of musical intervals remains a matter for discussion. These appear to be quite different views of the same phenomenon, with some of the difference being associated with the musical expertise of listeners. It has been suggested, for example, that the mel scale might be better interpreted as a scale of *brightness* rather than one of pitch [12]. With periodic sounds brightness and pitch are closely related, but there are sounds, such as bells, hisses, and clicks, that do not have all the properties of periodic sounds and yet convey enough of a sense of pitch to enable tunes to be played with them, even though they cannot be heard as combining into chords or harmony. In these cases, the impressions of brightness and pitch seem to be associated with a prominence of sound energy in a band of frequencies rather than with any of the spectral components (partials, overtones, or harmonics) that may be present in the sound. A separate confirmation of this concept of brightness is found in subjective assessments of reproduced sound quality, where there appears to be a perceptual dimension along a continuum of "darkness" to "brightness" in which brightness is associated with a frequency response that rises toward the high frequencies or in which there are peaks in the treble [13]. At this, we reach a point in the discussion where it is more relevant to move into a different but related domain.

3.3.3e   Timbre, Sound Quality, and Perceptual Dimensions

Sounds may be judged to have the same subjective dimensions of loudness and pitch and yet sound very different from one another. This difference in *sound quality*, known as *timbre* in musical terminology, can relate to the tonal quality of sounds from specific musical instruments as they are played in live performance, to the character of tone imparted to all sounds processed through a system of recording and reproduction, and to the tonal modifications added by the architectural space within which the original performance or a reproduction takes place. Timbre is, therefore, a matter of fundamental importance in audio, since it can be affected by almost anything that occurs in the production, processing, storage, and reproduction of sounds.

Timbre has many dimensions, not all of which have been confidently identified and few of which have been related with any certainty to the corresponding physical attributes of sound. There is, for example, no doubt that the shape and composition of the frequency spectrum of the sound are major factors, as are the temporal behaviors of individual elements comprising the spectrum, but progress has been slow in identifying those measurable aspects of the signal that correlate with specific perceived dimensions, mainly because there are so many interactions between the dimensions themselves and between the physical and psychological factors underlying them.

The field of electronic sound synthesis has contributed much to the understanding of why certain musical instruments sound the way they do, and from this understanding have followed devices that permit continuous variations of many of the sound parameters. The result has been progressively better imitations of acoustical instruments in electronic simulations, as well as an infinite array of new "instruments" exhibiting tonal colors, dynamics, and emotional connotations that are beyond the capability of traditional instruments. At the same time as this expansion of timbral variety is occurring on one front of technical progress, there is an effort on another front to faithfully preserve the timbre of real and synthesized instruments through the complex process of recording and reproduction.

A fundamental problem in coming to grips with the relationship between the technical descriptions of sounds and the perception of timbre is in establishing some order in the choice and quantitative evaluation of words and phrases used by listeners to describe aspects of sound quality. Some of the descriptors are fairly general in their application and seem to fall naturally to quantification on a continuous scale from say, "dull" to "bright" or from "full" to "thin." Others, though, are specific to particular instruments or lapse into poetic portrayals of the evoked emotions.

From carefully conducted assessments of reproduced sound quality involving forms of multivariate statistical analysis, it has become clear that the extensive list can be reduced to a few relatively independent dimensions. As might be expected, many of the descriptors are simply different ways of saying the same thing, or they are responses to different perceptual manifestations of the same physical phenomenon.

From such analyses can come useful clarifications of apparently anomalous results since these responses need not be unidirectional. For example, a relatively innocent rise in the high-frequency response of a sound reproducer might be perceived as causing violins to sound unpleasantly strident but cymbals to sound unusually clear and articulate. A nice sense of air and space might be somewhat offset by an accentuation of background hiss and vocal sibilants, and so on.

Inexperienced listeners tend to concentrate unduly on a few of the many descriptors that come to mind while listening, while slightly more sophisticated subjects may become confused by the numerous contradictory indications. Both groups, for different reasons, may fail to note that there is but a single underlying technical flaw. The task of critical listening is one that requires a broad perspective and an understanding of the meaning and relative importance of the many timbral clues that a varied musical program can reveal. Trained and experienced listeners tend to combine timbral clues in a quest for logical technical explanations for the perceived effects. However, with proper experimental controls and the necessary prompting through carefully prepared instructions and a questionnaire, listeners with little prior experience can arrive at similar evaluations of accuracy without understanding the technical explanations [14].

The following list of perceptual dimensions is derived from the work of Gabrielsson and various colleagues [13, 15]. The descriptions are slightly modified from the original [13].

- *Clarity, or definition*: This dimension is characterized by adjectives such as clear, well defined, distinct, clean or pure, and rich in details or detailed, as opposed to adjectives such as diffuse, muddy or confused, unclear, blurred, noisy, rough, harsh, or sometimes rumbling, dull, and faint. High ratings in this dimension seem to require that the reproduction system perform well in several respects, exhibiting a wide frequency range, flat frequency response, and low nonlinear distortion. Systems with limited bandwidth, spectral irregularities due to resonances, or audible distortion receive lower ratings. Low-frequency spectral emphasis seems also to be detrimental to performance in this dimension, resulting in descriptions of rumbling, for the obvious reason, and dullness, probably due to the upward masking effects of the strong low frequencies. Increased sound levels result in increased clarity and definition.

- *Sharpness, or hardness, versus softness*: Adjectives such as sharp, hard, shrill, screaming, pointed, and clashing are associated with this dimension, contrasted with the opposite qualities of soft, mild, calm or quiet, dull, and subdued. A rising high-frequency response or prominent resonances in the high-frequency region can elicit high ratings in this dimension, as can certain forms of distortion. A higher or lower sound level also contributes to movement within this dimension, with reduced levels enhancing the aspect of softness.

- *Brightness versus darkness*: This dimension is characterized by the adjective bright, as opposed to dark, rumbling, dull, and emphasized bass. There appears to be a similar relationship between this dimension and the physical attributes of the sound system as exists with the preceding dimension, sharpness, or hardness, versus softness. In experiments, the two dimensions sometimes appear together and sometimes separately. The sense of pitch associated with brightness might be a factor in distinguishing between these two dimensions.

- *Fullness versus thinness*: This dimension also can appear in combination with brightness versus darkness, and there are again certain similarities in the relationship to measured spectrum balance and smoothness. There appears to be an association with the bandwidth of the system, especially at the low frequencies, and with sound level. It seems possible that this dimension is a representation of one encountered elsewhere as volume, which has been found to increase with increasing sound level but to decrease with increasing frequency.

- *Spaciousness*: Almost self-explanatory, this dimension elicits expressions of spacious, airy, wide, and open, as opposed to closed or shut up, narrow, and dry. The phenomenon appears to be related to poorly correlated sounds at the two ears of the listener. Other aspects of spaciousness are related to the spectrum of the reproduced sound. Gabrielsson points out that increased treble response enhances spaciousness, while reducing the bandwidth encourages a closed or shut-up impression. It is well known that the directional properties of the external ear (Figure 3.3.5) encode incoming sounds with spectral cues that can be significant influences in sound localization [16]. One such cue is a moving spectral notch and an increase in the sound level reaching the eardrum over the band from 5 to 10 kHz for progressively elevated sources (Figure 3.3.6). The appropriate manipulation of the sound spectrum in this frequency region can alone create impressions of height [17, 18] and, in this sense, alter the impression of spaciousness. It is worthy of note that the dimension of spaciousness is clearly observed in monophonic as well as stereophonic reproductions, indicating that it is a rather fundamental aspect of sound quality [13, 19].

- *Nearness*: Differences in the apparent proximity of sound sources are regularly observed in listening tests. It is clear that sound level affects perception of distance, especially for sounds such as the human voice that are familiar to listeners. Evidence from other studies indicates that impressions of distance are also influenced by the relationship between the direct, early-reflected, and reverberant sounds and the degree of coherence that exists in these sounds as they appear at the listener's ears [17].

- *Absence of extraneous sounds*: This dimension refers to nonmusical sounds that either exist in the original program material and are accentuated by aspects of the reproducer (such as tape hiss being aggravated by a treble boost) or are generated within the device itself (such as electronic amplifier clipping or mechanical noises from a loudspeaker).

- *Loudness*: This self-explanatory dimension is a useful check on the accuracy with which the sound levels of comparison sounds have been matched. It should, however, be noted that some listeners seem to regard the adjective loud as a synonym for sharp, hard, or painful.

The relative importance of these dimensions in describing overall sound quality changes slightly according to the specific nature of the devices under test, the form of the listener questionnaire, the program material, and, to some extent, the listeners themselves. In general, Gabrielsson and colleagues [13, 15] have found that clarity, or definition, brightness versus darkness, and sharpness, or hardness, versus softness are major contributors to the overall impression of sound quality.

3.3.3f   Audibility of Variations in Amplitude and Phase

Other things being equal, very small differences in sound level can be heard: down to a fraction of a decibel in direct A/B comparisons. Level differences that exist over only a small part of the spectrum tend to be less audible than differences that occupy a greater bandwidth. In other words, a small difference that extends over several octaves may be as significant as a much larger difference that is localized in a narrow band of frequencies. Spectral tilts of as little as 0.1 dB per octave are audible. For simple sounds the only audible difference may be loudness, but for com-

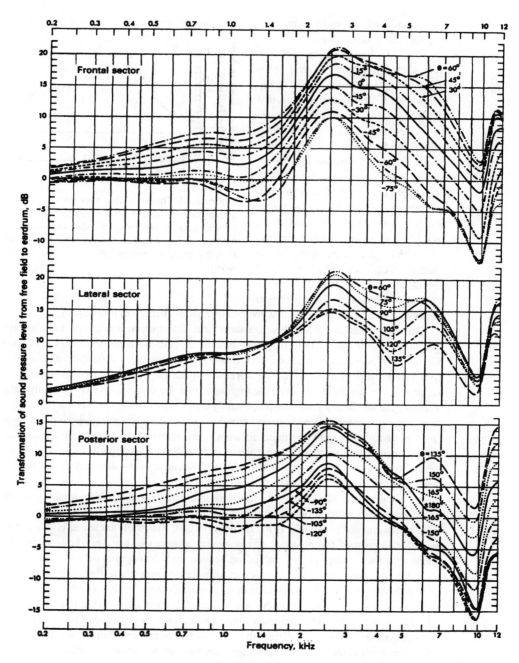

Figure 3.3.5 Family of curves showing the transformation of sound pressure level from the free field to the eardrum in the horizontal plane as a function of frequency, averaged over many listeners in several independent studies. The horizontal angles are referred to zero (the forward direction) and increase positively toward the ear in which the measurement is made and negatively away from it. (*From* [27]. *Used with permission.*)

plex sounds differences in timbre may be more easily detectable.

The audibility of phase shift is a very different matter. Several independent investigations over many years have led to the conclusion that while there are some special signals and listening situations where phase effects can be heard, their importance when listening to music in conventional environments is small [19]. Psychophysical studies indicate that, in general, sensitivity to phase is small compared with sensitivity to the amplitude spectrum and that sensitivity to phase decreases as the fundamental frequency of the signal increases. At the same time, it appears to be phase shifts in the upper harmonics of a complex signal that contribute most to changes in timbre [20].

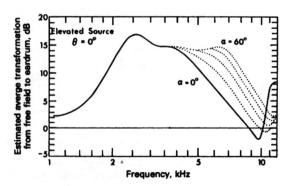

Figure 3.3.6 The estimated average transformation of sound pressure level from the free field to the eardrum as a function of frequency, showing the variations as a function of the angle of elevation for sounds arriving from the forward direction. (*From* [1]. *Used with permission.*)

The notion that phase, and therefore waveform, information is relatively unimportant is consistent with some observations of normal hearing. Sounds from real sources (voices and musical instruments) generally arrive at our ears after traveling over many different paths, some of which may involve several reflections. The waveform at the ear therefore depends on various factors other than the source itself. Even the argument that the direct sound is especially selected for audition and that later arrivals are perceptually suppressed does not substantially change the situation because sources themselves do not radiate waveforms that are invariably distinctive. With musical instruments radiating quite different components of their sound in different directions (consider the complexity of a grand piano or the cello, for example), the sum of these components—the waveform at issue—will itself be different at every different angle and distance; a recording microphone is in just such a situation.

The fact that the ear seems to be relatively insensitive to phase shifts would therefore appear to be simply a condition born of necessity. It would be incorrect to assume, however, that the phase performance of devices is *totally* unimportant. Spectrally localized phase anomalies are useful indicators of the presence of resonances in systems, and very large accumulations of phase shift over a range of frequencies can become audible as group delays.

While the presence of resonances can be inferred from phase fluctuations, their audibility may be better predicted from evidence in the amplitude domain [19]. It should be added that resonances of low Q in sound reproduction systems are more easily heard than those of higher Q [21–23]. This has the additional interesting ramification that evidence of sustained ringing in the time domain may be less significant than ringing that is rapidly damped; waveform features and other measured evidence that attract visual attention do not always correspond directly with the sound colorations that are audible in typical listening situations.

3.3.3g Perception of Direction and Space

Sounds are commonly perceived as arriving from specific directions, usually coinciding with the physical location of the sound source. This perception may also carry with it a strong impression of the acoustical setting of the sound event, which normally is related to the dimensions, locations, and sound-reflecting properties of the structures surrounding the listener and the sound source as well as objects in the intervening path.

Blauert, in his thorough review of this field [17], defines *spatial hearing* as embracing "the relationships between the locations of auditory events and other parameters—particularly those of sound events, but also others such as those that are related to the physiology of the brain." This statement introduces terms and concepts that may require some explanation. The adjective *sound*, as in *sound event*, refers to a physical source of sound, while the adjective *auditory* identifies a perception. Thus, the perceived location of an auditory event usually coincides with the physical location of the source of sound. Under certain circumstances, however, the two locations may differ slightly or even substantially. The difference is then attributed to other parameters having nothing whatever to do with the physical direction of the sound waves impinging on the ears of the listener, such as subtle aspects of a complex sound event or the processing of the sound signals within the brain.

Thus have developed the parallel studies of *monaural*, or one-eared, hearing and *binaural*, or two-eared, hearing. Commercial sound reproduction has stimulated a corresponding interest in the auditory events associated with sounds emanating from a single source and from multiple sources that may be caused to differ in various ways. In common usage it is assumed that stereophonic reproduction involves only two loudspeakers, but there are many other possible configurations. In stereophonic reproduction the objective is to create many more auditory events than the number of real sound sources would seem to permit. This is accomplished by presenting to the listener combinations of sounds that take advantage of certain inbuilt perceptual processes in the brain to create auditory events in locations other than those of the sound events and in auditory spaces that may differ from the space within which the reproduction occurs.

Understanding the processes that create auditory events would ideally permit the construction of predictable auditory spatial illusions in domestic stereophonic reproduction, in cinemas, in concert halls, and in auditoria. Although this ideal is far from being completely realized, there are some important patterns of auditory behavior that can be used as guides for the processing of sound signals reproduced through loudspeakers as well as for certain aspects of listening room, concert hall, and auditorium design.

3.3.3h Monaural Transfer Functions of the Ear

Sounds arriving at the ears of the listener are subject to modification by sound reflection, diffraction, and resonances in the structures of the external ear, head, shoulders, and torso. The amount and form of the modification are dependent on the frequency of the sound and the direction and distance of the source from which the sound emanates. In addition to the effect that this has on the sensitivity of the hearing process, which affects signal detection, there are modifications that amount to a kind of directional encoding, wherein sounds arriving from specific directions are subject to changes characteristic of those directions.

Each ear is partially sheltered from sounds arriving from the other side of the head. The effect of diffraction is such that low-frequency sounds, with wavelengths that are large compared with the dimensions of the head, pass around the head with little or no attenuation, while higher frequencies are progressively more greatly affected by the directional effects of diffraction. There is, in addition, the acoustical interference that occurs among the components of sound that have traveled over paths of slightly different length around the front and back and over the top of the head.

Superimposed on these effects are those of the pinna, or external ear. The intriguingly complex shape of this structure has prompted a number of theories of its behavior, but only relatively recently have some of its important functions been properly put into perspective. According to one view, the folds of the pinna form reflecting surfaces, the effect of which is to create, at the entrance to the ear canal, a system of interferences between the direct and these locally reflected sounds that depends on the direction and distance of the incoming sound [24]. The small size of the structures involved compared with the wavelengths of audible sounds indicates that dispersive scattering, rather than simple reflection, is likely to be the dominant effect. Nevertheless, measurements have identified some acoustical interferences resembling those that such a view would predict, and these have been found to correlate with some aspects of localization [18, 25].

In the end, however, the utility of the theory must be judged on the basis of how effectively it explains the physical functions of the device and how well it predicts the perceptual consequences of the process. From this point of view, time-domain descriptions would appear to be at a disadvantage since the hearing process is demonstrably insensitive to the fine structure of signals at frequencies above about 1.5 kHz [17]. Partly for this reason most workers have favored descriptions in terms of spectral cues.

It is therefore convenient that the most nearly complete picture of external-ear function has resulted from examinations of the behavior of the external ear in the frequency domain. By carefully measuring the pressure distributions in the standing-wave patterns, the dominant resonances in the external ear have been identified [26.] These have been related to the physical structures and to the measured acoustical performance of the external ear [1].

A particularly informative view of the factors involved in this discussion comes from an examination of curves showing the transformation of SPL from the free field to the eardrum [27]. These curves reveal, as a function of frequency, the amplitude modifications imposed on incident sounds by the external hearing apparatus. Figure 3.3.5 shows the family of curves representing this transformation for sounds arriving from different directions in the horizontal plane. Figure 3.3.6 shows the estimated transformations for sound sources at different elevations.

An interesting perspective on these data is shown in Figure 3.3.7, where it is possible to see the contributions of the various acoustical elements to the total acoustical gain of the ear. It should be emphasized that there is substantial acoustical interaction among these components, so that the sum of any combination of them is not a simple arithmetic addition. Nevertheless, this presentation is a useful means of acquiring a feel for the importance of the various components.

It is clear from these curves that there are substantial direction-dependent spectral changes, some rather narrowband in influence and others amounting to significant broadband tilts. Several studies in localization have found that, especially with pure tones and narrowband signals, listeners could attribute direction to auditory events resulting from sounds presented through only one ear (monaural localization) or presented identically in two ears, resulting in localization in the *median plane* (the plane bisecting the head vertically into symmetrical left-right halves). So strong are some of these effects that they can cause auditory events to appear in places different from the sound event, depending only on the spectral content of the sound. Fortunately such confusing effects are not common in the panorama of sounds we normally encounter, partly because of familiarity with the sounds themselves, but the process is almost certainly a part of the mechanism by which we are able to distinguish between front and back and between up an

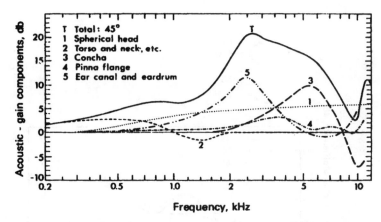

Figure 3.3.7 Contributions of various body parts to the total acoustic gain of the external hearing system for a sound source at a horizontal angle of 45°. Note that the interactions between these components prevent simple arithmetic addition of their individual contributions. (*From* [1]. *Used with permission.*)

down, directions that otherwise would be ambiguous because of the symmetrical locations of the two ears.

Interaural Differences

As useful as the monaural cues are, it is sound localization in the horizontal plane that is dominant, and for this the major cues come from the comparison of the sounds at the two ears and the analysis of the differences between them. From the data shown in Figure 3.3.5 it is evident that there is a substantial frequency-dependent *interaural amplitude difference* (IAD) that characterizes sounds arriving from different horizontal angles. Because of the path length differences there will also be an associated *interaural time difference* (ITD) that is similarly dependent on horizontal angle.

Figure 3.3.8 shows IADs as a function of frequency for three angles of incidence in the horizontal plane. These have been derived from the numerical data in [28], from which many other such curves can be calculated.

The variations in IAD as a function of both frequency and horizontal angle are natural consequences of the complex acoustical processes in the external hearing apparatus. Less obvious is the fact that there is frequency dependency in the ITDs. Figure 3.3.9 shows the relationship between ITD and horizontal angle for various pure tones and for broadband clicks. Also shown are the predictive curves for low-frequency sounds, based on diffraction theory, and for high-frequency sounds, based on the assumption that the sound reaches the more remote ear by traveling as a creeping wave that follows the contour of the head. At intermediate frequencies (0.5 to 2 kHz) the system is dispersive, and the temporal differences become very much dependent on the specific nature of the signal [29, 30].

It is evident from these data that at different frequencies, especially the higher frequencies, there are different combinations of ITD and IAD associated with each horizontal angle of incidence. Attempts at artificially manipulating the localization of auditory events by means of fre-

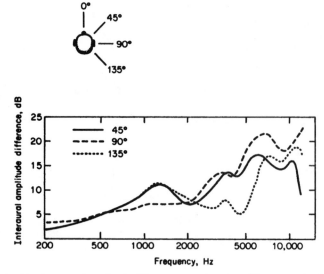

Figure 3.3.8 The interaural amplitude difference as a function of frequency for three angles of incidence. (*After* [28].)

quency-independent variations of these parameters are therefore unlikely to achieve the image size and positional precision associated with natural sound events.

Localization Blur

In normal hearing the precision with which we are able to identify the direction of sounds depends on a number of factors. The measure of this precision is called *localization blur*, the smallest displacement of the sound event that produces a just-noticeable difference in the corresponding auditory event. The concept of localization blur characterizes the fact that auditory space (the perception) is less precisely resolved than physical space and the measures we have of it.

The most precise localization is in the horizontal forward direction with broadband sounds preferably having some impulsive content. The lower limit of localization blur appears to be about 1°, with typical values ranging from 1 to 3°, though for some types of sound values of 10° or more are possible. Moving away from the forward axis, localization blur increases, with typical values for sources on either side of the head and to the rear being around 10 to 20°. Vertically, localization blur is generally rather large, ranging from about 5 to 20° in the forward direction to 30 to 40° behind and overhead [17].

Lateralization versus Localization

In exploring the various ways listeners react to interaural signal differences, it is natural that headphones be used, since the sounds presented to the two ears can then be independently controlled. The auditory events that result from this process are distinctive, however, in that the perceived images occur inside or very close to the head and image movement is predominantly

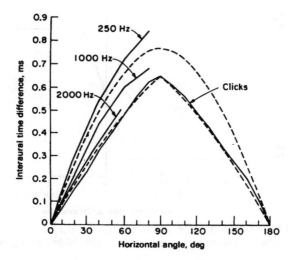

Figure 3.3.9 Interaural time difference as a function of horizontal angle. The curves show measured data for clicks and pure tones (solid lines) and predictive curves for low frequencies (top dashed curve), based on diffraction theory, and for high frequencies (bottom dashed curve), based on creeping-wave concepts. (*From* [41]. *Used with permission.*)

lateral. Hence, this phenomenon has come to be known as *lateralization*, as opposed to *localization*, which refers to auditory events perceived to be external and at a distance. Overcoming the in-head localization characteristic of headphone listening has been a major difficulty, inhibiting the widespread use of these devices for critical listening.

In headphone listening it is possible to move the auditory event by independently varying the interaural time or amplitude difference. Manipulating interaural time alone yields auditory image trajectories of the kind shown in Figure 3.3.10, indicating that the ITD required to displace the auditory image from center completely to one side is about 0.6 ms, a value that coincides with the maximum ITD occurring in natural hearing (Figure 3.3.9). Although most listeners would normally be aware of a single dominant auditory image even when the ITD exceeds this normal maximum value, it is possible for there to be multiple auditory images of lesser magnitude, each with a distinctive tonal character and each occupying a different position in perceptual space. With complex periodic signals, experienced listeners indicate that some of these images follow trajectories appropriate to the individual harmonics for frequencies that are below about 1 kHz [31]. This spatial complexity would not be expected in normal listening to a simple sound source, except when there are delayed versions of the direct sounds caused by strong reflections or introduced electronically. The result, if there are several such delayed-sound components, is a confused and spatially dispersed array of images, coming and going with the changing spectral and temporal structure of the sound. It seems probable that this is the origin of the often highly desirable sense of spaciousness in live and reproduced musical performances.

The sensitivity of the auditory system to changes in ITD in the lateralization of auditory images, or *lateralization blur* is dependent on both the frequency and the amplitude of the signal. According to various experimenters, lateralization blur varies from around 2 μs to about 60 μs, increasing as a function of signal frequency and sound level, and is at a minimum point around ITD = 0.

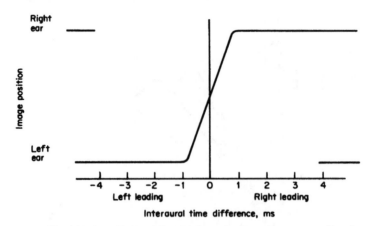

Figure 3.3.10 Perceived positions of the dominant auditory images resulting from impulsive signals (clicks) presented through headphones when the interaural time difference is varied.

Introducing an IAD displaces the auditory event toward the ear receiving the louder sound. An IAD of between 10 and 20 dB seems to be sufficient to cause the image to be moved completely to one side. The precise figure is difficult to ascertain because of the rapid increase in lateralization blur as a function of IAD; the auditory event becomes wider as it approaches the side of the head. Close to center, however, the lateralization blur is consistently in the vicinity of 1 to 2 dB.

Spatial Impression

Accompanying the auditory impression of images in any normal environment is a clear impression of the type and size of the listening environment itself. Two aspects appear to be distinguishable: *reverberance*, associated with the temporal stretching and blurring of auditory events caused by reverberation and late reflections; and *spaciousness*, often described as a spreading of auditory events so that they occupy more space than the physical ensemble of sound sources. Other descriptors such as ambience, width, or envelopment also apply. Spaciousness is a major determinant of listener preference in concert halls and as such has been the subject of considerable study.

In general, the impression of spaciousness is closely related to a lack of correlation between the input signals to the two ears. This appears to be most effectively generated by strong early lateral reflections (those arriving within about the first 80 ms after the direct sound). While all spectral components appear to add positively to the effect and to listener preference, they can contribute differently. Frequencies below about 3 kHz seem to contribute mainly to a sense of depth and envelopment, while high frequencies contribute to a broadening of the auditory event [32].

The acoustical interaction of several time-delayed and directionally displaced sounds at the ears results in a reduced interaural cross correlation; the sense of spaciousness is inversely proportional to this correlation. In other terms, there is a spectral and temporal incoherence in the sounds at the ears, leading to the fragmentation of auditory events as a function of both frequency and time. The fragments are dispersed throughout the perceptual space, contributing to the impression of a spatially extended auditory event.

3.3.3i   Distance Hearing

To identify the distance of a sound source listeners appear to rely on a variety of cues, depending on the nature of the sound and the environment. In the absence of strong reflections, as a sound source is moved farther from a listener, the sound level diminishes. It is possible to make judgments of distance on this factor alone, but only for sounds that are familiar, where there is a memory of absolute sound levels to use as a reference. With any sound, however, this cue provides a good sense of relative distance.

In an enclosed space the listener has more information to work with, because as a sound source is moved away, there will be a change in the relationship between the direct sound and the reflected and reverberant sounds in the room. The hearing mechanism appears to take note of the relative strengths of the direct and indirect sounds in establishing the distance of the auditory event. When the sound source is close, the direct sound is dominant and the auditory image is very compact; at greater distances, the indirect sounds grow proportionately stronger until eventually they dominate. The size of the auditory event increases with distance, as does the localization blur.

3.3.3j   Stereophonic Imaging

Consider the conventional stereophonic arrangement shown in Figure 3.3.11. If the two loudspeakers are radiating coherent sounds with identical levels and timing, the listener should perceive a single auditory event midway between the loudspeakers. This phantom, or virtual, sound source is the result of *summing localization*, the basis for the present system of two-channel stereophonic recording and reproduction.

Progressively increasing the time difference between the signals in the channels displaces the auditory event, or image, toward the side radiating the earlier sound until, at about 1 ms, the auditory image is coincident with the source of the earlier sound. At time differences greater than about 1 ms the perception may become spatially more dispersed, but the principal auditory event is

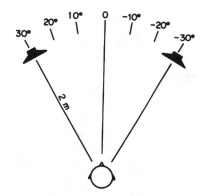

Figure 3.3.11 Standard stereophonic listening configuration.

generally perceived to remain at the position of the earlier sound event until, above some rather larger time difference, there will be two auditory events occurring separately in both time and space, the later of which is called an *echo*.

The region of time difference between that within which simple summing localization occurs and that above which echoes are perceived is one of considerable interest and complexity. In this region the position of the dominant auditory event is usually determined by the sound source that radiates the first sound to arrive at the listener's location. However, depending on the nature of

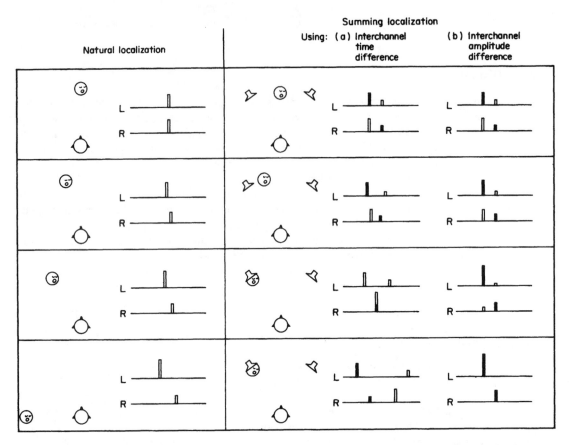

Figure 3.3.12 Comparison between sound localization in natural listening and localization in stereophonic listening within the range of simple summing. For the purposes of this simplified illustration, the sound waveform is an impulse. To the right of the pictoral diagrams showing a listener receiving sound from either a single source (natural localization) or a stereo pair of loudspeakers (summing localization) are shown the sounds received by the left and right ears of the listener. In the stereo illustrations, sounds from the left loudspeaker are indicated by dark bars and sounds from the right loudspeaker by light bars.

the signal, simple summing localization can break down and there can be subsidiary auditory images at other locations as well. The later sound arrivals also influence loudness, timbre, and intelligibility in ways that are not always obvious.

The cause of this complexity can be seen in Figure 3.3.12, showing the sounds arriving at the two ears when the sound is symbolically represented by a brief impulse. It is immediately clear that the fundamental difference between the situation of summing localization and that of natural localization is the presence of four sound components at the ears instead of just two.

In all cases the listener responds to identical ear input signals by indicating a single auditory event in the forward direction. Note, however, that in both stereo situations the signals at the two ears are not the same as the signals in normal localization. Thus, even though the spatial aspects have been simulated in stereo, the sounds at the two ears are modified by the *acoustical crosstalk* from each speaker to the opposite ear, meaning that perfectly accurate timbral reproduction for these sounds is not possible. This aspect of stereo persists through all conditions for time-difference manipulation of the auditory image, but with amplitude-difference manipulation the effect diminishes with increasing amplitude difference until, in the extreme, the listener hears only sound from a single speaker, a monophonic presentation.

Although impressions of image movement between the loudspeakers can be convincingly demonstrated by using either interchannel time or amplitude differences, there is an inherent limitation in the amount of movement: in both cases the lateral displacement of the principal auditory event is bounded by the loudspeakers themselves.

With time differences, temporal masking inhibits the contributions of the later arrivals, and the localization is dominated by the first sound to arrive at each ear. With small time differences the image can be moved between the loudspeakers, the first arrivals are from different loudspeakers, and it can be seen that an interchannel time difference is perceived as an ITD. At larger values of interchannel time difference the first arrivals are from the same loudspeaker, and the dominant auditory image remains at that location. This is because of the *law of the first wavefront*, also known as the precedence effect, according to which the dominant auditory event is perceived to be coincident with the loudspeaker radiating the earlier sound. The other sound components are still there nonetheless, and they can contribute to complexity in the spatial illusion as well as to changes in timbre.

With amplitude differences (also known as *intensity stereo*), the temporal pattern of events in the two ears is unchanged until the difference approaches infinity. At this point, the ears receive signals appropriate to a simple sound source with the attendant sound and localization accuracy. It is a real (monophonic) sound source generating a correspondingly real auditory event.

3.3.3k   Summing Localization with Interchannel Time and Amplitude Differences

Figure 3.3.13 shows the position of the auditory image as a function of interchannel time difference for the conventional stereophonic listening situation shown in Figure 3.3.11. The curves shown are but a few of the many that are possible since, as is apparent, the trajectory of the auditory image is strongly influenced by signal type and spectral composition.

In contrast, the curves in Figure 3.3.14, showing the position of the auditory image as a function of interchannel amplitude difference, are somewhat more orderly. Even so, there are significant differences in the slopes of the curves for different signals.

With a signal like music that is complex in all respects, it is to be expected that, at a fixed time or amplitude difference, the auditory event will not always be spatially well defined or positionally stable. There are situations where experienced listeners can sometimes identify and independently localize several coexisting auditory images. Generally, however, listeners are inclined to respond with a single compromise localization, representing either the "center of gravity" of a spatially complex image display or the dominant component of the array. If the spatial display is ambiguous, there can be a strong flywheel effect in which occasional clear spatial indications from specific components of the sound engender the perception that all of that sound is continuously originating from a specific region of space. This is especially noticeable with the onset of transient or any small mechanical sounds that are easily localized compared with the sustained portion of the sounds.

The blur in stereo localization, as in natural localizaton, is least for an image localized in the forward direction, where, depending on the type of sound, the *stereo localization blur* is typically about 3 to 7°. With the image fully displaced by amplitude difference (IAD = 30 dB), the blur

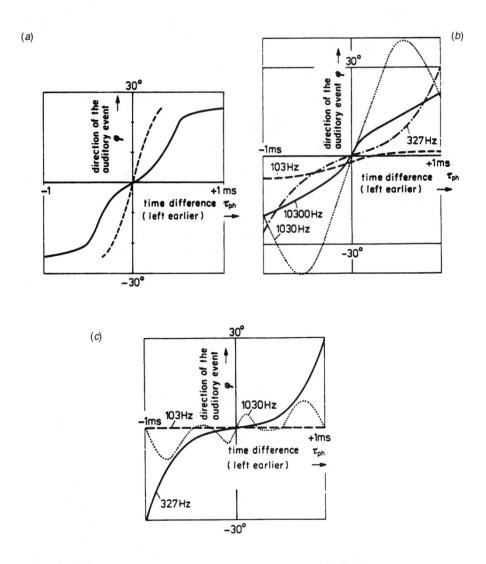

Figure 3.3.13 Direction of auditory images perceived by a listener in the situation of Figure 3.3.11 when the interchannel time difference is varied from 0 to +1 ms (left channel earlier) and −1 ms (right channel earlier). The curves show the results using different sounds: (*a*) dashed line = speech, solid line = impulses; (*b*) tone bursts; (*c*) continuous tones. (*From* [17]. *Used with permission.*)

increases to typical values of 5 to 11°. With the image fully displaced by means of time difference (ITD = 1 ms), the blur increases to typical values of 10 to 16° [17].

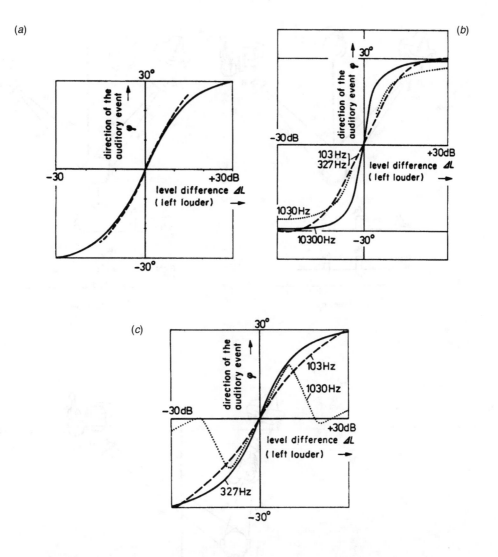

Figure 3.3.14 Direction of auditory images perceived by a listener in the situation of Figure 3.3.11 when the interchannel amplitude difference is varied from 0 to +30 dB (left louder) and −30 dB (right louder). The curves show the results with different sounds: (*a*) dashed line = speech, solid line = impulses; (*b*) tone bursts; (*c*) continuous tones. (*From* [17]. *Used with permission.*)

Effect of Listener Position

Sitting away from the line of symmetry between the speakers causes the central auditory images to be displaced toward the nearer loudspeaker. Interaural time differences between the sound arrivals at the ears are introduced as the path lengths from the two speakers change. Within the first several inches of movement away from the axis of symmetry, the sound components from the left and right loudspeakers remain the first arrivals at the respective ears. In this narrow

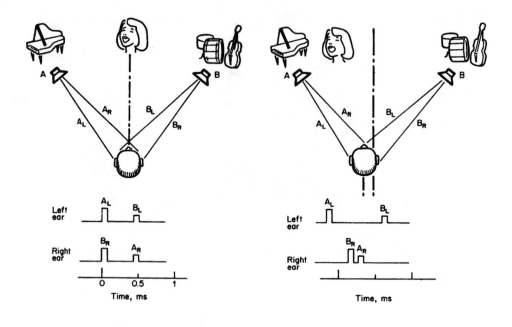

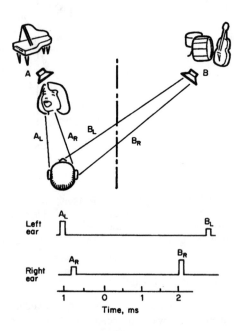

Figure 3.3.15 Sequence of events as a listener moves progressively away from the axis of symmetry in stereophonic listening.

region it is possible to compensate for the effect of the ITD by adding the appropriate opposite bias of interchannel amplitude difference (see Figure 3.3.15). This process is known as *time-intensity trading*, and it is the justification for the balance control on home stereo systems, supposedly allowing the listener to sit off the axis of symmetry and to compensate for it by introducing an interchannel amplitude bias. There are some problems, however, the first one being that the trading ratio is different for different sounds, so that the centering compensations do not work equally for all components of a complex signal; the image becomes blurred. The second problem arises when the listener moves beyond the limited range discussed previously, the simple form of summing localization breaks down, and the more complicated precedence effect comes into effect. In this region, it is to be expected that the auditory image will become rather muddled, increasing in size and spaciousness. Localization will tend to be related to the center of gravity of a spatially diffuse auditory event rather than of a specific compact event. Nevertheless, in recordings of ensembles with natural ambience the trading may be judged to be satisfactory, since the initial effect is, by design, rather diffuse. As the listener moves about, there will also be progressive changes to the timbre due to the directional properties of the loudspeakers and wave interference between essentially similar sounds arriving at the ears with various time delays.

Stereo Image Quality and Spaciousness

The position of auditory events is but a part of the total spatial impression. In stereo reproduction as in live performances, listeners appreciate the aspect of spaciousness as long as it creates a realistic impression. The process by which an impression of spaciousness is generated in stereo is much the same as in normal hearing—a reduction in interaural cross correlation. The tradeoff is also similar, in that as the feeling of space increases, the width of the auditory images also increases [33]. The extent to which the interchannel cross-correlation coefficient is altered to manipulate these effects is, therefore, a matter of artistic judgment depending on the type of music involved.

Special Role of the Loudspeakers

In the production of stereophonic recordings the impressions of image position, size, and spaciousness are controlled by manipulating the two-channel signals. However, the impressions received by the listener are also affected by the loudspeakers used for reproduction and their interaction with the listening room.

The directionality of the loudspeakers and the location of reflecting room boundaries together determine the relative strengths of the direct, early-reflected, and reverberant sounds that impinge on the listener. To the extent that the reflected sounds can reduce the correlation between the sounds at the two ears, it is clear that loudspeakers with substantial off-axis sound radiation can enhance the sense of spaciousness. For this to be effective, however, the listening room boundaries must be sound-reflecting at least at the points of the first reflections, especially the wall (lateral) reflections.

There is evidence that listeners in domestic situations prefer a certain amount of locally generated spaciousness [19, 34, 35]. In part this may be due to the more natural spatial distribution of the reflected sounds in the listening room as opposed to the recorded ambient sounds which are reproduced only as direct sounds from the loudspeakers. Loudspeakers placed in a room where the early reflections have been absorbed or directional loudspeakers placed in any type of room would be expected to yield a reproduction lacking spaciousness. This, it seems, is preferred by some listeners at home and many audio professionals in the control room [35, 36] especially with popular music. The fact that opinions are influenced by the type of music, individual preferences, and whether the listening is done for production or for pleasure makes this a matter for careful consideration. Once selected, the loudspeaker and the room tend to remain as fixed elements in a listening situation.

3.3.4 Sound in Rooms: The General Case

Taking the broadest view of complex sound sources, we can consider the combination of real sources and their reflected images as multiple sources. In this way, it is possible to deal with situations other than the special case of stereophonic reproduction.

3.3.4a Precedence Effect and the Law of the First Wavefront

For well over 100 years it has been known that the first sound arrival dominates sound localizaton. The phenomenon is known as the *law of the first wavefront* or the *precedence effect*. With time delays between the first and second arrivals of less than about 1 ms we are in the realm of simple summing localization. At longer delays the location of the auditory event is dictated by the location of the source of the first sound, but the presence of the later arrival is indicated by a distinctive timbre and a change in the spatial extent of the auditory event; it may be smeared toward the source of the second sound. At still longer time delays the second event is perceived as a discrete echo.

These interactions are physically complex, with many parametric variations possible. The perceived effects are correspondingly complex, and—as a consequence—the literature on the subject is extensive and not entirely unambiguous.

One of the best-known studies of the interaction of two sound events is that by Haas [37], who was concerned with the perception and intelligibility of speech in rooms, especially where there is sound reinforcement. He formed a number of conclusions, the most prominent of which is that for delays in the range of 1 to 30 ms, the delayed sound can be up to 10 dB higher in level than the direct sound before it is perceived as an echo. Within this range, there is an increase in loudness of the speech accompanied by "a pleasant modification of the quality of the sound (and) an apparent enlargement of the sound source." Over a wide range of delays the second sound was judged not to disturb the perception of speech, but this was found to depend on the syllabic rate. This has come to be known as the *Haas effect*, although the term has been extensively misused because of improper interpretation.

Examining the phenomenon more closely reveals a number of effects related to sound quality and to the localization dominance of the first-arrived sound. In general, the precedence effect is dependent on the presence of transient information in the sounds, but even this cannot prevent some interference from reflections in rooms. Several researchers have noted that high frequencies in delayed sounds were more disturbing than low frequencies, not only because of their relative audibility but because they were inclined to displace the localization. In fact, the situation in rooms is so complicated that it is to be expected that interaural difference cues will frequently be contradictory, depending on the frequency and temporal envelope of the sound. There are suggestions that the hearing process deals with the problem by means of a running plausibility analysis that pieces together evidence from the eyes and ears [38]. That this is true for normal listening where the sound sources are visible underlines the need in multi-channel reproduction to provide unambiguous directional cues for those auditory events that are intended to occupy specific locations.

3.3.4b Binaural Discrimination

The *cocktail-party effect*, in which it is demonstrably easier to carry on a conversation in a crowded noisy room when listening with two ears than with one, is an example of *binaural discrimination*. The spatial concentration that is possible with two ears has several other ramifications in audio. Reverberation is much less obtrusive in two-eared listening, as are certain effects of isolated reflections that arrive from directions away from that of the direct sound. For example, the timbral modifications that normally accompany the addition of a signal to a time-delayed duplicate (comb filtering) are substantially reduced when the delayed component arrives at the listener from a different direction [39]. This helps to explain the finding that listeners frequently enjoy the spaciousness from lateral reflections without complaining about the coloration. In this connection it has been observed that the disturbing effects of delayed sounds are reduced in the presence of room reverberation [37] and that reverberation tends to reduce the ability of listeners to discriminate differences in the timbre of sustained sounds like organ stops and vowels [20].

3.3.5 References

1. Shaw, E. A. G.: "The Acoustics of the External Ear," in W. D. Keidel and W. D. Neff (eds.), *Handbook* of *Sensory Physiology,* vol. V/I, *Auditory System,* Springer-Verlag, Berlin, 1974.

2. Fletcher, H., and W. A. Munson: "Loudness, Its Definition, Measurement and Calculation," *J. Acoust. Soc. Am.,* vol. 5, pp. 82–108, 1933.

3. Robinson, D. W., and R. S. Dadson: "A Redetermination of the Equal-Loudness Relations for Pure Tones," *Br. J. Appl. Physics,* vol. 7, pp. 166–181, 1956.

4. International Organization for Standardization: *Normal Equal-Loudness Contours for Pure Tones and Normal Threshold for Hearing under Free Field Listening Conditions,* Recommendation R226, December 1961.

5. Tonic, F. E.: "Loudness—Applications and Implications to Audio," *dB,* Part 1, vol. 7, no. 5, pp. 27–30; Part 2, vol. 7, no. 6, pp. 25–28, 1973.

6. Scharf, B.: "Loudness," in E. C. Carterette and M. P. Friedman (eds.), *Handbook* of *Perception,* vol. 4, *Hearing,* chapter 6, Academic, New York, N.Y., 1978.

7. Jones, B. L., and E. L. Torick: "A New Loudness Indicator for Use in Broadcasting," *J. SMPTE,* Society of Motion Picture and Television Engineers, White Plains, N.Y., vol. 90, pp. 772–777, 1981.

8. International Electrotechnical Commission: *Sound System Equipment,* part 10, *Programme Level Meters,* Publication 268-1 0A, 1978.

9. Zwislocki, J. J.: "Masking—Experimental and Theoretical Aspects of Simultaneous, Forward, Backward and Central Masking," in E. C. Carterette and M. P. Friedman (eds.), *Handbook of Perception,* vol. 4, *Hearing,* chapter 8, Academic, New York, N.Y., 1978.

10. Ward, W. D.: "Subjective Musical Pitch," *J. Acoust. Soc. Am.*, vol. 26, pp. 369–380, 1954.

11. Backus, John: *The Acoustical Foundations of Music,* Norton, New York, N.Y., 1969.

12. Pierce, John R.: *The Science of Musical Sound,* Scientific American Library, New York, N.Y., 1983.

13. Gabrielsson, A., and H. Siogren: "Perceived Sound Quality of Sound-Reproducing Systems," *J. Aoust. Soc. Am.*, vol. 65, pp. 1019–1033, 1979.

14. Toole, F. E.: "Subjective Measurements of Loudspeaker Sound Quality and Listener Performance," *J. Audio Eng. Soc.*, vol. 33, pp. 2–32, 1985.

15. Gabrielsson, A., and B. Lindstrom: "Perceived Sound Quality of High-Fidelity Loudspeakers." *J. Audio Eng. Soc.*, vol. 33, pp. 33–53, 1985.

16. Shaw, E. A. G.: "External Ear Response and Sound Localization," in R. W. Gatehouse (ed.), *Localization of Sound: Theory and Applications*, Amphora Press, Groton, Conn., 1982.

17. Blauert, J: *Spatial Hearing*, translation by J. S. Allen, M.I.T., Cambridge. Mass., 1983.

18. Bloom, P. J.: "Creating Source Elevation Illusions by Spectral Manipulations," *J. Audio Eng. Soc.*, vol. 25, pp. 560–565, 1977.

19. Toole, F. E.: "Loudspeaker Measurements and Their Relationship to Listener Preferences," *J. Audio Eng. Soc.*, vol. 34, part 1, pp. 227–235, part 2, pp. 323–348, 1986.

20. Plomp, R.: *Aspects of Tone Sensation—A Psychophysical Study,*" Academic, New York, N.Y., 1976.

21. Buchlein, R.: "The Audibility of Frequency Response Irregularities" (1962), reprinted in English translation in *J. Audio Eng. Soc.*, vol. 29, pp. 126–131, 1981.

22. Stevens, W. R.: "Loudspeakers—Cabinet Effects," *Hi-Fi News Record Rev.*, vol. 21, pp. 87–93, 1976.

23. Fryer, P.: "Loudspeaker Distortions—Can We Rear Them?," *Hi-Fi News Record Rev.*, vol. 22, pp. 51–56, 1977.

24. Batteau, D. W.: "The Role of the Pinna in Human Localization," *Proc. R. Soc. London*, B168, pp. 158–180, 1967.

25. Rasch, R. A., and R. Plomp: "The Listener and the Acoustic Environment," in D. Deutsch (ed.), *The Psychology of Music*, Academic, New York, N.Y., 1982.

26. Shaw, E. A. G., and R. Teranishi: "Sound Pressure Generated in an External-Ear Replica and Real Human Ears by a Nearby Sound Source," *J. Acoust. Soc. Am.*, vol. 44, pp. 240–249, 1968.

27. Shaw, E. A. G.: "Transformation of Sound Pressure Level from the Free Field to the Eardrum in the Horizontal Plane," *J. Acoust. Soc. Am.*, vol. 56, pp. 1848–1861, 1974.

28. Shaw, E. A. G., and M. M. Vaillancourt: "Transformation of Sound-Pressure Level from the Free Field to the Eardrum Presented in Numerical Form," *J. Acoust. Soc. Am.*, vol. 78, pp. 1120–1123, 1985.

29. Kuhn, G. F.: "Model for the Interaural Time Differences in the Azimuthal Plane," *J. Acoust. Soc. Am.*, vol. 62, pp. 157–167, 1977.

30. Shaw, E. A. G.: "Aural Reception," in A. Lara Saenz and R. W. B. Stevens (eds.), *Noise Pollution*, Wiley, New York, N.Y., 1986.

31. Toole, F. E., and B. McA. Sayers: "Lateralization Judgments and the Nature of Binaural Acoustic Images," *J. Acoust. Soc. Am.*, vol. 37, pp. 319–324, 1965.

32. Blauert, J., and W. Lindemann: "Auditory Spaciousness: Some Further Psychoacoustic Studies," *J. Acoust. Soc. Am.*, vol. 80, 533–542, 1986.

33. Kurozumi, K., and K. Ohgushi: "The Relationship between the Cross-Correlation Coefficient of Two-Channel Acoustic Signals and Sound Image Quality," *J. Acoust. Soc. Am.*, vol. 74, pp. 1726–1733, 1983.

34. Bose, A. G.: "On the Design, Measurement and Evaluation of Loudspeakers," presented at the 35th convention of the Audio Engineering Society, preprint 622, 1962.

35. Kuhl, W., and R. Plantz: "The Significance of the Diffuse Sound Radiated from Loudspeakers for the Subjective Hearing Event," *Acustica*, vol. 40, pp. 182–190, 1978.

36. Voelker, E. J.: "Control Rooms for Music Monitoring," *J. Audio Eng. Soc.*, vol. 33, pp. 452–462, 1985.

37. Haas, H.: "The Influence of a Single Echo on the Audibility of Speech," *Acustica*, vol. I, pp. 49–58, 1951; English translation reprinted in *J. Audio Eng. Soc.*, vol. 20, pp. 146–159, 1972.

38. Rakerd, B., and W. M. Hartmann: "Localization of Sound in Rooms, II—The Effects of a Single Reflecting Surface," *J. Acoust. Soc. Am.*, vol. 78, pp. 524–533, 1985.

39. Zurek, P. M.: "Measurements of Binaural Echo Suppression," *J. Acoust. Soc. Am.*, vol. 66, pp. 1750–1757, 1979.

40. Hall, Donald: *Musical Acoustics—An Introduction*, Wadsworth, Belmont, Calif., 1980.

41. Durlach, N. I., and H. S. Colburn: "Binaural Phenemena," in *Handbook of Perception*, E. C. Carterette and M. P. Friedman (eds.), vol. 4, Academic, New York, N.Y., 1978.

Audio Compression Systems

Fred Wylie, Jerry C. Whitaker[1]

3.4.1 Introduction

A number of real-time audio compression (bit rate reduction) coding methods have been developed that can significantly lower the data bandwidth and storage requirements for the transmission, distribution, and exchange of high-quality audio signals. The introduction in 1983 of the compact disc (CD) digital audio format set a quality benchmark that the manufacturers of subsequent professional audio equipment strive to match or improve upon. The discerning consumer expects the same quality from radio and television receivers.

3.4.1a PCM Versus Compression

It can be an expensive and complex technical exercise to fully implement a linear *pulse code modulation* (PCM) infrastructure, except over very short distances and within studio areas [1]. To demonstrate the advantages of distributing compressed digital audio over wireless or wired systems and networks, consider again the CD format as a reference. The CD is a 16 bit linear PCM process, but has one major handicap: the amount of circuit bandwidth the digital signal occupies in a transmission system. A stereo CD transfers information (data) at 1.411 Mbits/s, which would require a circuit with a bandwidth of approximately 700 kHz to avoid distortion of the digital signal. In practice, additional bits are added to the signal for channel coding, synchronization, and error correction; this increases the bandwidth demands yet again. 1.5 MHz is the commonly quoted bandwidth figure for a circuit capable of carrying a CD or similarly coded linear PCM digital stereo signal. This can be compared with the 20 kHz needed for each of two circuits to distribute the same stereo audio in the analog format, a 75-fold increase in bandwidth requirements.

---

1. From *Standard Handbook of Audio and Radio Engineering*, 2nd. ed., Jerry C. Whitaker and K. Blair Benson (eds.), McGraw-Hill, New York, N.Y., 2002. Used with permission.

3.4.2   Audio Bit Rate Reduction

In general, analog audio transmission requires fixed input and output bandwidths [2]. This condition implies that in a real-time compression system, the quality, bandwidth, and distortion/noise level of both the original and the decoded output sound should not be *subjectively* different, thus giving the appearance of a lossless and real-time process.

In a technical sense, all practical real-time bit-rate-reduction systems can be referred to as "lossy." In other words, the digital audio signal at the output is not identical to the input signal data stream. However, some compression algorithms are, for all intents and purposes, lossless; they lose as little as 2 percent of the original signal. Others remove much more of the original signal.

3.4.2a   Redundancy and Irrelevancy

A complex audio signal contains a great deal of information, some of which, because the human ear cannot hear it, is deemed irrelevant. [2]. The same signal, depending on its complexity, also contains information that is highly predictable and, therefore, can be made redundant.

Redundancy, measurable and quantifiable, can be removed in the coder and replaced in the decoder; this process often is referred to as *statistical compression. Irrelevancy*, on the other hand, referred to as *perceptual coding*, once removed from the signal cannot be replaced and is lost, irretrievably. This is entirely a subjective process, with each algorithm using a different psychoacoustic model.

Critically perceived signals, such as pure tones, are high in redundancy and low in irrelevancy. They compress quite easily, almost totally a statistical compression process. Conversely, noncritically perceived signals, such as complex audio or noisy signals, are low in redundancy and high in irrelevancy. These compress easily in the perceptual coder, but with the total loss of all the irrelevancy content.

3.4.2b   Human Auditory System

The sensitivity of the human ear is biased toward the lower end of the audible frequency spectrum, around 3 kHz [2]. At 50 Hz, the bottom end of the spectrum, and at 17 kHz at the top end, the sensitivity of the ear is down by approximately 50 dB relative to its sensitivity at 3 kHz (Figure 3.4.1). Additionally, very few audio signals—music- or speech-based—carry fundamental frequencies above 4 kHz. Taking advantage of these characteristics of the ear, the structure of audible sounds, and the redundancy content of the PCM signal is the basis used by the designers of the *predictive* range of compression algorithms.

Another well-known feature of the hearing process is that loud sounds mask out quieter sounds at a similar or nearby frequency. This compares with the action of an automatic gain control, turning the gain down when subjected to loud sounds, thus making quieter sounds less likely to be heard. For example, as illustrated in Figure 3.4.2, if we assume a 1 kHz tone at a level of 70 dBu, levels of greater than 40 dBu at 750 Hz and 2 kHz would be required for those frequencies to be heard. The ear also exercises a degree of temporal masking, being exceptionally tolerant of sharp transient sounds.

It is by mimicking these additional psychoacoustic features of the human ear and identifying the irrelevancy content of the input signal that the *transform* range of low bit-rate algorithms

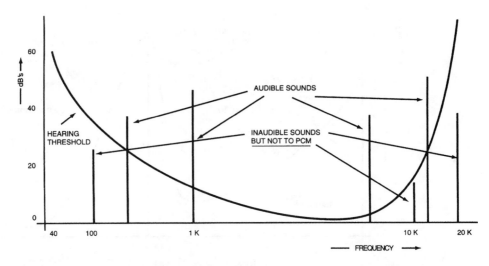

Figure 3.4.1 Generalized frequency response of the human ear. Note how the PCM process captures signals that the ear cannot distinguish. (*From* [2]. *Used with permission.*)

operate, adopting the principle that if the ear is unable to hear the sound then there is no point in transmitting it in the first place.

3.4.2c Quantization

Quantization is the process of converting an analog signal to its representative digital format or, as in the case with compression, the requantizing of an already converted signal [2]. This process is the limiting of a finite level measurement of a signal sample to a specific preset integer value. This means that the *actual* level of the sample may be greater or smaller than the preset *reference* level it is being compared with. The difference between these two levels, called the *quantization error*, is compounded in the decoded signal as *quantization noise*.

Quantization noise, therefore, will be injected into the audio signal after each A/D and D/A conversion, the level of that noise being governed by the bit allocation associated with the coding process (i.e., the number of bits allocated to represent the level of each sample taken of the analog signal). For linear PCM, the bit allocation is commonly 16. The level of each audio sample, therefore, will be compared with one of 2^{16} or 65,536 discrete levels or steps.

Compression or bit-rate reduction of the PCM signal leads to the requantizing of an already quantized signal, which will unavoidably inject further quantization noise. It always has been good operating practice to restrict the number of A/D and D/A conversions in an audio chain. Nothing has changed in this regard, and now the number of compression stages also should be kept to a minimum. Additionally, the bit rates of these stages should be set as high as practical; put another way, the compression ratio should be as low as possible.

Sooner or later—after a finite number of A/D, D/A conversions and passes of compression coding, of whatever type—the accumulation of quantization noise and other unpredictable signal

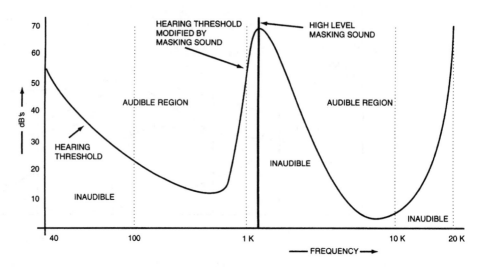

Figure 3.4.2 Example of the masking effect of a high-level sound. (*From* [2]. *Used with permission.*)

degradations eventually will break through the noise/signal threshold, be interpreted as part of the audio signal, be processed as such, and be heard by the listener.

3.4.2d   Sampling Frequency and Bit Rate

The bit rate of a digital signal is defined by

sampling frequency × bit resolution × number of audio channels

The rules regarding the selection of a sampling frequency are based on Nyquist's theorem [2]. This ensures that, in particular, the lower sideband of the sampling frequency does not encroach into the baseband audio. Objectionable and audible aliasing effects would occur if the two bands were to overlap. In practice, the sampling rate is set slightly above twice the highest audible frequency, which makes the filter designs less complex and less expensive.

In the case of a stereo CD with the audio signal having been sampled at 44.1 kHz, this sampling rate produces audio bandwidths of approximately 20 kHz for each channel. The resulting audio bit rate = 44.1 kHz × 16 × 2 = 1.411 Mbits/s, as discussed previously.

3.4.2e   Prediction and Transform Algorithms

Most audio-compression systems are based upon one of two basic technologies [2]:

- Predictive or *adaptive differential* PCM (ADPCM) time-domain coding
- Transform or *adaptive* PCM (APCM) frequency-domain coding

It is in their approaches to dealing with the redundancy and irrelevancy of the PCM signal that these techniques differ.

Subband Coding

In subband coding the PCM signal is split into it into a number of frequency subbands—in one case as few as two, and in another as many as 1024 [1]. Subband coding enables the frequency domain redundancies within the audio signals to be exploited. This permits a reduction in the coded bit rate, compared to PCM, for a given signal fidelity. Spectral redundancies are also present as a result of the signal energies in the various frequency bands being unequal at any instant in time. By altering the bit allocation for each subband, either by dynamically adapting it according to the energy of the contained signal or by fixing it for each subband, the quantization noise can be reduced across all bands. This process compares favorably with the noise characteristics of a PCM coder performing at the same overall bit rate.

Subband Gain

On its own, subband coding, incorporating PCM in each band, is capable of providing a performance improvement or *gain* compared with that of full band PCM coding, both being fed with the same complex, constant level input signal [1]. The improvement is defined as *subband gain* and is the ratio of the variations in quantization errors generated in each case while both are operating at the same transmission rate. The gain increases as the number of subbands increase, and with the complexity of the input signal. However, the implementation of the algorithm also becomes more complex.

Quantization noise generated during the coding process is constrained within each subband and cannot inter-

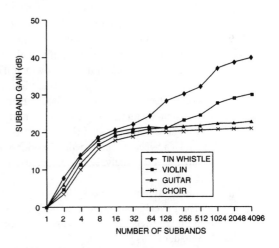

Figure 3.4.3 Variation of subband gain as a function of t number of subbands. (*From* [2]. *Used with permission.*)

fere with any other band. The advantage of this approach is that the masking by each of the subband dominant signals is much more effective because of the reduction in the noise bandwidth. Figure 3.4.3 charts subband gain as a function of the number of subbands for four essentially stationary, but differing, complex audio signals.

In practical implementations of compression codecs, several factors tend to limit the number of subbands employed. The primary considerations include:

- The level variation of normal audio signals leading to an averaging of the energy across bands and a subsequent reduction in the coding gain

- The coding or processing delay introduced by additional subbands

- The overall computational complexity of the system

The two key issues in the analysis of a subband framework are:

- Determining the likely improvement associated with additional subbands

- Determining the relationships between subband gain, the number of subbands, and the response of the filter bank used to create those subbands

APCM Coding

The APCM processor acts in a similar fashion to an automatic gain control system, continually making adjustments in response to the dynamics—at all frequencies—of the incoming audio signal [1]. Transform coding takes a time block of signal, analyzes it for frequency and energy, and identifies irrelevant content. Again, to exploit the spectral response of the ear, the frequency spectrum of the signal is divided into a number of subbands, and the most important criteria are coded with a bias toward the more sensitive low frequencies. At the same time, through the use of psychoacoustic masking techniques, those frequencies which it is assumed will be masked by the ear are also identified and removed. The data generated, therefore, describes the frequency content and the energy level at those frequencies, with more bits being allocated to the higher-energy frequencies than those with lower energy.

The larger the time block of signal being analyzed, the better the frequency resolution and the greater the amount of irrelevancy identified. The penalty, however, is an increase in coding delay and a decrease in temporal resolution. A balance has been struck with advances in perceptual coding techniques and psychoacoustic modeling leading to increased efficiency. It has been found that with this approach to compression, some 80 percent of the input audio can be removed with acceptable results [2].

This hybrid arrangement of working with time-domain subbands and simultaneously carrying out a spectral analysis can be achieved by using a *dynamic bit allocation* process for each subband.

Additionally, some systems exploit the significant redundancy between stereo channels by using a technique known as *joint stereo coding*. After the common information between left and right channels of a stereo signal has been identified, it is coded only once, thus reducing the bit-rate demands yet again.

Each of the subbands has its own defined *masking threshold*. The output data from each of the filtered subbands is requantized with just enough bit resolution to maintain adequate headroom between the quantization noise and the masking threshold for each band. In more complex coders, any spare bit capacity is utilized by those subbands with the greater need for increased masking threshold separation. The maintenance of these signal-to-masking threshold ratios is crucial if further compression is contemplated for any postproduction or transmission process.

3.4.2f   Processing and Propagation Delay

As noted previously, the current range of popular compression algorithms operate—for all intents and purposes—in real time [1]. However, this process does of necessity introduce some measurable delay into the audio chain. All algorithms take a finite time to analyze the incoming signal, which can range from a few milliseconds to tens of milliseconds. The amount of processing delay will be crucial if the equipment is to be used in any interactive or two-way application. As a rule of thumb, any more than 20 ms of delay in a two-way audio exchange is problematic.

Propagation delay in satellite and long terrestrial circuits is a fact of life. A two-way hook up over a 1000 km, full duplex, telecom digital link has a propagation delay of about 3 ms in each direction. This is comparable to having a conversation with someone standing 1 m away. It is obvious that even over a very short distance, the use of a codec with a long processing delay characteristic will have a dramatic effect on operation.

3.4.2g Bit Rate and Compression Ratio

With successive stages of compression, either or both the noise floor and the audio bandwidth will be set by the stage operating at the lowest bit rate. It is, therefore, worth emphasizing that after these platforms have been set by a low bit rate stage, they cannot be subsequently improved by using a following stage operating at a higher bit rate.

A stage of compression may well be followed in the audio chain by another digital stage, either of compression or linear, but—more importantly—operating at a different sampling frequency [1]. If another D/A conversion is to be avoided, a sample rate converter must be used. This can be a stand alone unit or it may already be installed as a module in existing equipment. Where a following stage of compression is operating at the same sampling frequency but a different compression ratio, the bit resolution will change by default.

If the stages have the same sampling frequencies, a direct digital link can be made, thus avoiding the conversion to the analog domain.

3.4.3 Objective Quality Measurements

Perceptual audio coding has revolutionized the processing and distribution of digital audio signals. One aspect of this technology, not often emphasized, is the difficulty of determining, *objectively*, the quality of perceptually coded signals. Audio professionals could greatly benefit from an objective approach to signal characterization because it would offer a simple but accurate approach for verification of good audio quality within a given facility.

Most of the discussions regarding this topic involve reference to the results of subjective evaluations of audio quality, where for example, groups of listeners compare reference audio material to coded audio material and then judge the *level of impairment* caused by the coding process. A procedure for this process has been documented in ITU-R Rec. BS.1116, and makes use of the ITU-R five grade impairment scale:

- 5.0—Imperceptible
- 4.0—Perceptible but not annoying
- 3.0—Slightly annoying
- 2.0—Annoying
- 10—Very annoying

Quality measurements made with properly executed subjective evaluations are widely accepted and have been used for a variety of purposes, from determining which of a group of perceptual coders performs best, to assessing the overall performance of an audio broadcasting system.

The problem with subjective evaluations is that, while accurate, they are time consuming and expensive to undertake. Traditional objective benchmarks of audio performance, such as signal-to-noise ratio or total harmonic distortion, are not reliable measures of perceived audio quality, especially when perceptually coded signals are being considered.

3.4.3a Perspective on Audio Compression

A balance must be struck between the degree of compression available and the level of distortion that can be tolerated, whether the result of a single coding pass or the result of a number of passes, as would be experienced in a complex audio chain or network [1]. Any compression system involves a number of tradeoffs. Quality must be measured against the coding algorithm being used, the compression ratio, bit rate, and coding delay resulting from the process.

There is continued progress in expanding the arithmetical capabilities of digital signal processors, and the supporting hardware developments usually follow a parallel course.

3.4.4 References

1. Wylie, Fred: "Audio Compression Technologies," *NAB Engineering Handbook*, 9th ed., Jerry C. Whitaker (ed.), National Association of Broadcasters, Washington, D.C., 1998.

2. Wylie, Fred: "Audio Compression Techniques," *The Electronics Handbook*, Jerry C. Whitaker (ed.), CRC Press, Boca Raton, Fla., pp. 1260–1272, 1996.

3.4.5 Bibliography

Brandenburg, K., and Gerhard Stoll: "ISO-MPEG-1 Audio: A Generic Standard for Coding of High Quality Digital Audio," *92nd AES Convention Proceedings*, Audio Engineering Society, New York, N.Y., 1992, revised 1994.

Smyth, Stephen: "Digital Audio Data Compression," *Broadcast Engineering*, Intertec Publishing, Overland Park, Kan., February 1992.

Fundamental Transmission Principles

The usable spectrum of electromagnetic radiation frequencies extends over a range from below 100 Hz for power distribution to 10^{20} Hz for the shortest X rays. The lower frequencies are used primarily for terrestrial broadcasting and communications. The higher frequencies include visible and near-visible infrared and ultraviolet light, and X rays. The frequencies typically of interest to RF engineers range from 30 kHz to 30 GHz:

- **Low Frequency** (LF): 30 to 300 kHz: The LF band is used for around-the-clock communications services over long distances and where adequate power is available to overcome high levels of atmospheric noise.

- **Medium Frequency** (MF): 300 kHz to 3 MHz: The low-frequency portion of this band is used for around-the-clock communication services over moderately long distances. The upper portion of the MF band is used principally for moderate-distance voice communications.

- **High Frequency** (HF): 3 to 30 MHz: This band provides reliable medium-range coverage during daylight and, when the transmission path is in total darkness, worldwide long-distance service. The reliability and signal quality of long-distance service depends to a large degree upon ionospheric conditions and related long-term variations in sunspot activity affecting skywave propagation.

- **Very High Frequency** (VHF): 30 to 300 MHz: The VHF band is characterized by reliable transmission over medium distances. At the higher portion of the VHF band, communication is limited by the horizon.

- **Ultrahigh Frequency** (UHF): 300 MHz to 3 GHz: Transmissions in this band are typically line of sight. Short wavelengths at the upper end of the band permit the use of highly directional parabolic or multielement antennas.

- **Superhigh Frequency** (SHF): 3 to 30 GHz: Communication in this band is strictly line of sight. Very short wavelengths permit the use of parabolic transmit and receive antennas of exceptional gain.

This section examines the fundamental principles of RF transmission, with special emphasis on applications relating to digital television.

In This Section:

4.1

The Electromagnetic Spectrum

John Norgard[1]

4.1.1 Introduction

The electromagnetic (EM) spectrum consists of all forms of EM radiation—EM waves (radiant energy) propagating through space, from dc to light to gamma rays. The EM spectrum can be arranged in order of frequency and/or wavelength into a number of regions, usually wide in extent, within which the EM waves have some specified common characteristics, such as characteristics relating to the production or detection of the radiation. A common example is the spectrum of the radiant energy in white light, as dispersed by a prism, to produce a "rainbow" of its constituent colors. Specific frequency ranges are often called *bands*; several contiguous frequency bands are usually called *spectrums*; and sub-frequency ranges within a band are sometimes called *segments*.

The EM spectrum can be displayed as a function of frequency (or wavelength). In air, frequency and wavelength are inversely proportional, $f = c/\lambda$ (where $c \approx 3 \times 10^8$ m/s, the speed of light in a vacuum). The MKS unit of frequency is the Hertz and the MKS unit of wavelength is the meter. Frequency is also measured in the following sub-units:

- Kilohertz, 1 kHz = 10^3 Hz

- Megahertz, 1 MHz = 10^6 Hz

- Gigahertz, 1 GHz = 10^9 Hz

- Terahertz, 1 THz = 10^{12} Hz

- Petahertz, 1 PHz = 10^{15} Hz

- Exahertz, 1 EHz = 10^{18} Hz

Or for very high frequencies, *electron volts*, 1 ev ~ 2.41×10^{14} Hz. Wavelength is also measured in the following sub-units:

- Centimeters, 1 cm = 10^{-2} m

---

1. From *Standard Handbook of Broadcast Engineering*, Jerry C. Whitaker (ed.), McGraw-Hill, New York, N.Y., 2005. Used with permission.

- Millimeters, 1 mm = 10^{-3} m

- Micrometers, 1 μm = 10^{-6} m (microns)

- Nanometers, 1 nm = 10^{-9} m

- Ångstroms, 1 Å = 10^{-10} m

- Picometers, 1 pm = 10^{-12} m

- Femtometers, 1 fm = 10^{-15} m

- Attometers, 1 am = 10^{-18} m

4.1.2   Spectral Sub-Regions

For convenience, the overall EM spectrum can be divided into three main sub-regions:

- *Optical spectrum*

- *DC to light spectrum*

- *Light to gamma ray spectrum*

These main sub-regions of the EM spectrum are next discussed. Note that the boundaries between some of the spectral regions are somewhat arbitrary. Certain spectral bands have no sharp edges and merge into each other, while other spectral segments overlap each other slightly.

4.1.2a   Optical Spectrum

The optical spectrum is the "middle" frequency/wavelength region of the EM spectrum. It is defined here as the visible and near-visible regions of the EM spectrum and includes:

- The *infrared (IR)* band, circa 300 μm–0.7 μm (circa 1 THz–429 THz)

- The *visible light* band, 0.7 μm–0.4 μm (429 THz–750 THz)

- The *ultraviolet (UV)* band, 0.4 μm–circa 10 nm (750 THz–circa 30 PHz), approximately 100 ev

These regions of the EM spectrum are usually described in terms of their wavelengths.

Atomic and molecular radiation produce radiant light energy. Molecular radiation and radiation from hot bodies produce EM waves in the IR band. Atomic radiation (outer shell electrons) and radiation from arcs and sparks produce EM waves in the UV band.

Visible Light Band

In the "middle" of the optical spectrum is the visible light band, extending approximately from 0.4 μm (violet) up to 0.7 μm (red); i.e., from 750 THz (violet) down to 429 THz (red). EM radiation in this region of the EM spectrum, when entering the eye, gives rise to visual sensations (colors), according to the spectral response of the eye, which responds only to radiant energy in the visible light band extending from the extreme long wavelength edge of red to the extreme short wavelength edge of violet. (The spectral response of the eye is sometimes quoted as extending from 0.38 μm (violet) up to 0.75 or 0.78 μm (red); i.e., from 789 THz down to 400 or 385

THz.) This visible light band is further subdivided into the various colors of the rainbow, in decreasing wavelength/increasing frequency:

- Red, a primary color, peak intensity at 700.0 nm (429 THz)

- Orange

- Yellow

- Green, a primary color, peak intensity at 546.1 nm (549 THz)

- Cyan

- Blue, a primary color, peak intensity at 435.8 nm (688 THz)

- Indigo

- Violet

IR Band

The IR band is the region of the EM spectrum lying immediately below the visible light band. The IR band consists of EM radiation with wavelengths extending between the longest visible red (circa 0.7 μm) and the shortest microwaves (300 μm–1 mm); i.e., from circa 429 THz down to 1 THz–300 GHz.

The IR band is further subdivided into the "near" (shortwave), "intermediate" (midwave), and "far" (longwave) IR segments as follows [2]:

- *Near* IR segment, 0.7 μm up to 3 μm (429 THz down to 100 THz)

- *Intermediate* IR segment, 3 μm up to 7 μm (100 THz down to 42.9 THz)

- *Far* IR segment, 7 μm up to 300 μm (42.9 THz down to 1 THz)

- Sub-millimeter band, 100 μm up to 1 mm (3 THz down to 300 GHz). Note that the sub-millimeter region of wavelengths is sometimes included in the very far region of the IR band.

EM radiation is produced by oscillating and rotating molecules and atoms. Therefore, all objects at temperatures above absolute zero emit EM radiation by virtue of their thermal motion (warmth) alone. Objects near room temperature emit most of their radiation in the IR band. However, even relatively cool objects emit some IR radiation; hot objects, such as incandescent filaments, emit strong IR radiation.

IR radiation is sometimes incorrectly called "radiant heat" because warm bodies emit IR radiation and bodies that absorb IR radiation are warmed. However, IR radiation is not itself "heat". This radiant energy is called "black body" radiation. Such waves are emitted by all material objects. For example, the background cosmic radiation (2.7K) emits microwaves; room temperature objects (293K) emit IR rays; the Sun (6000K) emits yellow light; the Solar Corona (1 million K) emits X rays.

---

2. Some reference texts use 2.5 mm (120 THz) as the breakpoint between the near and the intermediate IR bands, and 10 mm (30 THz) as the breakpoint between the intermediate and the far IR bands. Also, 15 mm (20 Thz) is sometimes considered as the long wavelength end of the far IR band.

IR astronomy uses the 1 μm to 1 mm part of the IR band to study celestial objects by their IR emissions. IR detectors are used in night vision systems, intruder alarm systems, weather forecasting, and missile guidance systems. IR photography uses multilayered color film, with an IR sensitive emulsion in the wavelengths between 700–900 nm, for medical and forensic applications, and for aerial surveying.

UV Band

The UV band is the region of the EM spectrum lying immediately above the visible light band. The UV band consists of EM radiation with wavelengths extending between the shortest visible violet (circa 0.4 μm) and the longest X rays (circa 10 nm); i.e., from 750 THz—approximately 3 ev—up to circa 30 PHz—approximately 100 ev.[3]

The UV band is further subdivided into the "near" and the "far" UV segments as follows:

- *Near* UV segment, circa 0.4 μm down to 100 nm (circa 750 THz up to 3 PHz, approximately 3 ev up to 10 ev)

- *Far* UV segment, 100 nm down to circa 10 nm, (3 PHz up to circa 30 PHz, approximately 10 ev up to 100 ev)

The far UV band is also referred to as the *vacuum UV band*, since air is opaque to all UV radiation in this region.

UV radiation is produced by electron transitions in atoms and molecules, as in a mercury discharge lamp. Radiation in the UV range is easily detected and can cause florescence in some substances, and can produce photographic and ionizing effects.

In UV astronomy, the emissions of celestial bodies in the wavelength band between 50–320 nm are detected and analyzed to study the heavens. The hottest stars emit most of their radiation in the UV band.

4.1.2b   DC to Light

Below the IR band are the lower frequency (longer wavelength) regions of the EM spectrum, subdivided generally into the following spectral bands (by frequency/wavelength):

- *Microwave* band, 300 GHz down to 300 MHz (1 mm up to 1 m). Some reference works define the lower edge of the microwave spectrum at 1 GHz.

- *Radio frequency* band, 300 MHz down to 10 kHz (1 m up to 30 Km)

- *Power/telephony* band, 10 kHz down to dc (30 Km up to ∞)

These regions of the EM spectrum are usually described in terms of their frequencies.

Radiations whose wavelengths are of the order of millimeters and centimeters are called *microwaves*, and those still longer are called radio frequency (RF) waves (or *Hertzian waves*).

Radiation from electronic devices produces EM waves in both the microwave and RF bands. Power frequency energy is generated by rotating machinery. Direct current (dc) is produced by batteries or rectified alternating current (ac).

---

3.   Some references use 4, 5, or 6 nm as the upper edge of the UV band.

Microwave Band

The microwave band is the region of wavelengths lying between the far IR/sub-millimeter region and the conventional RF region. The boundaries of the microwave band have not been definitely fixed, but it is commonly regarded as the region of the EM spectrum extending from about 1 mm up to 1 m in wavelengths; i.e., from 300 GHz down to 300 MHz. The microwave band is further sub-divided into the following segments:

- *Millimeter* waves, 300 GHz down to 30 GHz (1 mm up to 1 cm); the Extremely High Frequency band. (Some references consider the top edge of the millimeter region to stop at 100 GHz.)

- *Centimeter* waves, 30 GHz down to 3 GHz (1 cm up to 10 cm); the Super High Frequency band.

The microwave band usually includes the Ultra High Frequency band from 3 GHz down to 300 MHz (from 10 cm up to 1 m). Microwaves are used in radar, space communication, terrestrial links spanning moderate distances, as radio carrier waves in television broadcasting, for mechanical heating, and cooking in microwave ovens.

Radio Frequency (RF) Band

The RF range of the EM spectrum is the wavelength band suitable for utilization in radio communications extending from 10 kHz up to 300 MHz (from 30 Km down to 1 m). (Some references consider the RF band as extending from 10 kHz to 300 GHz, with the microwave band as a subset of the RF band from 300 MHz to 300 GHz.) Some of the radio waves in this band serve as the carriers of low-frequency audio signals; other radio waves are modulated by video and digital information.

In the U.S., the Federal Communications Commission (FCC) is responsible for assigning a range of frequencies to specific services. The International Telecommunications Union (ITU) coordinates frequency band allocation and cooperation on a worldwide basis.

Radio astronomy uses radio telescopes to receive and study radio waves naturally emitted by objects in space. Radio waves are emitted from hot gases (*thermal radiation*), from charged particles spiraling in magnetic fields (*synchrotron radiation*), and from excited atoms and molecules in space (*spectral lines*), such as the 21 cm line emitted by hydrogen gas.

Power Frequency (PF)/Telephone Band

The PF range of the EM spectrum is the wavelength band suitable for generating, transmitting, and consuming low frequency power, extending from 10 kHz down to dc (zero frequency); i.e., from 30 Km up in wavelength. In the US, most power is generated at 60 Hz (some military and computer applications use 400 Hz); in other countries, including Europe, power is generated at 50 Hz.

Frequency Band Designations

The combined microwave, RF (Hertzian Waves), and power/telephone spectra are subdivided into the specific bands given in Table 4.1.1, which lists the international radio frequency band designations and the numerical designations. Note that the band designated (12) has no commonly used name or abbreviation.

Table 4.1.1 Frequency Band Designations

Description	Band Designation	Frequency	Wavelength
Extremely Low Frequency	ELF (1) Band	3 Hz up to 30 Hz	100 Mm down to 10 Mm
Super Low Frequency	SLF (2) Band	30 Hz up to 300 Hz	10 Mm down to 1 Mm
Ultra Low Frequency	ULF (3) Band	300 Hz up to 3 kHz	1 Mm down to 100 Km
Very Low Frequency	VLF (4) Band	3 kHz up to 30 kHz	100 Km down to 10 Km
Low Frequency	LF (5) Band	30 kHz up to 300 kHz	10 Km down to 1 Km
Medium Frequency	MF (6) Band	300 kHz up to 3 MHz	1 Km down to 100 m
High Frequency	HF (7) Band	3 MHz up to 30 MHz	100 m down to 10 m
Very High Frequency	VHF (8) Band	30 MHz up to 300 MHz	10 m down to 1 m
Ultra High Frequency	UHF (9) Band	300 MHz up to 3 GHz	1 m down to 10 cm
Super High Frequency	SHF (10) Band	3 GHz up to 30 GHz	10 cm down to 1 cm
Extremely High Frequency	EHF (11) Band	30 GHz up to 300 GHz	1 cm down to 1 mm
—	(12) Band	300 GHz up to 3 THz	1 mm down to 100 µm

The radar band often is considered to extend from the middle of the High Frequency (7) band to the end of the EHF (11) band. The current U.S. Tri-Service radar band designations are listed in Table 4.1.2. An alternate and more detailed sub-division of the UHF (9), SHF (10), and EHF (11) bands is given in Table 4.1.3. Several other frequency bands of interest (not exclusive) are listed in Tables 4.1.4–4.1.6.

4.1.2c   Light to Gamma Rays

Above the UV spectrum are the higher frequency (shorter wavelength) regions of the EM spectrum, subdivided generally into the following spectral bands (by frequency/wavelength):

- *X ray* band, approximately 10 ev up to 1 Mev (circa 10 nm down to circa 1 pm), circa 3 PHz up to circa 300 EHz

Table 4.1.2 Radar Band Designations

Band	Frequency	Wavelength
A Band	0 Hz up to 250 MHz	∞ down to 1.2 m
B Band	250 MHz up to 500 MHz	1.2 m down to 60 cm
C Band	500 MHz up to 1 GHz	60 cm down to 30 cm
D Band	1 GHz up to 2 GHz	30 cm down to 15 cm
E Band	2 GHz up to 3 GHz	15 cm down to 10 cm
F Band	3 GHz up to 4 GHz	10 cm down to 7.5 cm
G Band	4 GHz up to 6 GHz	7.5 cm down to 5 cm
H Band	6 GHz up to 8 GHz	5 cm down to 3.75 cm
I Band	8 GHz up to 10 GHz	3.75 cm down to 3 cm
J Band	10 GHz up to 20 GHz	3 cm down to 1.5 cm
K Band	20 GHz up to 40 GHz	1.5 cm down to 7.5 mm
L Band	40 GHz up to 60 GHz	7.5 mm down to 5 mm)
M Band	60 GHz up to 100 GHz	5 mm down to 3 mm
N Band	100 GHz up to 200 GHz	3 mm down to 1.5 mm
O Band	200 GHz up to 300 GHz	1.5 mm down to 1 mm

Table 4.1.3 Detail of UHF, SHF, and EHF Band Designations

Band	Frequency	Wavelength
L Band	1.12 GHz up to 1.7 GHz	26.8 cm down to 17.6 cm
LS Band	1.7 GHz up to 2.6 GHz	17.6 cm down to 11.5 cm
S Band	2.6 GHz up to 3.95 GHz	11.5 cm down to 7.59 cm
C(G) Band	3.95 GHz up to 5.85 GHz	7.59 cm down to 5.13 cm
XN(J, XC) Band	5.85 GHz up to 8.2 GHz	5.13 cm down to 3.66 cm
XB(H, BL) Band	7.05 GHz up to 10 GHz	4.26 cm down to 3 cm
X Band	8.2 GHz up to 12.4 GHz	3.66 cm down to 2.42 cm
Ku(P) Band	12.4 GHz up to 18 GHz	2.42 cm down to 1.67 cm
K Band	18 GHz up to 26.5 GHz	1.67 cm down to 1.13 cm
V(R, Ka) Band	26.5 GHz up to 40 GHz	1.13 cm down to 7.5 mm
Q(V) Band	33 GHz up to 50 GHz	9.09 mm down to 6 mm
M(W) Band	50 GHz up to 75 GHz	6 mm down to 4 mm
E(Y) Band	60 GHz up to 90 GHz	5 mm down to 3.33 mm
F(N) Band	90 GHz up to 140 GHz	3.33 mm down to 2.14 mm
G(A)	140 GHz p to 220 GHz	2.14 mm down to 1.36 mm
R Band	220 GHz up to 325 GHz	1.36 mm down to 0.923 mm

- *Gamma ray* band, approximately 1 Kev up to ∞ (circa 300 pm down to 0 m), circa 1 EHz up to ∞

These regions of the EM spectrum are usually described in terms of their photon energies in electron volts. Note that the bottom of the gamma ray band overlaps the top of the X ray band.

It should be pointed out that *cosmic "rays"* (from astronomical sources) are not EM waves (rays) and, therefore, are not part of the EM spectrum. Cosmic rays are high energy charged particles (electrons, protons, and ions) of extraterrestrial origin moving through space, which may have energies as high as 10^{20} ev. Cosmic rays have been traced to cataclysmic astrophysical/cosmological events, such as exploding stars and black holes. Cosmic rays are emitted by supernova remnants, pulsars, quasars, and radio galaxies. Comic rays that collide with molecules in the Earth's upper atmosphere produce secondary cosmic rays and gamma rays of high energy that also contribute to natural background radiation. These gamma rays are sometimes called cosmic or

Table 4.1.4 Low Frequency Bands of Interest

Band	Frequency
Sub-sonic band	0 Hz–10 Hz
Audio band	10 Hz–10 kHz
Ultra-sonic band	10 kHz and up

Table 4.1.5 Applications of Interest in the RF Band

Band	Frequency
Longwave broadcasting band	150–290 kHz
AM broadcasting band	535–1705 kHz (1.640 MHz), 107 channels, 10 kHz separation
International broadcasting band	3–30 MHz
Shortwave broadcasting band	5.95–26.1 MHz (8 bands)
VHF TV (Channels 2 - 4)	54–72 MHz
VHF TV (Channels 5 - 6)	76–88 MHz
FM broadcasting band	88–108 MHz
VHF TV (Channels 7 - 13)	174–216 MHz
UHF TV (Channels 14 - 69)	512–806 MHz

Table 4.1.6 Applications of Interest in the Microwave Band

Application	Frequency
Aero Navigation	0.96–1.215 GHz
GPS Down Link	1.2276 GHz
Military COM/Radar	1.35–1.40 GHz
Miscellaneous COM/Radar	1.40–1.71 GHz
L-Band Telemetry	1.435–1.535 GHz
GPS Down Link	1.57 GHz
Military COM (Troposcatter/Telemetry)	1.71–1.85 GHz
Commercial COM & Private LOS	1.85–2.20 GHz
Microwave Ovens	2.45 GHz
Commercial COM/Radar	2.45–2.69 GHz
Instructional TV	2.50–2.69 GHz
Military Radar (Airport Surveillance)	2.70–2.90 GHz
Maritime Navigation Radar	2.90–3.10 GHz
Miscellaneous Radars	2.90–3.70 GHz
Commercial C-Band SAT COM Down Link	3.70–4.20 GHz
Radar Altimeter	4.20–4.40 GHz
Military COM (Troposcatter)	4.40–4.99 GHz
Commercial Microwave Landing System	5.00–5.25 GHz
Miscellaneous Radars	5.25–5.925 GHz
C-Band Weather Radar	5.35–5.47 GHz
Commercial C-Band SAT COM Up Link	5.925–6.425 GHz
Commercial COM	6.425–7.125 GHz
Mobile TV Links	6.875–7.125 GHz
Military LOS COM	7.125–7.25 GHz
Military SAT COM Down Link	7.25–7.75 GHz
Military LOS COM	7.75–7.9 GHz
Military SAT COM Up Link	7.90–8.40 GHz
Miscellaneous Radars	8.50–10.55 GHz
Precision Approach Radar	9.00–9.20 GHz
X-Band Weather Radar (& Maritime Navigation Radar)	9.30–9.50 GHz
Police Radar	10.525 GHz
Commercial Mobile COM (LOS & ENG)	10.55–10.68 GHz
Common Carrier LOS COM	10.70–11.70 GHz
Commercial COM	10.70–13.25 GHz
Commercial Ku-Band SAT COM Down Link	11.70–12.20 GHz
DBS Down Link & Private LOS COM	12.20–12.70 GHz
ENG & LOS COM	12.75–13.25 GHz
Miscellaneous Radars & SAT COM	13.25–14.00 GHz
Commercial Ku-Band SAT COM Up Link	14.00–14.50 GHz
Military COM (LOS, Mobile, &Tactical)	14.50–15.35 GHz
Aero Navigation	15.40–15.70 GHz
Miscellaneous Radars	15.70–17.70 GHz
DBS Up Link	17.30–17.80 GHz
Common Carrier LOS COM	17.70–19.70 GHz
Commercial COM (SAT COM & LOS)	17.70–20.20 GHz
Private LOS COM	18.36–19.04 GHz
Military SAT COM	20.20–21.20 GHz
Miscellaneous COM	21.20–24.00 GHz
Police Radar	24.15 GHz
Navigation Radar	24.25–25.25 GHz
Military COM	25.25–27.50 GHz
Commercial COM	27.50–30.00 GHz
Military SAT COM	30.00–31.00 GHz
Commercial COM	31.00–31.20 GHz

secondary gamma rays. Cosmic rays are a useful source of high-energy particles for certain scientific experiments.

Radiation from atomic inner shell excitations produces EM waves in the X ray band. Radiation from naturally radioactive nuclei produces EM waves in the gamma ray band.

X Ray Band

The X ray band is further sub-divided into the following segments:

* *Soft* X rays, approximately 10 ev up to 10 Kev (circa 10 nm down to 100 pm), circa 3 PHz up to 3 EHz

* *Hard* X rays, approximately 10 Kev up to 1 Mev (100 pm down to circa 1 pm), 3 EHz up to circa 300 EHz

Because the physical nature of these rays was at first unknown, this radiation was called "X rays." The designation continues to this day. The more powerful X rays are called *hard* X rays and are of high frequencies and, therefore, are more energetic; less powerful X rays are called *soft* X rays and have lower energies.

X rays are produced by transitions of electrons in the inner levels of excited atoms or by rapid deceleration of charged particles (*Brehmsstrahlung* or breaking radiation). An important source of X rays is *synchrotron radiation*. X rays can also be produced when high energy electrons from a heated filament cathode strike the surface of a target anode (usually tungsten) between which a high alternating voltage (approximately 100 kV) is applied.

X rays are a highly penetrating form of EM radiation and applications of X rays are based on their short wavelengths and their ability to easily pass through matter. X rays are very useful in crystallography for determining crystalline structure and in medicine for photographing the body. Because different parts of the body absorb X rays to a different extent, X rays passing through the body provide a visual image of its interior structure when striking a photographic plate. X rays are dangerous and can destroy living tissue. They can also cause severe skin burns. X rays are useful in the diagnosis and non-destructive testing of products for defects.

Gamma Ray Band

The gamma ray band is sub-divided into the following segments:

* *Primary* gamma rays, approximately 1 Kev up to 1 Mev (circa 300 pm down to 300 fm), circa 1 EHz up to 1000 EHz

* *Secondary* gamma rays, approximately 1 Mev up to ∞ (300 fm down to 0 m), 1000 EHz up to ∞

Secondary gamma rays are created from collisions of high energy cosmic rays with particles in the Earth's upper atmosphere.

The primary gamma rays are further sub-divided into the following segments:

* *Soft* gamma rays, approximately 1 Kev up to circa 300 Kev (circa 300 pm down to circa 3 pm), circa 1 EHz up to circa 100 EHz

* *Hard* gamma rays, approximately 300 Kev up to 1 Mev (circa 3 pm down to 300 fm), circa 100 EHz up to 1000 EHz

Gamma rays are essentially very energetic X rays. The distinction between the two is based on their origin. X rays are emitted during atomic processes involving energetic electrons; gamma rays are emitted by excited nuclei or other processes involving sub-atomic particles.

Gamma rays are emitted by the nucleus of radioactive material during the process of natural radioactive decay as a result of transitions from high energy excited states to low energy states in atomic nuclei. Cobalt 90 is a common gamma ray source (with a half-life of 5.26 years). Gamma rays are also produced by the interaction of high energy electrons with matter. "Cosmic" gamma rays cannot penetrate the Earth's atmosphere.

Applications of gamma rays are found both in medicine and in industry. In medicine, gamma rays are used for cancer treatment and diagnoses. Gamma ray emitting radioisotopes are used as tracers. In industry, gamma rays are used in the inspection of castings, seams, and welds.

4.1.3 Bibliography

Collocott, T. C., A. B. Dobson, and W. R. Chambers (eds.): *Dictionary of Science & Technology.*

Handbook of Physics, McGraw-Hill, New York, N.Y., 1958.

Judd, D. B., and G. Wyszecki: *Color in Business, Science and Industry*, 3rd ed., John Wiley and Sons, New York, N.Y.

Kaufman, Ed: *IES Illumination Handbook*, Illumination Engineering Society.

Lapedes, D. N. (ed.): *The McGraw-Hill Encyclopedia of Science & Technology*, 2nd ed., McGraw-Hill, New York, N.Y.

Norgard, John: "Electromagnetic Spectrum," *NAB Engineering Handbook*, 9th ed., Jerry C. Whitaker (ed.), National Association of Broadcasters, Washington, D.C., 1999.

Norgard, John: "Electromagnetic Spectrum," *The Electronics Handbook*, Jerry C. Whitaker (ed.), CRC Press, Boca Raton, Fla., 1996.

Stemson, A: *Photometry and Radiometry for Engineers*, John Wiley and Sons, New York, N.Y.

The Cambridge Encyclopedia, Cambridge University Press, 1990.

The Columbia Encyclopedia, Columbia University Press, 1993.

Webster's New World Encyclopedia, Prentice Hall, 1992.

Wyszecki, G., and W. S. Stiles: *Color Science, Concepts and Methods, Quantitative Data and Formulae*, 2nd ed., John Wiley and Sons, New York, N.Y.

William Daniel, Edward W. Allen, Donald G. Fink[1]

4.2.1 Introduction

The portion of the electromagnetic spectrum commonly used for radio transmissions lies between approximately 10 kHz and 40 GHz. The influence on radio waves of the medium through which they propagate is frequency-dependent. The lower frequencies are greatly influenced by the characteristics of the earth's surface and the ionosphere, while the highest frequencies are greatly affected by the atmosphere, especially rain. There are no clear-cut boundaries between frequency ranges but instead considerable overlap in propagation modes and effects of the path medium.

In the U.S., those frequencies allocated for broadcast-related use include the following:

- 550–1640 kHz: AM radio

- 54–72 MHz: TV channels 2–4

- 76–88 MHz: TV channels 5–6

- 88–108 MHz: FM radio

- 174–216 MHz: TV channels 7–13

- 470–806 MHz: TV channels 14–69

- 0.9–12.2 GHz: nonexclusive TV terrestrial and satellite ancillary services

- 12.2–12.7 GHz: direct satellite broadcasting

- 12.7–40 GHz: nonexclusive direct satellite broadcasting

---

1. From *Standard Handbook of Broadcast Engineering*, Jerry C. Whitaker (ed.), McGraw-Hill, New York, N.Y., 2005. Used with permission.

4.2.2 Propagation in Free Space

For simplicity and ease of explanation, propagation in space and under certain conditions involving simple geometry, in which the wave fronts remain coherent, may be treated as *ray propagation*. It. should be kept in mind that this assumption may not hold in the presence of obstructions, surface roughness, and other conditions which are often encountered in practice.

For the simplest case of propagation in space, namely that of uniform radiation in all directions from a point source, or *isotropic radiator*, it is useful to consider the analogy to a point source of light. The radiant energy passes with uniform intensity through all portions of an imaginary spherical surface located at a radius r from the source. The area of such a surface is $4\pi r^2$ and the power flow per unit area $W = P_t / 4\pi r^2$, where P_t is the total power radiated by the source and W is represented in W/m^2. In the engineering of broadcasting and of some other radio services, it is conventional to measure the intensity of radiation in terms of the strength of the electric field E_o rather than in terms of power density W. The power density is equal to the square of the field strength divided by the impedance of the medium, so for free space

$$W = \frac{E_o^2}{120\pi}$$

(4.2.1)

and

$$P_t = \frac{4\pi r^2 E_o^2}{120\pi}$$

(4.2.2)

or

$$P_t = \frac{r^2 E_o^2}{30}$$

(4.2.3)

Where:
P_t = watts radiated
E_o = the free space field in volts per meter
r = the radius in meters

A more conventional and useful form of this equation, which applies also to antennas other than isotropic radiators, is

$$E_o = \frac{\sqrt{30 g_t P_t}}{r}$$

(4.2.4)

where g_t is the power gain of the antenna in the pertinent direction compared to an isotropic radiator.

An isotropic antenna is useful as a reference for specifying the radiation patterns for more complex antennas but does not in fact exist. The simplest forms of practical antennas are the *electric doublet* and the *magnetic doublet*, the former a straight conductor that is short compared with the wavelength and the latter a conducting loop of short radius compared with the wavelength. For the doublet radiator, the gain is 1.5 and the field strength in the equatorial plane is

$$E_o = \frac{\sqrt{45P_t}}{r} \tag{4.2.5}$$

For a half-wave dipole, namely, a straight conductor one-half wave in length, the power gain is 1.64 and

$$E_o = \frac{7\sqrt{P_t}}{r} \tag{4.2.6}$$

From the foregoing equations it can be seen that for free space:

- The radiation intensity in watts per square meter is proportional to the radiated power and inversely proportional to the square of the radius or distance from the radiator.

- The electric field strength is proportional to the square root of the radiated power and inversely proportional to the distance from the radiator.

4.2.2a   Transmission Loss Between Antennas in Free Space

The maximum useful power P_r that can be delivered to a matched receiver is given by [1]

$$P_r = \left(\frac{E\lambda}{2\pi}\right)^2 \frac{g_r}{120} \text{ W} \tag{4.2.7}$$

Where:
E = received field strength in volts per meter
λ = wavelength in meters, $300/F$
F = frequency in MHz
g_r = receiving antenna power gain over an isotropic radiator

This relationship between received power and the received field strength is shown by scales 2, 3, and 4 in Figure 4.2.1 for a half-wave dipole. For example, the maximum useful power at 100 MHz that can be delivered by a half-wave dipole in a field of 50 dB above 1 μV/m is 95 dB below 1 W.

A general relation for the ratio of the received power to the radiated power obtained from Equations (4.2.4) and (4.2.7) is

$$\frac{P_r}{P_t} = \left(\frac{\lambda}{4\pi r}\right)^2 g_t g_r \left(\frac{E}{E_o}\right)^2 \tag{4.2.8}$$

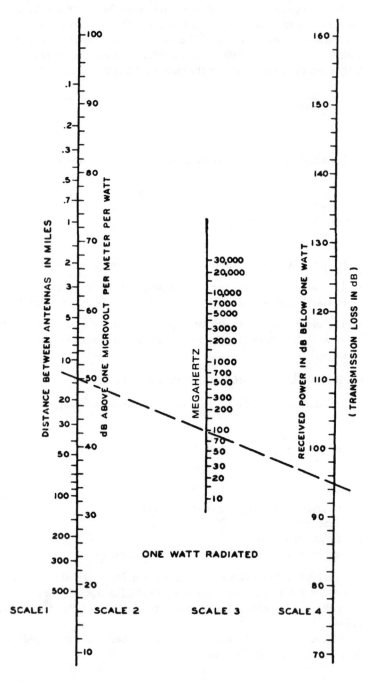

Figure 4.2.1 Free-space field intensity and received power between half-wave dipoles. (*From* [2]. *Used with permission.*)

When both antennas are half-wave dipoles, the power-transfer ratio is

$$\frac{P_r}{P_t} = \left(\frac{1.64\lambda}{4\pi r}\right)^2 \left(\frac{E}{E_o}\right)^2 = \left(\frac{0.13\lambda}{r}\right)^2 \left(\frac{E}{E_o}\right)^2 \tag{4.2.9}$$

and is shown on scales 1 to 4 of Figure 4.2.2. For free-space transmission, $E/E_o = 1$.

When the antennas are horns, paraboloids, or multielement arrays, a more convenient expression for the ratio of the received power to the radiated power is given by

$$\frac{P_r}{P_t} = \frac{B_t B_r}{(\lambda r)^2} \left(\frac{E}{E_o}\right)^2 \tag{4.2.10}$$

where B_t and B_r are the effective areas of the transmitting and receiving antennas, respectively. This relation is obtained from Equation (4.2.8) by substituting as follows

$$g = \frac{4\pi B}{\lambda^2} \tag{4.2.11}$$

This is shown in Figure 4.2.2 for free-space transmission when $B_t = B_r$. For example, the free-space loss at 4000 MHz between two antennas of 10 ft^2 (0.93 m^2) effective area is about 72 dB for a distance of 30 mi (48 km).

4.2.3 Propagation Over Plane Earth

The presence of the ground modifies the generation and propagation of radio waves so that the received field strength is ordinarily different than would be expected in free space [3, 4]. The ground acts as a partial reflector and as a partial absorber, and both of these properties affect the distribution of energy in the region above the earth.

4.2.3a Field Strengths Over Plane Earth

The geometry of the simple case of propagation between two antennas each placed several wavelengths above a plane earth is shown in Figure 4.2.3. For isotropic antennas, for simple magnetic-doublet antennas with vertical polarization, or for simple electric-doublet antennas with horizontal polarization the resultant received field is [4, 5]

$$E = \frac{E_o d}{r_1} + \frac{E_o d R e^{j\Delta}}{r_2} = E_o(\cos\theta_1 + R\cos\theta_2 e^{j\Delta}) \tag{4.2.12}$$

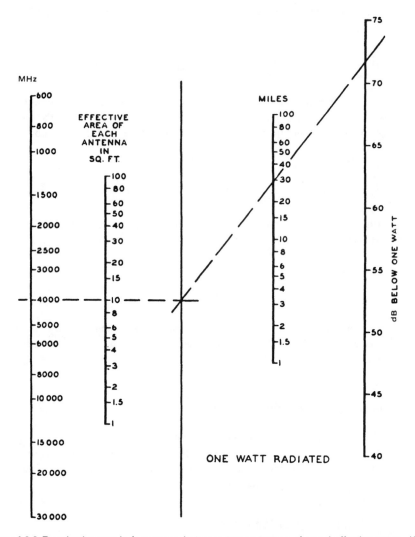

Figure 4.2.2 Received power in free space between two antennas of equal effective areas. (*From* [2]. *Used with permission.*)

For simple magnetic-doublet antennas with horizontal polarization or electric-doublet antennas with vertical polarization at both the transmitter and receiver, it is necessary to correct for the cosine radiation and absorption patterns in the plane of propagation. The received field is

$$E = E_o(\cos^3\theta_1 + R\cos^3\theta_2\, e^{\,j\Delta})$$

(4.2.13)

Where:

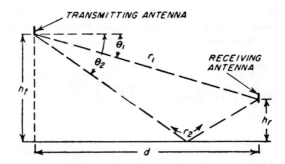

Figure 4.2.3 Ray paths for antennas above plane earth. (*From* [2]. *Used with permission.*)

E_o = the free-space field at distance d in the equatorial plane of the doublet
R = the complex reflection coefficient of the earth
j = the square root of -1
$e^{j\Delta} = \cos \Delta + j \sin \Delta$
Δ = the phase difference between the direct wave received over path r_1 and the ground-reflected wave received over path r_2, which is due to the difference in path lengths

For distances such that θ is small and the differences between d and r_1 and r_2 can be neglected, Equations (4.2.12) and (4.2.13) become

$$E = E_o(1 + Re^{j\Delta}) \qquad\qquad (4.2.14)$$

When the angle θ is very small, R is approximately equal to -1. For the case of two antennas, one or both of which may be relatively close to the earth, a surface-wave term must be added and Equation (4.2.14) becomes [3, 6]

$$E = E_o[1 + Re^{j\Delta} + (1 - R)Ae^{j\Delta}] \qquad\qquad (4.2.15)$$

The quantity A is the *surface-wave attenuation factor*, which depends upon the frequency, ground constants, and type of polarization. It is never greater than unity and decreases with increasing distance and frequency, as indicated by the following approximate equation [1]

$$A \cong \frac{-1}{1 + j\left(\dfrac{2\pi d}{\lambda}\right)(\sin\theta + z)^2} \qquad\qquad (4.2.16)$$

This approximate expression is sufficiently accurate as long as $A < 0.1$, and it gives the magnitude of A within about 2 dB for all values of A. However, as A approaches unity, the error in phase approaches 180°. More accurate values are given by Norton [3] where, in his nomenclature, $A = f(P,B)\, e^{i\phi}$.

The equation (4.2.15) for the absolute value of field strength has been developed from the successive consideration of the various components that make up the ground wave, but the following equivalent expressions may be found more convenient for rapid calculation

$$E = E_o \left\{ 2\sin\frac{\Delta}{2} + j\left[(1+R)+(1-R)A\right]e^{j\Delta/2} \right\}$$

(4.2.17)

When the distance d between antennas is greater than about five times the sum of the two antenna heights h_t and h_r, the phase difference angle Δ (rad) is

$$\Delta = \frac{4\pi h_t h_r}{\lambda d}$$

(4.2.18)

Also, when the angle Δ is greater than about 0.5 rad, the terms inside the brackets of Equation (4.2.17)—which include the surface wave—are usually negligible, and a sufficiently accurate expression is given by

$$E = E_o \left(2\sin\frac{2\pi h_t h_r}{\lambda d} \right)$$

(4.2.19)

In this case, the principal effect of the ground is to produce interference fringes or *lobes*, so that the field strength oscillates about the free-space field as the distance between antennas or the height of either antenna is varied.

When the angle Δ is less than about 0.5 rad, there is a region in which the surface wave may be important but not controlling. In this region, sin $\Delta/2$ is approximately equal to $\Delta/2$ and

$$E = E_o \frac{4\pi h'_t h'_r}{\lambda d}$$

(4.2.20)

In this equation $h' = h + jh_o$, where h is the actual antenna height and $h_o = \lambda/2\pi z$ has been designated as the minimum effective antenna height. The magnitude of the minimum effective height h_o is shown in Figure 4.2.4 for seawater and for "good" and "poor" soil. "Good" soil corresponds roughly to clay, loam, marsh, or swamp, while "poor" soil means rocky or sandy ground [1].

The surface wave is controlling for antenna heights less than the minimum effective height, and in this region the received field or power is not affected appreciably by changes in the antenna height. For antenna heights that are greater than the minimum effective height, the received field or power is increased approximately 6 dB every time the antenna height is doubled, until free-space transmission is reached. It is ordinarily sufficiently accurate to assume that h' is equal to the actual antenna height or the minimum effective antenna height, whichever is the larger.

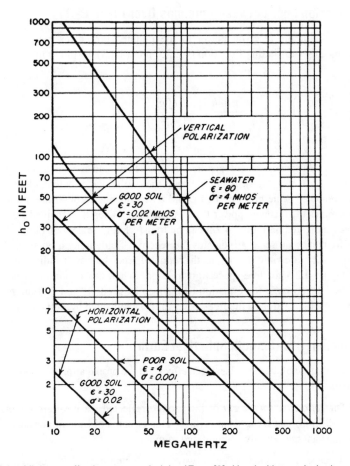

Figure 4.2.4 Minimum effective antenna height. (*From* [2]. *Used with permission.*)

When translated into terms of antenna heights in feet, distance in miles, effective power in kilowatts radiated from a half-wave dipole, and frequency F in megahertz, Equation (4.2.20) becomes the following very useful formula for the rapid calculation of approximate values of field strength for purposes of prediction or for comparison with measured values

$$E \cong F \frac{h_t' \, h_r' \, \sqrt{P_t}}{3 \, d^2}$$

(4.2.21)

4.2.3b Transmission Loss Between Antennas Over Plane Earth

The ratio of the received power to the radiated power for transmission over plane earth is obtained by substituting Equation (4.2.20) into (4.2.8), resulting in

$$
\frac{P_r}{P_t} = \left(\frac{\lambda}{4\pi d} \right)^2 g_t \, g_r \left(\frac{4\pi h_t' h_r'}{\lambda d} \right) = \left(\frac{h_t' h_r'}{d^2} \right)^2 g_t \, g_r
$$

(4.2.22)

This relationship is independent of frequency, and is shown on Figure 4.2.5 for half-wave dipoles ($g_t = g_r = 1.64$). A line through the two scales of antenna height determines a point on the unlabeled scale between them, and a second line through this point and the distance scale determines the received power for 1 W radiated. When the received field strength is desired, the power indicated on Figure 4.2.5 can be transferred to scale 4 of Figure 4.2.1, and a line through the frequency on scale 3 indicates the received field strength on scale 2. The results shown on Figure 4.2.5 are valid as long as the value of received power indicated is lower than that shown on Figure 4.2.3 for free-space transmission. When this condition is not met, it means that the angle Δ is too large for Equation (4.2.20) to be accurate and that the received field strength or power oscillates around the free-space value as indicated by Equation (4.2.19) [1].

4.2.3c Propagation Over Smooth Spherical Earth

The curvature of the earth has three effects on the propagation of radio waves at points within the line of sight:

- The *reflection coefficient* of the ground-reflected wave differs for the curved surface of the earth from that for a plane surface. This effect is of little importance, however, under the circumstances normally encountered in practice.

- Because the ground-reflected wave is reflected against the curved surface of the earth, its energy diverges more than would be indicated by the inverse distance-squared law, and the ground-reflected wave must be multiplied by a divergence factor D.

- The heights of the transmitting and receiving antennas h_t' and h_r', above the plane that is tangent to the surface of the earth at the point of reflection of the ground-reflected wave, are less than the antenna heights h_t and h_r above the surface of the earth, as shown in Figure 4.2.6.

Under these conditions, Equation (4.2.14), which applies to larger distances within the line of sight and to antennas of sufficient height that the surface component may be neglected, becomes

$$
E = E_o (1 + DR' e^{j\Delta})
$$

(4.2.23)

Similar substitutions of the values that correspond in Figures 4.2.3 and 4.2.6 can be made in Equations (4.2.15 through 4.2.22). However, under practical conditions, it is generally satisfactory to use the plane-earth formulas for the purpose of calculating smooth-earth values. An exception to this is usually made in the preparation of standard reference curves, which are generally calculated by the use of the more exact formulas [1, 4–9].

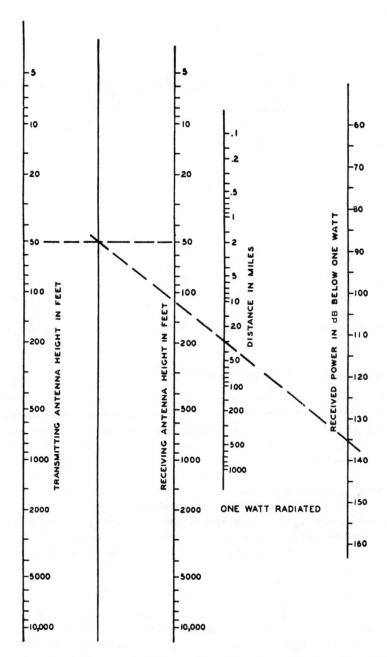

Figure 4.2.5 Received power over plane earth between half-wave dipoles. *Notes:* (1) This chart is not valid when the indicated received power is greater than the free space power shown in Figure 4.2.1. (2) Use the actual antenna height or the minimum effective height shown in Figure 4.2.4, whichever is the larger. (*From* [2]. *Used with permission.*)

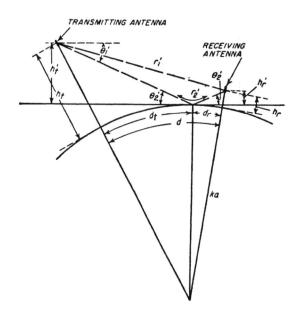

Figure 4.2.6 Ray paths for antennas above spherical earth. (*From* [2]. *Used with permission.*)

4.2.3d   Propagation Beyond the Line of Sight

Radio waves are bent around the earth by the phenomenon of *diffraction*, with the ease of bending decreasing as the frequency increases. Diffraction is a fundamental property of wave motion, and in optics it is the correction to apply to geometrical optics (*ray theory*) to obtain the more accurate *wave optics*. In wave optics, each point on the wave front is considered to act as a radiating source. When the wave front is coherent or undisturbed, the resultant is a progression of the front in a direction perpendicular thereto, along a path that constitutes the ray. When the front is disturbed, the resultant front can be changed in both magnitude and direction with resulting attenuation and bending of the ray. Thus, all shadows are somewhat "fuzzy" on the edges and the transition from "light" to "dark" areas is gradual, rather than infinitely sharp.

The effect of diffraction around the earth's curvature is to make possible transmission beyond the line of sight, with somewhat greater loss than is incurred in free space or over plane earth. The magnitude of this loss increases as either the distance or the frequency is increased and it depends to some extent on the antenna height.

The calculation of the field strength to be expected at any particular point in space beyond the line of sight around a spherical earth is rather complex, so that individual calculations are seldom made except with specially designed software. Rather, nomograms or families of curves are usually prepared for general application to large numbers of cases. The original wave equations of Van der Pol and Bremmer [6] have been modified by Burrows [7] and by Norton [3, 5] so as to make them more readily usable and particularly adaptable to the production of families of curves. Such curves have been prepared by a variety of organizations. These curves have not been included herein, in view of the large number of curves that are required to satisfy the possible variations in frequency, electrical characteristics of the earth, polarization, and antenna

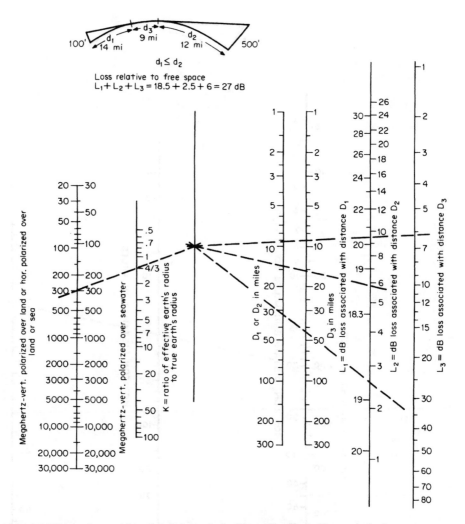

Figure 4.2.7 Loss beyond line of sight in decibels. (*From* [2]. *Used with permission.*)

height. Also, the values of field strength indicated by smooth-earth curves are subject to considerable modification under actual conditions found in practice.

Figure 4.2.7 is a nomogram to determine the additional loss caused by the curvature of the earth [1]. This loss must be added to the free-space loss found from Figure 4.2.1. A scale is included to provide for the effect of changes in the effective radius of the earth, caused by atmospheric refraction. Figure 4.2.7 gives the loss relative to free space as a function of three distances; d_1 is the distance to the horizon from the lower antenna, d_2 is the distance to the horizon from the higher antenna, and d_3 is the distance between the horizons. The total distance between antennas is $d = d_1 + d_2 + d_3$.

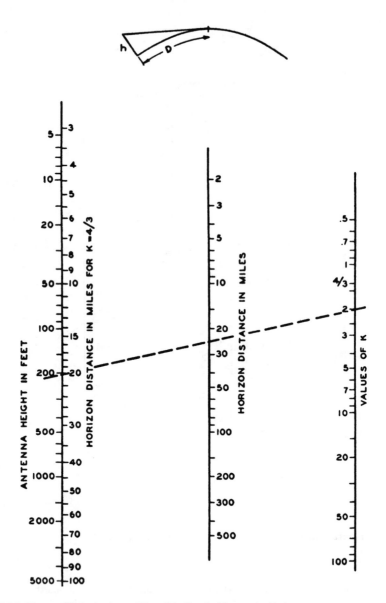

Figure 4.2.8 Distance to the horizon. (*From* [2]. *Used with permission.*)

The horizon distances d_1 and d_2 for the respective antenna heights h_1 and h_2 and for any assumed value of the earth's radius factor k can be determined from Figure 4.2.8 [1].

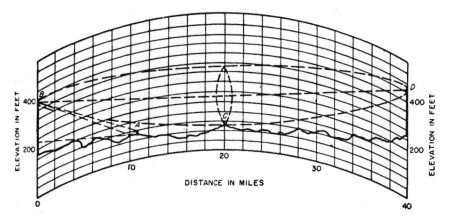

Figure 4.2.9 Ray paths for antennas over rough terrain. (*From* [2]. *Used with permission.*)

4.2.3e Effects of Hills, Buildings, Vegetation, and the Atmosphere

The preceding discussion assumes that the earth is a perfectly smooth sphere with a uniform or a simple atmosphere, for which condition calculations of expected field strengths or transmission losses can be computed for the regions within the line of sight and regions well beyond the line of sight, and interpolations can be made for intermediate distances. The presence of hills, buildings, and trees has such complex effects on propagation that it is impossible to compute in detail the field strengths to be expected at discrete points in the immediate vicinity of such obstructions or even the median values over very small areas. However, by the examination of the earth profile over the path of propagation and by the use of certain simplifying assumptions, predictions that are more accurate than smooth-earth calculations can be made of the median values to be expected over areas representative of the gross features of terrain.

Effects of Hills

The profile of the earth between the transmitting and receiving points is taken from available topographic maps and is plotted on a chart that provides for average air refraction by the use of a 4/3 earth radius, as shown in Figure 4.2.9. The vertical scale is greatly exaggerated for convenience in displaying significant angles and path differences. Under these conditions, vertical dimensions are measured along vertical parallel lines rather than along radii normal to the curved surface, and the propagation paths appear as straight lines. The field to be expected at a low receiving antenna at *A* from a high transmitting antenna at *B* can be predicted by plane-earth methods, by drawing a tangent to the profile at the point at which reflection appears to occur with equal incident and reflection angles. The heights of the transmitting and receiving antennas above the tangent are used in conjunction with Figure 4.2.5 to compute the transmission loss, or with Equation (4.2.21) to compute the field strength. A similar procedure can be used for more distantly spaced high antennas when the line of sight does not clear the profile by at least the first *Fresnel zone* [10].

Propagation over a sharp ridge, or over a hill when both the transmitting and receiving antenna locations are distant from the hill, may be treated as diffraction over a knife edge, shown

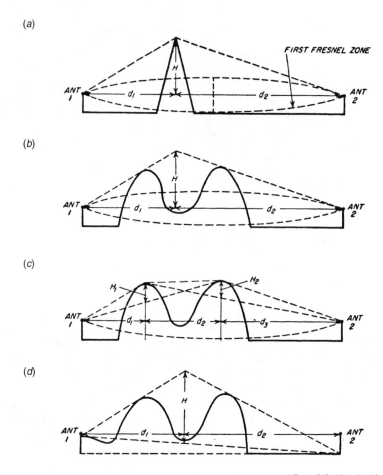

Figure 4.2.10 Ray paths for antennas behind hills: (*a–d*), see text. (*From* [2]. *Used with permission.*)

schematically in Figure 4.2.10*a* [1, 9–14]. The height of the obstruction *H* is measured from the line joining the centers of the two antennas to the top of the ridge. As shown in Figure 4.2.11, the shadow loss approaches 6 dB as *H* approaches 0—*grazing incidence*—and it increases with increasing positive values of *H*. When the direct ray clears the obstruction, *H* is negative, and the shadow loss approaches 0 dB in an oscillatory manner as the clearance is increased. Thus, a substantial clearance is required over line-of-sight paths in order to obtain free-space transmission. There is an optimum clearance, called the first Fresnel-zone clearance, for which the transmission is theoretically 1.2 dB better than in free space. Physically, this clearance is of such magnitude that the phase shift along a line from the antenna to the top of the obstruction and from there to the second antenna is about one-half wavelength greater than the phase shift of the direct path between antennas.

The locations of the first three Fresnel zones are indicated on the right-hand scale on Figure 4.2.11, and by means of this chart the required clearances can be obtained. At 3000 MHz, for example, the direct ray should clear all obstructions in the center of a 40 mi (64 km) path by

Note: When accuracy greater than ±1.5 dB is required, values on the d_1 scale should be:

$$d_1 \frac{\sqrt{2}}{1 + d_1/d_2}$$

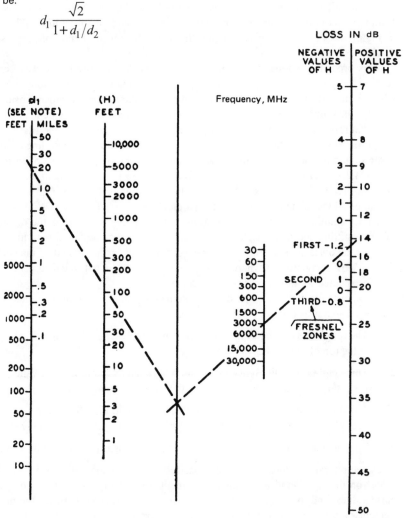

Figure 4.2.11 Shadow loss relative to free space. (*From* [2]. *Used with permission.*)

about 120 ft (36 m) to obtain full first-zone clearance, as shown at "C" in Figure 4.2.9. The corresponding clearance for a ridge 100 ft (30 m) in front of either antenna is 4 ft (1.2 m). The locus of all points that satisfy this condition for all distances is an ellipsoid of revolution with foci at the two antennas.

When there are two or more knife-edge obstructions or hills between the transmitting and receiving antennas, an equivalent knife edge can be represented by drawing a line from each antenna through the top of the peak that blocks the line of sight, as in Figure 4.2.10*b*.

Alternatively, the transmission loss can be computed by adding the losses incurred when passing over each of the successive hills, as in Figure 4.2.10c. The height H_1 is measured from the top of hill 1 to the line connecting antenna 1 and the top of hill 2. Similarly, H_2 is measured from the top of hill 2 to the line connecting antenna 2 and the top of hill 1. The nomogram given in Figure 4.2.11 is used for calculating the losses for terrain conditions represented by Figure 4.2.10a–c.

This procedure applies to conditions for which the earth-reflected wave can be neglected, such as the presence of rough earth, trees, or structures at locations along the profile at points where earth reflection would otherwise take place at the frequency under consideration; or where first Fresnel-zone clearance is obtained in the foreground of each antenna and the geometry is such that reflected components do not contribute to the field within the first Fresnel zone above the obstruction. If conditions are favorable to earth reflection, the base line of the diffraction triangle should not be drawn through the antennas, but through the points of earth reflection, as in Figure 4.2.10d. H is measured vertically from this base line to the top of the hill, while d_1 and d_2 are measured to the antennas as before. In this case, Figure 4.2.12 is used to estimate the shadow loss to be added to the plane-earth attenuation [1].

Under conditions where the earth-reflected components reinforce the direct components at the transmitting and receiving antenna locations, paths may be found for which the transmission loss over an obstacle is less than the loss over spherical earth. This effect may be useful in establishing VHF relay circuits where line-of-sight operation is not practical. Little utility, however, can be expected for mobile or broadcast services [14].

An alternative method for predicting the median value for all measurements in a completely shadowed area is as follows [15]:

1. The roughness of the terrain is assumed to be represented by height H, shown on the profile at the top of Figure 4.2.13.

2. This height is the difference in elevation between the bottom of the valley and the elevation necessary to obtain line of sight with the transmitting antenna.

3. The difference between the measured value of field intensity and the value to be expected over plane earth is computed for each point of measurement within the shadowed area.

4. The median value for each of several such locations is plotted as a function of $\sqrt{(H/\lambda)}$.

These empirical relationships are summarized in the nomogram shown in Figure 4.2.13. The scales on the right-hand line indicate the median value of shadow loss, compared with plane-earth values, and the difference in shadow loss to be expected between the median and the 90 percent values. For example, with variations in terrain of 500 ft (150 m), the estimated median shadow loss at 4500 MHz is about 20 dB and the shadow loss exceeded in 90 percent of the possible locations is about $20 + 15 = 35$ dB. This analysis is based on large-scale variations in field intensity, and does not include the standing-wave effects that sometimes cause the field intensity to vary considerably within a matter of a few feet.

Effects of Buildings

Built-up areas have little effect on radio transmission at frequencies below a few megahertz, since the size of any obstruction is usually small compared with the wavelength, and the shadows caused by steel buildings and bridges are not noticeable except immediately behind these obstructions. However, at 30 MHz and above, the absorption of a radio wave in going through an

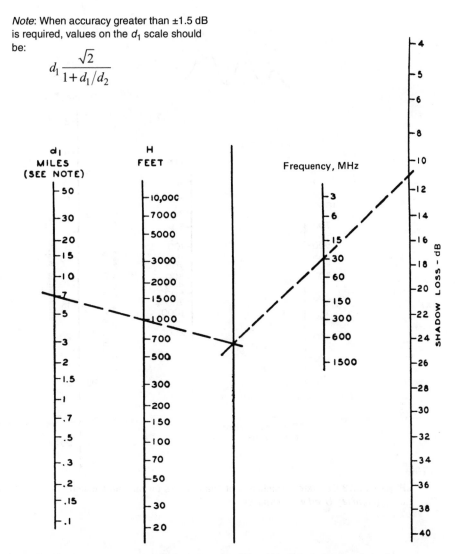

Note: When accuracy greater than ±1.5 dB is required, values on the d_1 scale should be:

$$d_1 \frac{\sqrt{2}}{1 + d_1/d_2}$$

Figure 4.2.12 Shadow loss relative to plane earth. (*From* [2]. *Used with permission.*)

obstruction and the shadow loss in going over it are not negligible, and both types of losses tend to increase as the frequency increases. The attenuation through a brick wall, for example, can vary from 2 to 5 dB at 30 MHz and from 10 to 40 dB at 3000 MHz, depending on whether the wall is dry or wet. Consequently, most buildings are rather opaque at frequencies of the order of thousands of megahertz.

For radio-relay purposes, it is the usual practice to select clear sites; but where this is not feasible the expected fields behind large buildings can be predicted by the preceding diffraction methods. In the engineering of mobile- and broadcast-radio systems it has not been found practi-

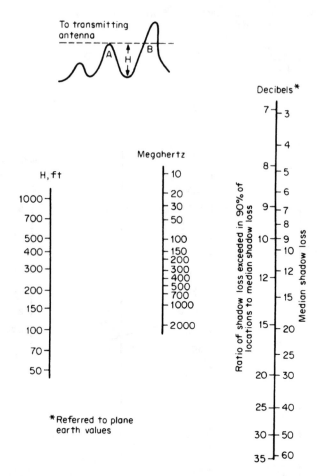

Figure 4.2.13 Estimated distribution of shadow loss for random locations (referred to plane-earth values). (*From* [2]. *Used with permission.*)

cal in general to relate measurements made in built-up areas to the particular geometry of buildings, so that it is conventional to treat them statistically. However, measurements have been divided according to general categories into which buildings can readily be classified, namely, the tall buildings typical of the centers of cities on the one hand, and typical two-story residential areas on the other.

Buildings are more transparent to radio waves than the solid earth, and there is ordinarily much more backscatter in the city than in the open country. Both of these factors tend to reduce the shadow losses caused by the buildings. On the other hand, the angles of diffraction over or around the buildings are usually greater than for natural terrain, and this factor tends to increase the loss resulting from the presence of buildings. Quantitative data on the effects of buildings indicate that in the range of 40 to 450 MHz there is no significant change with frequency, or at least the variation with frequency is somewhat less than the square-root relationship noted in the case of hills. The median field strength at street level for random locations in New York City is about 25 dB below the corresponding plane-earth value. The corresponding values for the 10 percent and 90 percent points are about –15 and –35 dB, respectively [1, 15]. Measurements in congested residential areas indicate somewhat less attenuation than among large buildings.

Effects of Trees and Other Vegetation

When an antenna is surrounded by moderately thick trees and below treetop level, the average loss at 30 MHz resulting from the trees is usually 2 or 3 dB for vertical polarization and negligible with horizontal polarization. However, large and rapid variations in the received field strength can exist within a small area, resulting from the standing-wave pattern set up by reflections from trees located at a distance of as much as 100 ft (30 m) or more from the antenna. Consequently, several nearby locations should be investigated for best results. At 100 MHz, the average loss from surrounding trees may be 5 to 10 dB for vertical polarization and 2 or 3 dB for horizontal polarization. The tree losses continue to increase as the frequency increases, and above 300 to 500 MHz they tend to be independent of the type of polarization. Above 1000 MHz, trees that are thick enough to block vision present an almost solid obstruction, and the diffraction loss over or around these obstructions can be obtained from Figures 4.2.9 or 4.2.11.

There is a pronounced seasonal effect in the case of deciduous trees, with less shadowing and absorption in the winter months when the leaves have fallen. However, when the path of travel through the trees is sufficiently long that it is obscured, losses of the above magnitudes can be incurred, and the principal mode of propagation may be by diffraction over the trees.

When the antenna is raised above trees and other forms of vegetation, the prediction of field strengths again depends upon the proper estimation of the height of the antenna above the areas of reflection and of the applicable reflection coefficients. For growth of fairly uniform height and for angles near grazing incidence, reflection coefficients will approach –1 at frequencies near 30 MHz. As indicated by Rayleigh's criterion of roughness, the apparent roughness for given conditions of geometry increases with frequency so that near 1000 MHz even such low and relatively uniform growth as farm crops or tall grass may have reflection coefficients of about –0.3 for small angles of reflection [17].

The distribution of losses in the immediate vicinity of trees does not follow normal probability law but is more accurately represented by Rayleigh's law, which is the distribution of the sum of a large number of equal vectors having random phases.

4.2.3f   Effects of the Lower Atmosphere (Troposphere)

Radio waves propagating through the lower atmosphere, or troposphere, are subject to absorption, scattering, and bending. Absorption is negligible in the VHF–UHF frequency range but becomes significant at frequencies above 10 GHz. The index of refraction of the atmosphere, n, is slightly greater than 1 and varies with temperature, pressure, and water vapor pressure, and therefore with height, climate, and local meteorological conditions. An exponential model showing a decrease with height to 37 to 43 mi (60 to 70 kin) is generally accepted [18, 19]. For this model, variation of n is approximately linear for the first kilometer above the surface in which most of the effect on radio waves traveling horizontally occurs. For average conditions, the effect of the atmosphere can be included in the expression of earth diffraction around the smooth earth without discarding the useful concept of straight-line propagation by multiplying the actual earth's radius by k to obtain an effective earth's radius, where

$$k = \frac{1}{1 + a\left(dn/dh\right)}$$

(4.2.24)

Where:
a = the actual radius of the earth
dn/dh = the rate of change of the refractive index with height

Through the use of average annual values of the refractive index gradient, k is found to be 4/3 for temperate climates.

Stratification and Ducts

As a result of climatological and weather processes such as *subsidence*, *advection*, and surface heating and radiative cooling, the lower atmosphere tends to be stratified in layers with contrasting refractivity gradients [20]. For convenience in evaluating the effect of this stratification, *radio refractivity N* is defined as $N = (n-1) \times 10^6$ and can be derived from

$$N = 77.6\frac{P}{T} + 3.73 \times 10^5 \, \frac{e}{T^2}$$

(4.2.25)

Where:
P = atmospheric pressure, mbar
T = absolute temperature, K
e = water vapor pressure, mbar

When the gradient of N is equal to –39 N-units per kilometer, normal propagation takes place, corresponding to the effective earth's radius ka, where $k = 4/3$.

When dN/dh is less than –39 N-units per kilometer, *subrefraction* occurs and the radio wave is bent strongly downward.

When dN/dh is less than –157 N-units per kilometer, the radio energy may be bent downward sufficiently to be reflected from the earth, after which the ray is again bent toward the earth, and so on. The radio energy thus is trapped in a duct or waveguide. The wave also may be trapped between two elevated layers, in which case energy is not lost at the ground reflection points and even greater enhancement occurs. Radio waves thus trapped or *ducted* can produce fields exceeding those for free-space propagation because the spread of energy in the vertical direction is eliminated as opposed to the free-space case, where the energy spreads out in two directions orthogonal to the direction of propagation. Ducting is responsible for abnormally high fields beyond the radio horizon. These enhanced fields occur for significant periods of time on overwater paths in areas where meteorological conditions are favorable. Such conditions exist for significant periods of time and over significant horizontal extent in the coastal areas of southern California and around the Gulf of Mexico. Over land, the effect is less pronounced because surface features of the earth tend to limit the horizontal dimension of ducting layers [20].

Tropospheric Scatter

The most consistent long-term mode of propagation beyond the radio horizon is that of scattering by small-scale fluctuations in the refractive index resulting from turbulence. Energy is scattered from multitudinous irregularities in the common volume which consists of that portion of troposphere visible to both the transmitting and receiving sites. There are some empirical data that show a correlation between the variations in the field beyond the horizon and ΔN, the difference between the reflectivity on the ground and at a height of 1 km [21]. Procedures have been developed for calculating scatter fields for beyond-the-horizon radio relay systems as a function of frequency and distance [22, 23]. These procedures, however, require detailed knowledge of path configuration and climate.

The effect of scatter propagation is incorporated in the statistical evaluation of propagation (considered previously in this chapter), where the attenuation of fields beyond the diffraction zone is based on empirical data and shows a linear decrease with distance of approximately 0.2 dB/mi (0.1 dB/km) for the VHF–UHF frequency band.

4.2.3g   Atmospheric Fading

Variations in the received field strengths around the median values are caused by changes in atmospheric conditions. Field strengths tend to be higher in summer than in winter, and higher at night than during the day, for paths over land beyond the line of sight. As a first approximation, the distribution of long-term variations in field strength in decibels follows a normal probability law.

Measurements indicate that the fading range reaches a maximum somewhat beyond the horizon and then decreases slowly with distance out to several hundred miles. Also, the fading range at the distance of maximum fading increases with frequency, while at the greater distances where the fading range decreases, the range is also less dependent on frequency. Thus, the slope of the graph N must be adjusted for both distance and frequency. This behavior does not lend itself to treatment as a function of the earth's radius factor k, since calculations based on the same range of k produce families of curves in which the fading range increases systematically with increasing distance and with increasing frequency.

Effects of the Upper Atmosphere (Ionosphere)

Four principal recognized layers or regions in the ionosphere are the E layer, the $F1$ layer, the $F2$ layer (centered at heights of about 100, 200, and 300 km, respectively), and the D region, which is less clearly defined but lies below the E layer. These *regular* layers are produced by radiation from the sun, so that the ion density—and hence the frequency of the radio waves that can be reflected thereby—is higher during the day than at night. The characteristics of the layers are different for different geographic locations and the geographic effects are not the same for all layers. The characteristics also differ with the seasons and with the intensity of the sun's radiation, as evidenced by the sunspot numbers, and the differences are generally more pronounced upon the $F2$ than upon the $F1$ and E layers. There are also certain random effects that are associated with solar and magnetic disturbances. Other effects that occur at or just below the E layer have been established as being caused by meteors [24].

The greatest potential for television interference by way of the ionosphere is from *sporadic E ionization*, which consists of occasional patches of intense ionization occurring 62 to 75 mi (100 to 120 km) above the earth's surface and apparently formed by the interaction of winds in the neutral atmosphere with the earth's magnetic field. Sporadic *E* ionization can reflect VHF signals back to earth at levels capable of causing interference to some services. In the U.S., VHF sporadic *E* propagation occurs a greater percentage of the time in the southern half of the country and during the May to August period [25].

4.2.4 References

1. Bullington, K.: "Radio Propagation at Frequencies above 30 Mc," *Proc. IRE*, pg. 1122, October 1947.

2. Fink, D. G., (ed.): *Television Engineering Handbook*, McGraw-Hill, New York, N.Y., 1957.

3. Eckersley, T. L.: "Ultra-Short-Wave Refraction and Diffraction," *J. Inst. Elec. Engrs.*, pg. 286, March 1937.

4. Norton, K. A.: "Ground Wave Intensity over a Finitely Conducting Spherical Earth," *Proc. IRE*, pg. 622, December 1941.

5. Norton, K. A.: "The Propagation of Radio Waves over a Finitely Conducting Spherical Earth," *Phil. Mag.*, June 1938.

6. van der Pol, Balth, and H. Bremmer: "The Diffraction of Electromagnetic Waves from an Electrical Point Source Round a Finitely Conducting Sphere, with Applications to Radio-telegraphy and to Theory of the Rainbow," pt. 1, *Phil. Mag.*, July, 1937; pt. 2, *Phil. Mag.*, November 1937.

7. Burrows, C. R., and M. C. Gray: "The Effect of the Earth's Curvature on Groundwave Propagation," *Proc. IRE*, pg. 16, January 1941.

8. "The Propagation of Radio Waves through the Standard Atmosphere," Summary Technical Report of the Committee on Propagation, vol. 3, National Defense Research Council, Washington, D.C., 1946, published by Academic Press, New York, N.Y.

9. "Radio Wave Propagation," Summary Technical Report of the Committee on Propagation of the National Defense Research Committee, Academic Press, New York, N.Y., 1949.

10. de Lisle, E. W.: "Computations of VHF and UHF Propagation for Radio Relay Applications," RCA, Report by International Division, New York, N.Y.

11. Selvidge, H.: "Diffraction Measurements at Ultra High Frequencies," *Proc. IRE*, pg. 10, January 1941.

12. McPetrie, J. S., and L. H. Ford: "An Experimental Investigation on the Propagation of Radio Waves over Bare Ridges in the Wavelength Range 10 cm to 10 m," *J. Inst. Elec. Engrs.*, pt. 3, vol. 93, pg. 527, 1946.

13. Megaw, E. C. S.: "Some Effects of Obstacles on the Propagation of Very Short Radio Waves," *J. Inst. Elec. Engrs.*, pt. 3, vol. 95, no. 34, pg. 97, March 1948.

14. Dickson, F. H., J. J. Egli, J. W. Herbstreit, and G. S. Wickizer: "Large Reductions of VHF Transmission Loss and Fading by the Presence of a Mountain Obstacle in Beyond-Line-of-Sight Paths," *Proc. IRE*, vol. 41, no. 8, pg. 96, August 1953.

15. Bullington, K.: "Radio Propagation Variations at VHF and UHF," *Proc. IRE*, pg. 27, January 1950.

16. "Report of the Ad Hoc Committee, Federal Communications Commission," vol. 1, May 1949; vol. 2, July 1950.

17. Epstein, J., and D. Peterson: "An Experimental Study of Wave Propagation at 850 Mc," *Proc. IRE*, pg. 595, May 1953.

18. "Documents of the XVth Plenary Assembly," CCIR Report 563, vol. 5, Geneva, 1982.

19. Bean, B. R., and E. J. Dutton: "Radio Meteorology," National Bureau of Standards Monograph 92, March 1, 1966.

20. Dougherty, H. T., and E. J. Dutton: "The Role of Elevated Ducting for Radio Service and Interference Fields," NTIA Report 81–69, March 1981.

21. "Documents of the XVth Plenary Assembly," CCIR Report 881, vol. 5, Geneva, 1982.

22. "Documents of the XVth Plenary Assembly," CCIR Report 238, vol. 5, Geneva, 1982.

23. Longley, A. G., and P. L. Rice: "Prediction of Tropospheric Radio Transmission over Irregular Terrain—A Computer Method," ESSA (Environmental Science Services Administration), U.S. Dept. of Commerce, Report ERL (Environment Research Laboratories) 79-ITS 67, July 1968.

24. National Bureau of Standards Circular 462, "Ionospheric Radio Propagation," June 1948.

25. Smith, E. E., and E. W. Davis: "Wind-induced Ions Thwart TV Reception," *IEEE Spectrum*, pp. 52—55, February 1981.

Longley-Rice Propagation Model

4.3.1 Introduction[1]

The Longley-Rice radio propagation model is used to make predictions of radio field strength at specific geographic points based on the elevation profile of terrain between the transmitter and each specific reception point [1]. A computer is needed to make these predictions because of the large number of reception points that must be individually examined. Computer code for the Longley-Rice point-to-point radio propagation model is published in [2].

This chapter describes the process used by the U.S. Federal Communications Commission (FCC) in determining the digital television (DTV) channel plan using the Longley-Rice propagation model.

4.3.2 Evaluation of Service

Under the FCC's rules, computation of service area or coverage using the Longley-Rice methodology is limited to the areas within certain specific geographic contours [1].

For analog television service, computations are made inside the conventional Grade B contour defined in Section 73.683 of the FCC rules, with the exception that the defining field for UHF channels is modified by a dipole factor equal to 20 log [615/(channel mid-frequency)]. Thus, the area subject to calculation for analog TV consists of the geographic points at which the field strength predicted for 50 percent of locations and 50 percent of time by FCC curves is at least as great as the values given in Table 4.3.1. The relevant curves for predicting these fields are the F(50, 50) curves found in Section 73.699 of FCC rules.

For digital television stations, service is evaluated inside contours determined by DTV planning factors in combination with field strength curves derived for 50 percent of locations and 90 percent of the time from curves which are also found in Section 73.699 of FCC rules. The family

---

1. This chapter is based on: FCC: "OET Bulletin No. 69—Longley-Rice Methodology for Evaluating TV Coverage and Interference," Federal Communications Commission, Washington, D.C., July 2, 1997.

Table 4.3.1 Field Strengths Defining the Area Subject to Calculation for Analog Stations (*After* [1].)

Channels	Defining Field Strength, dBu, to be Predicted Using F(50, 50) Curves
2 – 6	47
7 – 13	56
14 – 69	64 – 20 log[615 / (channel mid-frequency)]

Table 4.3.2 Field Strengths Defining the Area Subject to Calculation for DTV Stations (*After* [1].)

Channels	Defining Field Strength, dBu, to be Predicted Using F(50, 90) Curves
2 – 6	28
7 – 13	36
14 – 69	41– 20 log[615 / (channel mid-frequency)]

of FCC propagation curves for predicting field strength at 50 percent of locations 90 percent of the time is found from

$$F(50, 90) = F(50, 50) - [F(50, 10) - F(50, 50)] \qquad (4.3.1)$$

That is, the F(50, 90) value is lower than F(50, 50) by the same amount that F(50, 10) exceeds F(50, 50).

The defining field strengths for DTV service, contained in Section 73.622 of the FCC rules, are shown in Table 4.3.2. These values are determined from the DTV planning factors identified in Table 4.3.3. They are used first to determine the area subject to calculation using FCC curves, and subsequently to determine whether service is present at particular points within this area using Longley-Rice terrain-dependent prediction.

For digital TV, three different situations arise:

- For DTV stations of the initial allotment plan located at the initial reference coordinates, the area subject to calculation extends in each direction to the distance at which the field strength predicted by FCC curves falls to the value identified in Table 4.3.2. The bounding contour is identical, in most cases, to that of the analog station with which the initial allotment is paired. The initial allotment plan and reference coordinates are set forth in [3].

- For new DTV stations, the area subject to calculation extends from the transmitter site to the distance at which the field strength predicted by FCC curves falls to the value identified in Table 4.3.2.

- In the case where a DTV station of the initial allotment plan has moved, the area subject to calculation is the combination (logical union) of the area determined for the initial allotment and the area inside the contour which would apply in the case of a new DTV station.

4.3.2a   Planning Factors

The planning factors shown in Table 4.3.3 lead to the values of field strength given in Table 4.3.2 to define the area subject to calculation for DTV stations [1]. These planning factors are assumed to characterize the equipment, including antenna systems, used for home reception.

Table 4.3.3 Planning Factors for DTV Reception (*After* [1].)

Planning Factor	Symbol	Low VHF	High VHF	UHF
Geometric mean frequency (Mhz)	F	69	194	615
Dipole factor (dBm-dBu)	K_d	−111.8	−120.8	−130.8
Dipole factor adjustment	K_a	none	none	see text
Thermal noise (dBm)	N_t	−106.2	−106.2	−106.2
Antenna gain (dB)	G	4	6	10
Downlead line loss (dB)	L	1	2	4
System noise figure (dB)	N_s	10	10	7
Required carrier/noise ratio (dB)	C/N	15	15	15

They determine the minimum field strength for DTV reception as a function of frequency band and as a function of channel number in the UHF band.

The adjustment, $K_a = 20 \log[615/(\text{channel mid-frequency})]$, is added to K_d to account for the fact that field strength requirements are greater for UHF channels above the geometric mean frequency of the UHF band and smaller for UHF channels below that frequency. The geometric mean frequency, 615 MHz, is approximately the mid-frequency of channel 38.

The modified Grade B contour of analog UHF stations is determined by applying this same adjustment factor to the Grade B field strength given in 47 CFR §73.683. With this dipole factor modification, the field strength defining the Grade B of UHF channels becomes

$$64 - 20 \log[615/(\text{channel mid-frequency})] \text{ dBu}$$

in place of simply 64. Thus, the modified Grade B contour for channel 14 is determined by a median field strength of 61.7 dBu, and the value for channel 51 is 66.3 dBu. The modified values have been presented in Table 4.3.1. This modified Grade B contour bounds the area subject to Longley-Rice calculations for analog stations.

The values appearing in Table 4.3.2 follow from the planning factors. They were derived from Table 4.3.3 by solving the equation

$$\text{Field} + K_d + K_a + G - L - N_t - N_s = C/N \tag{4.3.2}$$

For a new DTV station with a particular authorized set of facilities, the values given in Table 4.3.2 determine the contour within which the FCC makes all subsequent calculations of service and interference.

4.3.2b   Reference Value of ERP for DTV Operation

The initial allotment plan established a reference value for the effective radiated power (ERP) of DTV stations [1]. This ERP is the maximum of the values needed to match the service contour of the paired analog station in each direction supposing that the new station operates at the same location with the same antenna height. The reference ERP was calculated using the following methodology.

The distance to the existing analog grade B contour is determined in each of 360 uniformly spaced compass directions starting from true north using linear interpolation of available data as necessary. This determination is made using information in the FCC Engineering Data Base of April 3, 1997, including directional antenna data, and from terrain elevation data at points separated by 3 arc-seconds of longitude and latitude. FCC curves (Section 73.699 of FCC rules) are applied in the usual way, as described in Section 73.684 of the rules, to find this grade B contour distance, with the exception that dipole factor considerations are applied to the field strength contour for UHF.

Height above average terrain is determined every 45° from terrain elevation data in combination with the height of the transmitter radiation center above mean sea level, and by linear interpolation for compass directions in between. In cases where the TV Engineering Data Base indicates that a directional antenna is employed, the ERP in each specific direction is determined through linear interpolation of the relative field values describing the directional pattern. The directional pattern stored in the FCC Directional Antenna Data Base provides relative field values at 10 degree intervals and may include additional values in special directions. The result of linear interpolation of these relative field values is squared and multiplied by the overall maximum ERP listed for the station in the TV Engineering Data Base to find the ERP in a specific direction.

The corresponding values of ERP for DTV signals in each direction is then calculated by a further application of FCC curves, with noise-limited DTV coverage defined as the presence of the field strengths identified in Table 4.3.2 at 50 percent of locations and 90 percent of the time. These ERP values are computed for all 360 azimuths using the same radial-specific height above average terrain as for the analog TV case, but now in conjunction with $F(50, 90)$ curves.

Finally, the ERP for DTV is modified so that it does not exceed 1 megawatt and is not less than 50 kilowatts. This is done by scaling the azimuthal power, pattern rather than by truncation. Thus, if replication by FCC curves as described above required an ERP of 2 megawatts, the power pattern is reduced by a factor of 2 in all directions. The resulting ERP is the reference value cited in Section 73.622 of the rules.

4.3.2c   DTV Transmitting Antenna Patterns

In general, these computations of DTV power to match the distance to the Grade B contour of an analog station result in ERP values which vary with azimuth [1]. For example, the azimuthal ERP pattern which replicates in UHF the Grade B contour of an omnidirectional VHF operation will be somewhat distorted because terrain has a different effect on propagation in the two bands. In addition, the 90 percent time variability allowance for DTV has an effect on the DTV pattern. Thus, the procedure described here effectively derives a new directional antenna pattern wherever necessary for a precise match according to FCC curves.

These DTV azimuthal patterns may be calculated using the procedure outlined above. In addition, they are available from the FCC's Internet site, www.fcc.gov, to supplement the information contained in Appendix B of the *Sixth Report and Order* [3]. The format for describing DTV transmitting antenna patterns is identical to that of the FCC Directional Antenna Data Base for analog stations. Relative field values are given at intervals of 10 degrees, and supplemental values are given at special azimuths. For DTV patterns, special azimuths are included where the pattern factor is unity, while both bracketing factors at 10-degree azimuths are less than unity.

Table 4.3.4 Parameter Values Used in FCC Implementation of the Longley-Rice Fortran Code (*After* [1].)

Parameter	Value	Meaning/Comment
EPS	15.0	Relative permittivity of ground.
SGM	0.005	Ground conductivity, Siemens per meter.
ZSYS	0.0	Coordinated with setting of EN0. See page 72 of NTIA Report.
EN0	301.0	Surface refractivity in N-units (parts per million).
IPOL	0	Denotes horizontal polarization.
MDVAR	3	Code 3 sets broadcast mode of variability calculations.
KLIM	5	Climate code 5 for continental temperate.
HG(1)	see text	Height of the radiation center above ground.
HG(2)	10 m	Height of TV receiving antenna above ground.

4.3.2d   Application of the Longley-Rice Methodology

The area subject to calculation is divided into rectangular cells, and the Longley-Rice point-to-point propagation model Version 1.2.2 is applied to a point in each cell to determine whether the predicted field strength is above the value found in Table 4.3.1 or Table 4.3.2, as appropriate [1]. The values identified in those tables are considered to be thresholds for reception in the absence of interference. For cells with population, the point chosen by the FCC computer program is the population centroid; otherwise it is the geometric center; and the point so determined represents the cell in all subsequent service and interference calculations. The station's directional transmitting antenna pattern, if any, is taken into account in determining the ERP in the direction of each cell. Cells 2 kilometers on a side were used to produce the service and interference data appearing in Appendix B of the *Sixth Report and Order*.

Parameter values set in the Longley-Rice Fortran code as implemented by the FCC are given in Table 4.3.4. In addition to these parameters, execution of the code requires a specification of the percent of time and locations at which the predicted fields will be realized or exceeded, and a third percentage identifying the degree of confidence desired in the results. To predict TV service at cells of the area subject to calculation, the FCC set the location variability at 50 percent and the time variability at 90 percent. The percent confidence is set at 50 percent, indicating that we are interested in median situations.

HG(1) in Table 4.3.4 is the height of the radiation center above ground. It is determined by subtracting the ground elevation above mean sea level (AMSL) at the transmitter location from the height of the radiation center AMSL. The latter is found in the TV Engineering Data Base while the former is retrieved from the terrain elevation data base as a function of the transmitter site coordinates also found in the TV Engineering Data Base.

Finally, terrain elevation data at uniformly spaced points between the between transmitter and receiver must be provided. The FCC computer program is linked to a terrain elevation data base with values every 3 arc-seconds of latitude and longitude. The program retrieves elevations from this data base at regular intervals with a spacing increment that is chosen at the time the program is compiled. The computer runs which evaluated service and interference for the *Sixth Report and Order* used a spacing increment of 1 kilometer. The elevation of a point of interest is determined by linear interpolation of the values retrieved for the corners of the coordinate rectangle in which the point of interest lies.

Evaluations of service coverage and interference using finer spacing increments were expected to be consistent with those using 1 kilometer. Evaluations using cells smaller than 2 km on a side were also expected to be consistent with the evaluations given in Appendix B of the *Sixth Report and Order*.

4.3.3   Evaluation of Interference

The presence or absence of interference in each grid cell of the area subject to calculation is determined by further application of Longley-Rice. Radio paths between undesired TV transmitters and the point representing each cell are examined [1]. The undesired transmitters included in the analysis of each cell are those which are possible sources of interference at that cell, considering their distance from the cell and channel offset relationships. For each such radio path, the Longley-Rice procedure is applied for median situations (that is, confidence 50 percent), and for 50 percent of locations, 10 percent of the time.

The interference analysis examines only those cells that have already been determined to have a desired field strength above the threshold for reception given in Table 4.3.1 for analog stations and Table 4.3.2 for DTV stations. A cell being examined is counted as having interference if the ratio of the desired field strength to that of any one of the possible interference sources is less than a certain critical minimum value. The comparison is made after applying the discrimination effect of the receiving antenna. The *critical value* is a function of the channel offset relationship.

Cells of the area subject to calculation for an analog station are examined first as to whether the desired signal is above the threshold for reception, second with regard to whether there is interference from another analog station, and finally as to whether there is interference from DTV stations. Thus, a DTV station does not cause interference to analog stations in places where there is no service because of a weak desired signal, or in places where interference from other analogue stations already exists.

4.3.3a   D/U Ratios

Criteria for the ratio of desired to undesired field strength are specified in Section 73.623 of FCC rules for interference involving DTV stations as *desired* or *undesired* [1]. These criteria are summarized in Tables 4.3.5a and 4.3.5b. The tables also include the criteria for interference between analog stations used in preparing the service and interference evaluation in Appendix B of the *Sixth Report and Order.*

The evaluation of service and interference in Appendix B of the *Sixth Report and Order* considered taboo channel relationships for interference into DTV. However, the D/U ratios (approximately –60 dB) were such that they rarely if ever had an effect on the results, and the FCC rules adopted in the *Sixth Report and Order* do not require attention to UHF taboo interference to DTV stations.

4.3.3b   Receiving Antenna Pattern

The receiving antenna is assumed to have a directional gain pattern which tends to discriminate against off-axis undesired stations [1]. This pattern is a planning factor affecting interference. The specific form of this pattern was chosen by a working group of the FCC Advisory Committee on Advanced Television. It is built into the service and interference computer program developed by the Broadcasters' Caucus and also used in the FCC program.

The discrimination, in relative volts, provided by the assumed receiving pattern is a fourth-power cosine function of the angle between the lines joining the desired and undesired stations to the reception point. One of these lines goes directly to the desired station, the other goes to the undesired station. The discrimination is calculated as the fourth power of the cosine of the angle

Table 4.3.5a Interference Criteria for Co- and Adjacent Channels (*After* [1].)

Channel Offset	D/U Ratio, dB			
	Analog into Analog	DTV into Analog	Analog into DTV	DTV into DTV
−1 (lower adjacent)	−3	−17	−48	−42
0 (co-channel)	+28	+34	+2	+15
+1 (upper adjacent)	−13	−12	−49	−43

Table 4.3.5b Interference Criteria for UHF Taboo Channels (*After* [1].)

Channel Offset Relative to Desired Channel N	D/U Ratio, dB			
	Analog into Analog	DTV into Analog	Analog into DTV	DTV into DTV
N − 8	−32	−32	NC	NC
N − 7	−30	−35	NC	NC
N − 4	NC	−34	NC	NC
N - 3	−33	−30	NC	NC
N − 2	−26	−24	NC	NC
N + 2	−29	−28	NC	NC
N + 3	−34	−34	NC	NC
N + 4	−23	−25	NC	NC
N + 7	−33	−34	NC	NC
N + 8	−41	−43	NC	NC
N+14	−25	−33	NC	NC
N+15	−9	−31	NC	NC
(NC means not considered)				

Table 4.3.6 Front-to-Back Ratios Assumed for Receiving Antennas (*After* [1].)

TV Service	Front-to-Back Ratios, dB		
	Low VHF	High VHF	UHF
Analog	6	6	6
DTV	10	12	14

between these lines but never more than represented by the front-to-back ratios identified in Table 4.3.6. When both desired and undesired stations are dead ahead, the angle is 0.0 giving a cosine of unity so that there is no discrimination. When the undesired station is somewhat off-axis, the cosine will be less than unity bringing discrimination into play; and when the undesired station is far off axis, the maximum discrimination given by the front-to-back ratio is attained.

4.3.4   References

1.   FCC: "OET Bulletin No. 69—Longley-Rice Methodology for Evaluating TV Coverage and Interference," Federal Communications Commission, Washington, D.C., July 2, 1997

2. Hufford, G. A., A. G. Longley, and W. A. Kissick: *A Guide to the Use of the ITS Irregular Terrain Model in the Area Prediction Mode*, U.S. Department of Commerce, Washington, D.C., NTIA Report 82-100, April 1982. (Note: some modifications to the code were described by G. A. Hufford in a memorandum to users of the model dated January 30, 1985. With these modifications, the code is referred to as Version 1.2.2 of the Longley-Rice model.)

3. FCC: "Appendix B, Sixth Report and Order," MM Docket 87-268, FCC 97-115, Federal Communications Commission, Washington, D.C., April 3, 1997.

Modulation Systems and Characteristics

4.4.1   Introduction

The primary purpose of most communications and signaling systems is to transfer information from one location to another. The message signals used in communication and control systems usually must be limited in frequency to provide for efficient transfer. This frequency may range from a few hertz for control systems to a few megahertz for video signals to many megahertz for multiplexed data signals. To facilitate efficient and controlled distribution of these components, an *encoder* generally is required between the source and the transmission channel. The encoder acts to *modulate* the signal, producing at its output the *modulated waveform*. Modulation is a process whereby the characteristics of a wave (the *carrier*) are varied in accordance with a message signal, the modulating waveform. Frequency translation is usually a by-product of this process. Modulation may be continuous, where the modulated wave is always present, or pulsed, where no signal is present between pulses.

There are a number of reasons for producing modulated waves, including:

- *Frequency translation*. The modulation process provides a vehicle to perform the necessary frequency translation required for distribution of information. An input signal may be translated to its assigned frequency band for transmission or radiation.

- *Signal processing*. It is often easier to amplify or process a signal in one frequency range as opposed to another.

- *Antenna efficiency*. Generally speaking, for an antenna to be efficient, it must be large compared with the signal wavelength. Frequency translation provided by modulation allows antenna gain and beamwidth to become part of the system design considerations. The use of higher frequencies permits antenna structures of reasonable size and cost.

- *Bandwidth modification*. The modulation process permits the bandwidth of the input signal to be increased or decreased as required by the application. Bandwidth reduction permits more efficient use of the spectrum, at the cost of signal fidelity. Increased bandwidth, on the other hand, provides increased immunity to transmission channel disturbances.

- *Signal multiplexing*. In a given transmission system, it may be necessary or desirable to combine several different signals into one baseband waveform for distribution. Modulation provides the vehicle for such *multiplexing*. Various modulation schemes allow separate signals to be combined at the transmission end and separated (*demultiplexed*) at the receiving end. Multiplexing may be accomplished by using, among other systems, *frequency-domain multiplexing* (FDM) or *time-domain multiplexing* (TDM).

Modulation of a signal does not come without the possible introduction of undesirable attributes. Bandwidth restriction or the addition of noise or other disturbances are the two primary problems faced by the transmission system designer.

4.4.2 Amplitude Modulation

In the simplest form of amplitude modulation, an analog carrier is controlled by an analog modulating signal. The desired result is an RF waveform whose amplitude is varied by the magnitude of the applied modulating signal and at a rate equal to the frequency of the applied signal. The resulting waveform consists of a carrier wave plus two additional signals:

- An upper-sideband signal, which is equal in frequency to the carrier *plus* the frequency of the modulating signal

- A lower-sideband signal, which is equal in frequency to the carrier *minus* the frequency of the modulating signal

This type of modulation system is referred to as *double-sideband amplitude modulation* (DSAM).

The radio carrier wave signal onto which the analog amplitude variations are to be impressed is expressed as

$$e(t) = AE_c \cos(\omega_c t) \tag{4.4.1}$$

Where:
$e(t)$ = instantaneous amplitude of carrier wave as a function of time (t)
A = a factor of amplitude modulation of the carrier wave
ω_c = angular frequency of carrier wave (radians per second)
E_c = peak amplitude of carrier wave

If A is a constant, the peak amplitude of the carrier wave is constant, and no modulation exists. Periodic modulation of the carrier wave results if the amplitude of A is caused to vary with respect to time, as in the case of a sinusoidal wave

$$A = 1 + \left(\frac{E_m}{E_c}\right) \cos(\omega_m t) \tag{4.4.2}$$

where E_m/E_c = the ratio of modulation amplitude to carrier amplitude.
The foregoing relationship leads to

$$e(t) = E_c \left[1 + \left(\frac{E_m}{E_c}\right) \cos(\omega_m t)\cos(\omega_c t)\right] \tag{4.4.3}$$

This is the basic equation for periodic (sinusoidal) amplitude modulation. When all multiplications and a simple trigonometric identity are performed, the result is

$$e(t) = E_c \cos(\omega_c t) + \frac{M}{2} \cos(\omega_c t + \omega_m t) + \frac{M}{2} \cos(\omega_c t - \omega_m t) \qquad (4.4.4)$$

where M = the amplitude modulation factor (E_m/E_c).

Amplitude modulation is, essentially, a multiplication process in which the time functions that describe the modulating signal and the carrier are multiplied to produce a modulated wave containing *intelligence* (information or data of some kind). The frequency components of the modulating signal are translated in this process to occupy a different position in the spectrum.

The bandwidth of an AM transmission is determined by the modulating frequency. The bandwidth required for full-fidelity reproduction in a receiver is equal to twice the applied modulating frequency.

The magnitude of the upper sideband and lower sideband will not normally exceed 50 percent of the carrier amplitude during modulation. This results in an upper-sideband power of one-fourth the carrier power. The same power exists in the lower sideband. As a result, up to one-half of the actual carrier power appears additionally in the sum of the sidebands of the modulated signal. A representation of the AM carrier and its sidebands is shown in Figure 4.4.1. The actual occupied bandwidth, assuming pure sinusoidal modulating signals and no distortion during the modulation process, is equal to twice the frequency of the modulating signal.

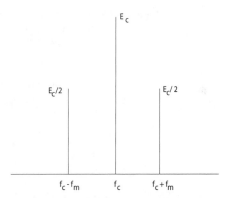

Figure 4.4.1 Frequency-domain representation of an amplitude-modulated signal at 100 percent modulation. E_c = carrier power, f_c = frequency of the carrier, and f_m = frequency of the modulating signal.

The extent of the amplitude variations in a modulated wave is expressed in terms of the *degree of modulation* or *percentage of modulation*. For sinusoidal variation, the degree of modulation m is determined from

$$m = \frac{E_{avg} - E_{min}}{E_{avg}} \qquad (4.4.5)$$

Where:
E_{avg} = average envelope amplitude
E_{min} = minimum envelope amplitude

Full (100 percent) modulation occurs when the peak value of the modulated envelope reaches twice the value of the unmodulated carrier, and the minimum value of the envelope is zero. The envelope of a modulated AM signal in the time domain is shown in Figure 4.4.2.

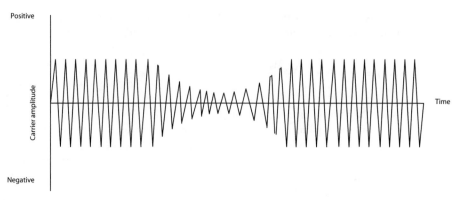

Figure 4.4.2 Time-domain representation of an amplitude-modulated signal. Modulation at 100 percent is defined as the point at which the peak of the waveform reaches twice the carrier level, and the minimum point of the waveform is zero.

When the envelope variation is not sinusoidal, it is necessary to define the degree of modulation separately for the peaks and troughs of the envelope

$$m_{pp} = \frac{E_{max} - E_{avg}}{E_{avg}} \times 100 \qquad (4.4.6)$$

$$m_{np} = \frac{E_{avg} - E_{min}}{E_{avg}} \times 100 \qquad (4.4.7)$$

Where:
m_{pp} = positive peak modulation (percent)
E_{max} = peak value of modulation envelope
m_{np} = negative peak modulation (percent)
E_{avg} = average envelope amplitude
E_{min} = minimum envelope amplitude

When modulation exceeds 100 percent on the negative swing of the carrier, spurious signals are emitted. It is possible to modulate an AM carrier asymmetrically; that is, to restrict modulation in the negative direction to 100 percent, but to allow modulation in the positive direction to exceed 100 percent without a significant loss of fidelity. In fact, many modulating signals normally exhibit asymmetry, most notably human speech waveforms.

The carrier wave represents the average amplitude of the envelope and, because it is the same regardless of the presence or absence of modulation, the carrier transmits no information. The information is carried by the sideband frequencies. The amplitude of the modulated envelope may be expressed as [1]

$$E = E_0 + E_1 \sin(2\pi f_1 t + \Phi_1) + E_2 \sin(2\pi f_2 + \Phi_2) \qquad (4.4.8)$$

Where:

E = envelope amplitude

E_0 = carrier wave crest value, V

$E_1 = 2 \times$ first sideband crest amplitude, V

f_1 = frequency difference between the carrier and the first upper/lower sidebands

$E_2 = 2 \times$ second sideband crest amplitude, V

f_2 = frequency difference between the carrier and the second upper/lower sidebands

Φ_1 = phase of the first sideband component

Φ_2 = phase of the second sideband component

The amplitude-modulated signal can be generated in the basic circuit shown in Figure 4.4.3

4.4.2a   Vestigial-Sideband Amplitude Modulation

Because the intelligence (modulating signal) of conventional AM transmission is identical in the upper *and* lower sidebands, it is possible to eliminate one sideband and still convey the required information. This scheme is implemented in *vestigial-sideband AM* (VSBAM). Complete elimination of one sideband (for example, the lower sideband) requires an ideal high-pass filter with infinitely sharp cutoff. Such a filter is quite difficult to implement in any practical design. VSBAM is a compromise technique wherein one sideband (typically the lower sideband) is attenuated significantly. The result is a savings in occupied bandwidth and transmitter power.

VSBAM is used for analog television broadcast transmission and other applications. A typical bandwidth trace for a VSBAM TV transmitter is shown in Figure 4.4.4.

4.4.2b   Single-Sideband Amplitude Modulation

The carrier in an AM signal does not convey any intelligence. All of the modulating information is in the sidebands. It is possible, therefore, to suppress the carrier upon transmission, radiating only one or both sidebands of the AM signal. The result is much greater efficiency at the transmitter (that is, a reduction in the required transmitter power). Suppression of the carrier may be accomplished with DSAM and SSBAM signals. *Single-sideband suppressed carrier* AM (SSB-SC) is the most spectrum- and energy-efficient mode of AM transmission. Figure 4.4.5 shows representative waveforms for suppressed carrier transmissions.

A waveform with carrier suppression differs from a modulated wave containing a carrier primarily in that the envelope varies at twice the modulating frequency. In addition, it will be noted that the SSB-SC wave has an apparent phase that reverses every time the modulating signal passes through zero. The wave representing a single sideband consists of a number of frequency components, one for each component in the original signal. Each of these components has an amplitude proportional to the amplitude of the corresponding modulating component and a frequency differing from that of the carrier by the modulating frequency. The result is that, in general, the envelope amplitude of the single sideband signal increases with the degree of modulation, and the envelope varies in amplitude in accordance with the difference frequencies formed by the various frequency components of the single sideband interacting with each other.

An SSB-SC system is capable of transmitting a given intelligence within a frequency band only half as wide as that required by a DSAM waveform. Furthermore, the SSB system saves more than two-thirds of the transmission power because of the elimination of one sideband and the carrier.

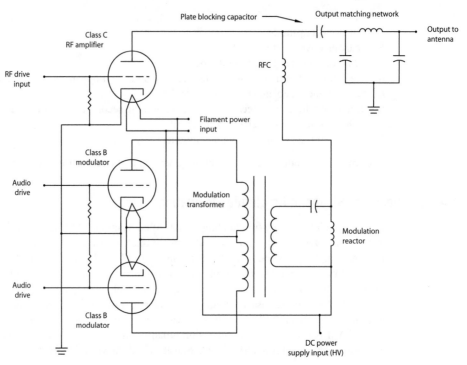

Figure 4.4.3 Simplified diagram of a high-level amplitude-modulated amplifier.

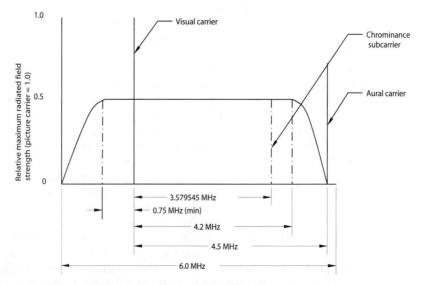

Figure 4.4.4 Idealized amplitude characteristics of the FCC standard waveform for monochrome and color TV transmission. (*Adapted from FCC Rules, Sec. 73.699.*)

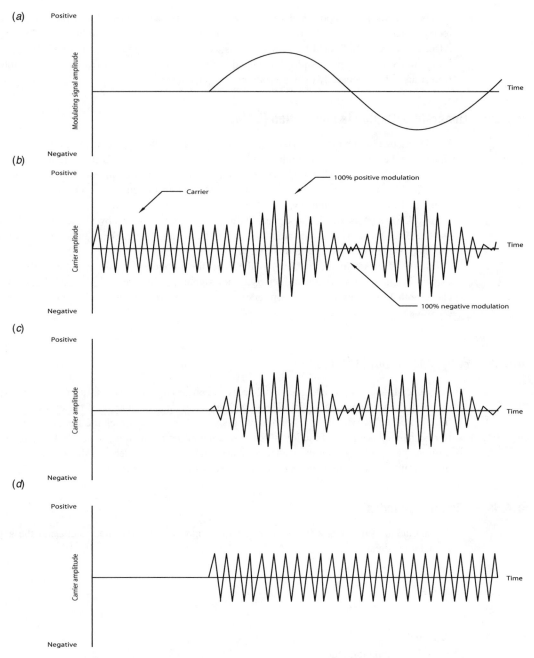

Figure 4.4.5 Types of suppressed carried amplitude modulation: (*a*) the modulating signal, (*b*) double-sideband AM, (*c*) double-sideband suppressed carrier AM, (*d*) single-sideband suppressed carrier AM.

The drawback to suppressed carrier systems is the requirement for a more complicated receiver. The carrier must be regenerated at the receiver to permit demodulation of the signal. Also, in the case of SSBAM transmitters, it is usually necessary to generate the SSB signal in a low-power stage and then amplify the signal with a linear power amplifier to drive the antenna. Linear amplifiers generally exhibit relatively low efficiency.

4.4.2c   Quadrature Amplitude Modulation (QAM)

Single sideband transmission makes very efficient use of the spectrum; for example, two SSB signals can be transmitted within the bandwidth normally required for a single DSB signal. However, DSB signals can achieve the same efficiency by means of *quadrature amplitude modulation* (QAM), which permits two DSB signals to be transmitted and received simultaneously using the same carrier frequency.

Two DSB signals coexist separately within the same bandwidth by virtue of the 90° phase shift between them. The signals are, thus, said to be in *quadrature*. Demodulation uses two local oscillator signals that are also in quadrature; i.e., a sine and a cosine signal.

The chief disadvantage of QAM is the need for a coherent local oscillator at the receiver exactly in phase with the transmitter oscillator signal. Slight errors in phase or frequency can cause both loss of signal and interference between the two signals (cochannel interference or crosstalk).

4.4.3   Frequency Modulation

Frequency modulation is a technique whereby the phase angle or phase shift of a carrier is varied by an applied modulating signal. The *magnitude* of frequency change of the carrier is a direct function of the *magnitude* of the modulating signal. The *rate* at which the frequency of the carrier is changed is a direct function of the *frequency* of the modulating signal. In FM modulation, multiple pairs of sidebands are produced. The actual number of sidebands that make up the modulated wave is determined by the *modulation index* (MI) of the system.

4.4.3a   Modulation Index

The modulation index is a function of the frequency deviation of the system and the applied modulating signal:

$$MI = \frac{F_d}{M_f}$$ (4.4.9)

Where:
MI = the modulation index
F_d = frequency deviation
M_f = modulating frequency

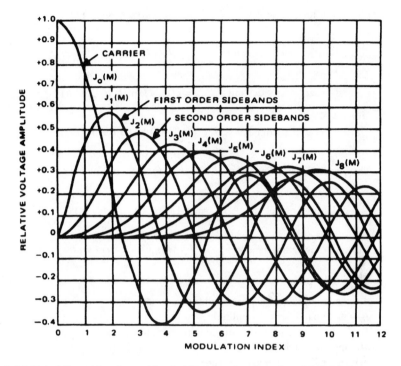

Figure 4.4.6 Plot of Bessel functions of the first kind as a function of modulation index.

The higher the MI, the more sidebands produced. It follows that the higher the modulating frequency for a given deviation, the fewer number of sidebands produced, but the greater their spacing.

To determine the frequency spectrum of a transmitted FM waveform, it is necessary to compute a Fourier series or Fourier expansion to show the actual signal components involved. This work is difficult for a waveform of this type, because the integrals that must be performed in the Fourier expansion or Fourier series are not easily solved. The result, however, is that the integral produces a particular class of solution that is identified as the *Bessel function*, illustrated in Figure 4.4.6.

The carrier amplitude and phase, plus the sidebands, can be expressed mathematically by making the modulation index the argument of a simplified Bessel function. The general expression is given from the following equations:

RF output voltage = $E_t \; = \; E_c + S_{1u} - S_{1l} + S_{2u} - S_{2l} + S_{3u} - S_{3l} + S_{nu} - S_{nl}$

Carrier amplitude = $E_c \; = \; A\,[J_0(M)\sin \omega c\,(t)]$

First-order upper sideband = $S_{1u} \; = \; J_1(M)\sin(\omega\,c + \omega\,m)\,t$

First-order lower sideband = $S_{1l} = J_1(M)\sin(\omega c - \omega m)t$

Second-order upper sideband = $S_{2u} = J_2(M)\sin(\omega c + 2\omega m)t$

Second-order lower sideband = $S_{2l} = J_2(M)\sin(\omega c - 2\omega m)t$

Third-order upper sideband = $S_{3u} = J_3(M)\sin(\omega c + 3\omega m)t$

Third-order lower sideband = $S_{3l} = J_3(M)\sin(\omega c - 3\omega m)t$

Nth-order upper sideband = $S_{nu} = J_n(M)\sin(\omega c + n\omega m)t$

Nth-order lower sideband = $S_{nl} = J_n(M)\sin(\omega c - n\omega m)t$

Where:
A = the unmodulated carrier amplitude constant
J_0 = modulated carrier amplitude
$J_1, J_2, J_3... J_n$ = amplitudes of the nth-order sidebands
M = modulation index
$\omega c = 2\pi F_c$, the carrier frequency
$\omega m = 2\pi F_m$, the modulating frequency

Further supporting mathematics will show that an FM signal using the modulation indices that occur in a wideband system will have a multitude of sidebands. From the purist point of view, *all* sidebands would have to be transmitted, received, and demodulated to reconstruct the modulating signal with complete accuracy. In practice, however, the channel bandwidths permitted practical FM systems usually are sufficient to reconstruct the modulating signal with little discernible loss in fidelity, or at least an acceptable loss in fidelity.

Figure 4.4.7 illustrates the frequency components present for a modulation index of 5. Figure 4.4.8 shows the components for an index of 15. Note that the number of significant sideband components becomes quite large with a high MI. This simple representation of a single-tone frequency-modulated spectrum is useful for understanding the general nature of FM, and for making tests and measurements. When typical modulation signals are applied, however, many more sideband components are generated. These components vary to the extent that sideband energy becomes distributed over the entire occupied bandwidth, rather than appearing at discrete frequencies.

Although complex modulation of an FM carrier greatly increases the number of frequency components present in the frequency-modulated wave, it does not, in general, widen the frequency band occupied by the energy of the wave. To a first approximation, this band is still roughly twice the sum of the maximum frequency deviation at the peak of the modulation cycle plus the highest modulating frequency involved.

FM is not a simple frequency translation, as with AM, but involves the generation of entirely new frequency components. In general, the new spectrum is much wider than the original modu-

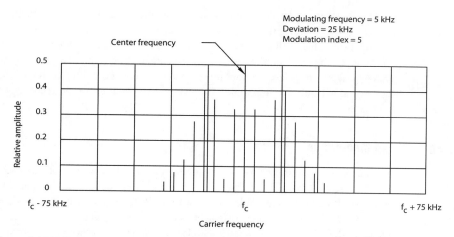

Figure 4.4.7 RF spectrum of a frequency-modulated signal with a modulation index of 5 and operating parameters as shown.

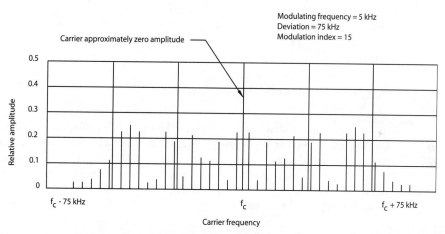

Figure 4.4.8 RF spectrum of a frequency-modulated signal with a modulation index of 15 and operating parameters as shown.

lating signal. This greater bandwidth may be used to improve the *signal-to-noise ratio* (S/N) of the transmission system. FM thereby makes it possible to exchange bandwidth for S/N enhancement.

The power in an FM system is constant throughout the modulation process. The output power is increased in the amplitude modulation system by the modulation process, but the FM system simply distributes the power throughout the various frequency components that are produced by modulation. During modulation, a wideband FM system does not have a high amount of energy present in the carrier. Most of the energy will be found in the sum of the sidebands.

The constant-amplitude characteristic of FM greatly assists in capitalizing on the low noise advantage of FM reception. Upon being received and amplified, the FM signal normally is clipped to eliminate all amplitude variations above a certain threshold. This removes noise picked up by the receiver as a result of man-made or atmospheric signals. It is not possible (generally speaking) for these random noise sources to change the frequency of the desired signal; they can affect only its amplitude. The use of *hard limiting* in the receiver will strip off such interference.

4.4.3b   Phase Modulation

In a phase modulation (PM) system, intelligence is conveyed by varying the phase of the RF wave. Phase modulation is similar in many respects to frequency modulation, except in the interpretation of the modulation index. In the case of PM, the modulation index depends only on the amplitude of the modulation; MI is independent of the frequency of the modulating signal. It is apparent, therefore, that the phase-modulated wave contains the same sideband components as the FM wave and, if the modulation indices in the two cases are the same, the relative amplitudes of these different components also will be the same.

The modulation parameters of a PM system relate as follows:

$$\Delta f = m_p \times f_m \qquad\qquad (4.4.10)$$

Where:
Δf = frequency deviation of the carrier
m_p = phase shift of the carrier
f_m = modulating frequency

In a phase-modulated wave, the phase shift m_p is independent of the modulating frequency; the frequency deviation Δf is proportional to the modulating frequency. In contrast, with a frequency-modulated wave, the frequency deviation is independent of modulating frequency. Therefore, a frequency-modulated wave can be obtained from a phase modulator by making the modulating voltage applied to the phase modulator inversely proportional to frequency. This can be readily achieved in hardware.

4.4.3c   Modifying FM Waves

When a frequency-modulated wave is passed through a harmonic generator, the effect is to increase the modulation index by a factor equal to the frequency multiplication involved. Similarly, if the frequency-modulated wave is passed through a frequency divider, the effect is to reduce the modulation index by the factor of frequency division. Thus, the frequency components contained in the wave—and, consequently, the bandwidth of the wave—will be increased or decreased, respectively, by frequency multiplication or division. No distortion in the nature of the modulation is introduced by the frequency change.

When an FM wave is translated in the frequency spectrum by heterodyne action, the modulation index—hence the relative positions of the sideband frequencies and the bandwidths occupied by them—remains unchanged.

4.4.3d Preemphasis and Deemphasis

The FM transmission/reception system offers significantly better noise-rejection characteristics than AM. However, FM noise rejection is more favorable at low modulating frequencies than at high frequencies because of the reduction in the number of sidebands at higher frequencies. To offset this problem, the input signal to the FM transmitter may be *preemphasized* to increase the amplitude of higher-frequency signal components in normal program material. FM receivers utilize complementary *deemphasis* to produce flat overall system frequency response.

4.4.4 Pulse Modulation

The growth of digital processing and communications has led to the development of modulation systems tailor-made for high-speed, spectrum-efficient transmission. In a *pulse modulation* system, the unmodulated carrier usually consists of a series of recurrent pulses. Information is conveyed by modulating some parameter of the pulses, such as amplitude, duration, time of occurrence, or shape. Pulse modulation is based on the *sampling principle*, which states that a message waveform with a spectrum of finite width can be recovered from a set of discrete samples if the sampling rate is higher than twice the highest sampled frequency (the Nyquist criteria). The samples of the input signal are used to modulate some characteristic of the carrier pulses.

4.4.4a Digital Modulation Systems

Because of the nature of digital signals (on or off), it follows that the amplitude of the signal in a pulse modulation system should be one of two heights (present or absent/positive or negative) for maximum efficiency. Noise immunity is a significant advantage of such a system. It is necessary for the receiving system to detect only the presence or absence (or polarity) of each transmitted pulse to allow complete reconstruction of the original intelligence. The pulse shape and noise level have minimal effect (to a point). Furthermore, if the waveform is to be transmitted over long distances, it is possible to regenerate the original signal exactly for retransmission to the next relay point. This feature is in striking contrast to analog modulation systems in which each modulation step introduces some amount of noise and signal corruption.

In any practical digital data system, some corruption of the intelligence is likely to occur over a sufficiently large span of time. Data encoding and manipulation schemes have been developed to detect and correct or conceal such errors. The addition of error-correction features comes at the expense of increased system overhead and (usually) slightly lower intelligence throughput.

4.4.4b Pulse Amplitude Modulation

Pulse amplitude modulation (PAM) is one of the simplest forms of data modulation. PAM departs from conventional modulation systems in that the carrier exists as a series of pulses, rather than as a continuous waveform. The amplitude of the pulse train is modified in accordance with the applied modulating signal to convey intelligence, as illustrated in Figure 4.4.9. There are two primary forms of PAM sampling:

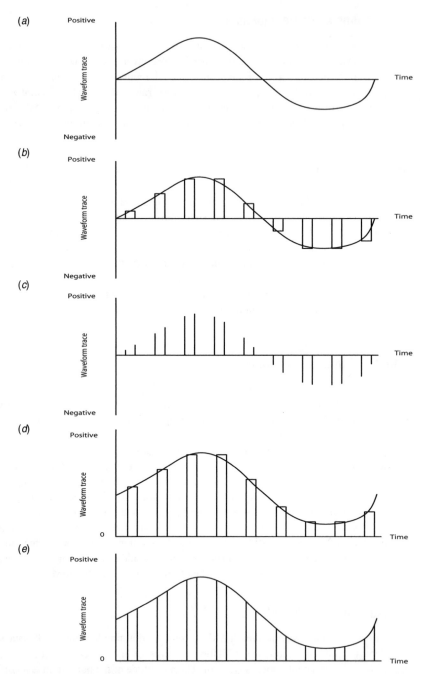

Figure 4.4.9 Pulse amplitude modulation waveforms: (*a*) modulating signal; (*b*) square-topped sampling, bipolar pulse train; (*c*) topped sampling, bipolar pulse train; (*d*) square-topped sampling, unipolar pulse train; (*e*) top sampling, unipolar pulse train.

- *Natural sampling* (or *top sampling*), where the modulated pulses follow the amplitude variation of the sampled time function during the sampling interval.

- *Instantaneous sampling* (or *square-topped sampling*), where the amplitude of the pulses is determined by the instantaneous value of the sampled time function corresponding to a single instant of the sampling interval. This "single instant" may be the center or edge of the sampling interval.

There are two common methods of generating a PAM signal:

- Variation of the amplitude of a pulse sequence about a fixed nonzero value (or *pedestal*). This approach constitutes double-sideband amplitude modulation.

- Double-polarity modulated pulses with no pedestal. This approach constitutes double-sideband suppressed carrier modulation.

4.4.4c   Pulse Time Modulation (PTM)

A number of modulating schemes have been developed to take advantage of the noise immunity afforded by a constant amplitude modulating system. *Pulse time modulation* (PTM) is one of those systems. In a PTM system, instantaneous samples of the intelligence are used to vary the time of occurrence of some parameter of the pulsed carrier. Subsets of the PTM process include:

- *Pulse duration modulation* (PDM), where the time of occurrence of either the leading or trailing edge of each pulse (or both pulses) is varied from its unmodulated position by samples of the input modulating waveform. PDM also may be described as *pulse length* or *pulse width* modulation (PWM).

- *Pulse position modulation* (PPM), where samples of the modulating input signal are used to vary the position in time of pulses, relative to the unmodulated waveform. Several types of pulse time modulation waveforms are shown in Figure 4.4.10.

- *Pulse frequency modulation* (PFM), where samples of the input signal are used to modulate the frequency of a series of carrier pulses. The PFM process is illustrated in Figure 4.4.11.

It should be emphasized that all of the pulse modulation systems discussed thus far may be used with both analog and digital input signals. Conversion is required for either signal into a form that can be accepted by the pulse modulator.

4.4.4d   Pulse Code Modulation

The pulse modulation systems discussed previously are *unencoded* systems. *Pulse code modulation* (PCM) is a scheme wherein the input signal is *quantized* into discrete steps and then sampled at regular intervals (as in conventional pulse modulation). In the *quantization* process, the input signal is sampled to produce a code representing the instantaneous value of the input within a predetermined range of values. Figure 4.4.12 illustrates the concept. Only certain discrete levels are allowed in the quantization process. The code is then transmitted over the communications system as a pattern of pulses.

Quantization inherently introduces an initial error in the amplitude of the samples taken. This *quantization error* is reduced as the number of quantization steps is increased. In system design, tradeoffs must be made regarding low quantization error, hardware complexity, and occupied

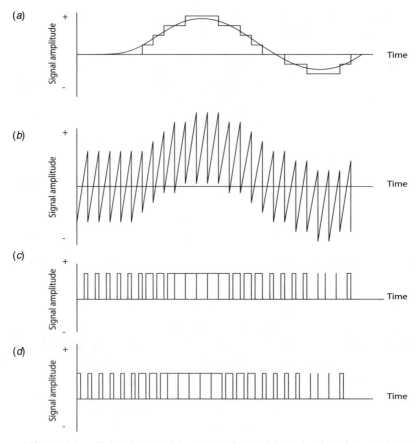

Figure 4.4.10 Pulse time modulation waveforms: (*a*) modulating signal and sample-and-hold (S/H) waveforms, (*b*) sawtooth waveform added to S/H, (*c*) leading-edge PTM, (*d*) trailing-edge PTM.

bandwidth. The greater the number of quantization steps, the wider the bandwidth required to transmit the intelligence or, in the case of some signal sources, the slower the intelligence must be transmitted.

In the classic design of a PCM encoder, the quantization steps are equal. The quantization error (or *quantization noise*) usually can be reduced, however, through the use of nonuniform spacing of levels. Smaller quantization steps are provided for weaker signals, and larger steps are provided near the peak of large signals. Quantization noise is reduced by providing an encoder that is matched to the *level distribution* (*probability density*) of the input signal.

Nonuniform quantization may be realized in an encoder through processing of the input (analog) signal to compress it to match the desired nonuniformity. After compression, the signal is fed to a uniform quantization stage.

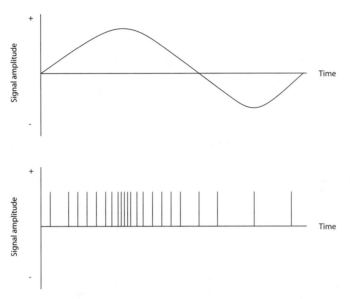

Figure 4.4.11 Pulse frequency modulation.

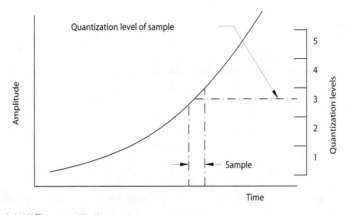

Figure 4.4.12 The quantization process.

4.4.4e Delta Modulation

Delta modulation (DM) is a coding system that measures changes in the direction of the input waveform, rather than the instantaneous value of the wave itself. Figure 4.4.13 illustrates the concept. The clock rate is assumed to be constant. Transmitted pulses from the pulse generator are positive if the signal is changing in a positive direction; they are negative if the signal is changing in a negative direction.

As with the PCM encoding system, quantization noise is a parameter of concern for DM. Quantization noise can be reduced by increasing the sampling frequency (the pulse generator fre-

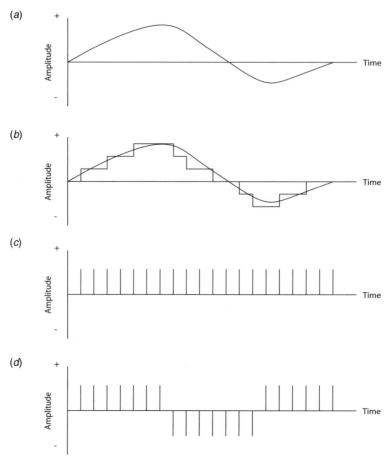

Figure 4.4.13 Delta modulation waveforms: (*a*) modulating signal, (*b*) quantized modulating signal, (*c*) pulse train, (*d*) resulting delta modulation waveform.

quency). The DM system has no fixed maximum (or minimum) signal amplitude. The limiting factor is the slope of the sampled signal, which must not change by more than one level or step during each pulse interval.

4.4.4f Digital Coding Systems

A number of methods exist to transmit digital signals over long distances in analog transmission channels. Some of the more common systems include:

- *Binary on-off keying* (BOOK), a method by which a high-frequency sinusoidal signal is switched on and off corresponding to 1 and 0 (on and off) periods in the input digital data stream. In practice, the transmitted sinusoidal waveform does not start or stop abruptly, but follows a predefined ramp up or down.

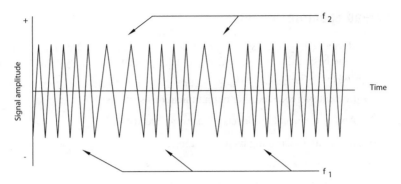

Figure 4.4.14 Binary FSK waveform

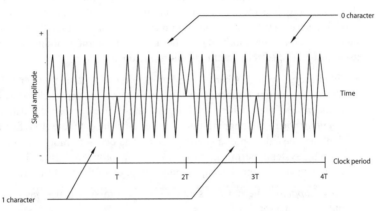

Figure 4.4.15 Binary PSK waveform.

- *Binary frequency-shift keying* (BFSK), a modulation method in which a continuous wave is transmitted that is shifted between two frequencies, representing 1s and 0s in the input data stream. The BFSK signal may be generated by switching between two oscillators (set to different operating frequencies) or by applying a binary baseband signal to the input of a voltage-controlled oscillator (VCO). The transmitted signals often are referred to as a *mark* (binary digit 1) or a *space* (binary digit 0). Figure 4.4.14 illustrates the transmitted waveform of a BFSK system.

- *Binary phase-shift keying* (BPSK), a modulating method in which the phase of the transmitted wave is shifted 180° in synchronism with the input digital signal. The phase of the RF carrier is shifted by $\pi/2$ radians or $-\pi/2$ radians, depending upon whether the data bit is a 0 or a 1. Figure 4.4.15 shows the BPSK transmitted waveform.

- *Quadriphase-shift keying* (QPSK), a modulation scheme similar to BPSK except that quaternary modulation is employed, rather than binary modulation. QPSK requires half the bandwidth of BPSK for the same transmitted data rate.

4.4.5   Spread Spectrum Systems

As the name implies, a *spread spectrum* system requires a frequency range substantially greater than the basic information-bearing signal. Spread spectrum systems have some or all of the following properties:

- Low interference to other communications systems

- Ability to reject high levels of external interference

- Immunity to jamming by hostile forces

- Provision for secure communications paths

- Operability over multiple RF paths

Spread spectrum systems operate with an entirely different set of requirements than transmission systems discussed previously. Conventional modulation methods are designed to provide for the easiest possible reception and demodulation of the transmitted intelligence. The goals of spread spectrum systems, on the other hand, are secure and reliable communications that cannot be intercepted by unauthorized persons. The most common modulating and encoding techniques used in spread spectrum communications include:

- *Frequency hopping*, where a random or *pseudorandom* number (PN) sequence is used to change the carrier frequency of the transmitter. This approach has two basic variations: *slow frequency hopping*, where the hopping rate is smaller than the data rate, and *fast frequency hopping*, where the hopping rate is larger than the data rate. In a fast frequency-hopping system, the transmission of a single piece of data occupies more than one frequency. Frequency-hopping systems permit multiple-access capability to a given band of frequencies because each transmitted signal occupies only a fraction of the total transmitted bandwidth.

- *Time hopping*, where a PN sequence is used to switch the position of a message-carrying pulse within a series of frames.

- *Message corruption*, where a PN sequence is added to the message before modulation.

- *Chirp spread spectrum*, where linear frequency modulation of the main carrier is used to spread the transmitted spectrum. This technique is commonly used in radar and also has been applied to communications systems.

In a spread spectrum system, the signal power is divided over a large bandwidth. The signal, therefore, has a small average power in any single narrowband slot. This means that a spread spectrum system can share a given frequency band with one or more narrowband systems. Furthermore, because of the low energy in any particular band, detection or interception of the transmission is difficult.

4.4.6   References

1. Terman, F. E.: *Radio Engineering*, 3rd ed., McGraw-Hill, New York, N.Y., pg. 468, 1947.

4.4.7   Bibliography

Benson, K. B., and Jerry. C. Whitaker: *Television and Audio Handbook for Technicians and Engineers*, McGraw-Hill, New York, N.Y., 1989.

Crutchfield, E. B. (ed.): *NAB Engineering Handbook*, 8th ed., National Association of Broadcasters, Washington, D.C., 1991.

Fink, D., and D. Christiansen (eds.): *Electronics Engineers' Handbook*, 3rd ed., McGraw-Hill, New York, N.Y., 1989.

Jordan, Edward C. (ed.): *Reference Data for Engineers: Radio, Electronics, Computer and Communications*, 7th ed., Howard W. Sams, Indianapolis, IN, 1985.

Kubichek, Robert: "Amplitude Modulation," in *The Electronics Handbook*, Jerry C. Whitaker (ed.), CRC Press, Boca Raton, Fla., pp. 1175–1187, 1996.

Seymour, Ken: "Frequency Modulation," in *The Electronics Handbook*, Jerry C. Whitaker (ed.), CRC Press, Boca Raton, Fla., pp. 1188–1200, 1996.

Whitaker, Jerry. C.: *Radio Frequency Transmission Systems: Design and Operation*, McGraw-Hill, New York, N.Y., 1991.

Ziemer, Rodger E.: "Pulse Modulation," in *The Electronics Handbook*, Jerry C. Whitaker (ed.), CRC Press, Boca Raton, Fla., pp. 1201–1212, 1996.

Digital Television Standards

This is an exciting time for the broadcast industry and for consumers. Digital television, the product of inspired work by scientists and engineers around the world, has emerged as a powerful force in the marketplace. High-definition to the home, made practical by DTV, has taken the consumer market by storm. In addition to beautiful pictures and stunning sound, DTV enables capabilities never before possible. Indeed, DTV offers features thought to be fantasy little more than a decade ago.

This section examines the DTV Standards developed by the Advanced Television Systems Committee, including supporting capabilities such as data broadcasting. Due to the practical limitations of page constraints in a printed book, other digital television systems—specifically DVB and ISDB—are not covered. Excellent reference books exist for these technologies and readers are encouraged to explore them appropriately.

In This Section:

5.1.1 Introduction[1]

The Digital Television (DTV) Standard has ushered in a new era in television broadcasting [1]. The impact of DTV is more significant than simply moving from an analog system to a digital system. Rather, DTV permits a level of flexibility wholly unattainable with analog broadcasting. An important element of this flexibility is the ability to expand system functions by building upon the technical foundations specified in ATSC standards A/53, "ATSC Digital Television Standard," and A/52, "Digital Audio Compression (AC-3) Standard."

With NTSC, and its PAL and SECAM counterparts, the video, audio, and some limited data information are conveyed by modulating an RF carrier in such a way that a receiver of relatively simple design can decode and reassemble the various elements of the signal to produce a program consisting of video and audio, and perhaps related data (e.g., closed captioning). As such, a complete program is transmitted by the broadcaster that is essentially in finished form. In the DTV system, however, additional levels of processing are required after the receiver demodulates the RF signal. The receiver processes the digital bit stream extracted from the received signal to yield a collection of program elements (video, audio, and/or data) that match the service(s) that the consumer has selected. This selection is made using system and service information that is also transmitted. Audio and video are delivered in digitally compressed form and must be decoded for presentation. Audio may be monophonic, stereo, or multi-channel. Data may supplement the main video/audio program (e.g., closed captioning, descriptive text, or commentary) or it may be a stand-alone service (e.g., a stock or news ticker).

1. This chapter is based on: ATSC, "Guide to the Use of the Digital Television Standard," Advanced Television Systems Committee, Washington, D.C., Doc. A/54A, December 4, 2003; and ATSC, "ATSC Digital Television Standard," Advanced Television Systems Committee, Washington, D.C., Doc. A/53E, December 27, 2005.

Editor's note: This chapter provides an overview of the ATSC DTV transmission system based on ATSC A/53E, and A/54A. For full details on this system, readers are encouraged to download the source documents from the ATSC Web site (http://www.atsc.org). All ATSC Standards, Recommended Practices, and Information Guides are available at no charge.

The nature of the DTV system is such that it is possible to provide new features that build upon the infrastructure within the broadcast plant and the receiver. One of the major enabling developments of digital television, in fact, is the integration of significant processing power in the receiving device itself. Historically, in the design of any broadcast system—be it radio or television—the goal has always been to concentrate technical sophistication (when needed) at the transmission end and thereby facilitate simpler receivers. Because there are far more receivers than transmitters, this approach has obvious business advantages. While this trend continues to be true, the complexity of the transmitted bit stream and compression of the audio and video components require a significant amount of processing power in the receiver, which is practical because of the enormous advancements made in computing technology. Once a receiver reaches a certain level of sophistication (and market success) additional processing power is essentially "free."

The Digital Television Standard describes a system designed to transmit high quality video and audio and ancillary data over a single 6 MHz channel. The system can deliver about 19 Mbits/s in a 6 MHz terrestrial broadcasting channel and about 38 Mbits/s in a 6 MHz cable television channel. This means that encoding HD video essence at 1.106 Gbits/s[2] (highest rate progressive input) or 1.244 Gbits/s[3] (highest rate interlaced picture input) requires a bit rate reduction by about a factor of 50 (when the overhead numbers are added, the rates become closer). To achieve this bit rate reduction, the system is designed to be efficient in utilizing available channel capacity by exploiting complex video and audio compression technology.

The compression scheme optimizes the throughput of the transmission channel by representing the video, audio, and data sources with as few bits as possible while preserving the level of quality required for the given application.

The RF/transmission subsystems described in the Digital Television Standard are designed specifically for terrestrial and cable applications. The structure is such that the video, audio, and service multiplex/transport subsystems are useful in other applications.

5.1.2 DTV System Overview

A basic block diagram representation of the ATSC DTV system is shown in Figure 5.1.1 [2]. The digital television system can be seen to consist of three subsystems [3].

- Source coding and compression

- Service multiplex and transport

- RF/transmission

Source coding and compression refers to the bit-rate reduction methods (data compression) appropriate for application to the video, audio, and ancillary digital data streams. The term ancillary data encompasses the following functions:

- Control data

---

2. $720 \times 1280 \times 60 \times 2 \times 10 = 1.105920$ Gbits/s (the 2 represents the factor needed for 4:2:2 color subsampling, and the 10 is for 10-bit systems).
3. $1080 \times 1920 \times 30 \times 2 \times 10 = 1.244160$ Gbits/s (the 2 represents the factor needed for 4:2:2 color subsampling, and the 10 is for 10-bit systems).

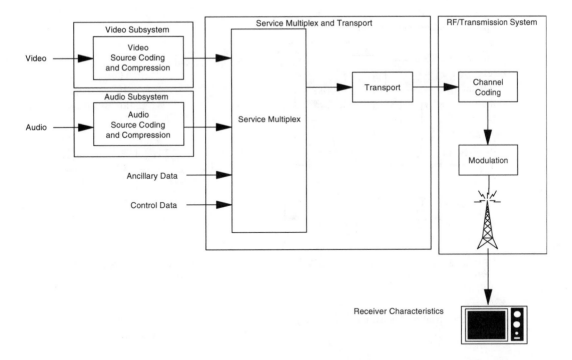

Figure 5.1.1 Digital terrestrial television broadcasting model. (*From* [2]. *Used with permission.*)

- Conditional-access control data

- Data associated with the program audio and video services, such as closed captioning

Ancillary data also can refer to independent program services. The purpose of the coder is to minimize the number of bits needed to represent the audio and video information.

Service multiplex and transport refers to the means of dividing the digital data stream into *packets* of information, the means of uniquely identifying each packet or packet type, and the appropriate methods of multiplexing video data-stream packets, audio data-stream packets, and ancillary data-stream packets into a single data stream. In developing the transport mechanism, interoperability among digital media—such as terrestrial broadcasting, cable distribution, satellite distribution, recording media, and computer interfaces—was a prime consideration. The DTV system employs the MPEG-2 transport-stream syntax for the packetization and multiplexing of video, audio, and data signals for digital broadcasting systems [4]. The MPEG-2 transport-stream syntax was developed for applications where channel bandwidth or recording media capacity is limited, and the requirement for an efficient transport mechanism is paramount.

RF/Transmission refers to channel coding and modulation. The channel coder takes the data bit stream and adds additional information that can be used by the receiver to reconstruct the data from the received signal which, because of transmission impairments, may not accurately represent the transmitted signal. The modulation (or *physical layer*) uses the digital data-stream information to modulate the transmitted signal. The modulation subsystem offers two modes:

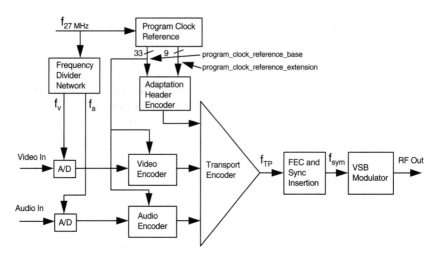

Figure 5.1.2 High-level view of the DTV encoding system. (*From* [2]. *Used with permission.*)

- Terrestrial broadcast mode (8-VSB)

- High-data-rate mode (16-VSB).

Figure 5.1.2 gives a high-level view of the encoding equipment. This view is not intended to be complete, but is used to illustrate the relationship of various clock frequencies within the encoder. There are two domains within the encoder where a set of frequencies are related: the source-coding domain and the channel-coding domain.

The source-coding domain, represented schematically by the video, audio, and transport encoders, uses a family of frequencies that are based on a 27 MHz clock (f_{27MHz}). This clock is used to generate a 42-bit sample of the frequency, which is partitioned into two elements defined by the MPEG-2 specification:

- The 33-bit *program clock reference base*

- The 9-bit *program clock reference extension*

The 33-bit program clock reference base is equivalent to a sample of a 90 kHz clock that is locked in frequency to the 27 MHz clock, and is used by the audio and video source encoders when encoding the *Presentation Time Stamp* (PTS) and the *Decode Time Stamp* (DTS). The audio and video sampling clocks, f_a and f_v, respectively, must be frequency-locked to the 27 MHz clock. This condition can be expressed as the requirement that there exist two pairs of integers, (n_a, m_a) and (n_v, m_v), such that

$$f_a = \frac{n_a}{m_a} \times 27 \ \text{MHz} \tag{5.1.1}$$

and

$$f_v = \frac{n_v}{m_v} \times 27 \text{ MHz} \tag{5.1.2}$$

The channel-coding domain is represented by the forward error correction/sync insertion subsystem and the vestigial sideband (VSB) modulator. The relevant frequencies in this domain are the VSB symbol frequency (f_{sym}) and the frequency of the transport stream (f_{TP}), which is the frequency of transmission of the encoded transport stream. These two frequencies must be locked, having the relation

$$f_{TP} = 2 \times \frac{188}{208} \frac{312}{313} f_{sym} \tag{5.1.3}$$

The signals in the two domains are not required to be frequency-locked to each other and, in many implementations, will operate asynchronously. In such systems, the frequency drift can necessitate the occasional insertion or deletion of a *null* packet from within the transport stream, thereby accommodating the frequency disparity.

A basic block diagram representation of the overall ATSC DTV system is shown in Figure 5.1.3.

5.1.2a   Basic Video Systems Characteristics

Table 5.1.1 lists the primary television production standards that define video formats relating to compression techniques applicable to the ATSC DTV standard. These picture formats may be derived from one or more appropriate video input formats.

The DTV video-compression algorithm conforms to the Main Profile syntax of ISO/IEC 13818-2 (MPEG-2). The allowable parameters are bounded by the upper limits specified for the Main Profile/High Level. Table 5.1.2 lists the allowed compression formats under the ATSC DTV standard.

5.1.2b   Receiver

The ATSC receiver must recover the bits representing the original video, audio and other data from the modulated signal [1]. In particular, the receiver must:

- Tune the selected 6 MHz channel

- Reject adjacent channels and other sources of interference

- Demodulate (equalize as necessary) the received signal, applying error correction to produce a transport bit stream

- Identify the elements of the bit stream using a transport layer processor

- Select each desired element and send it to its appropriate processor

- Decode and synchronize each element

- Present the programming

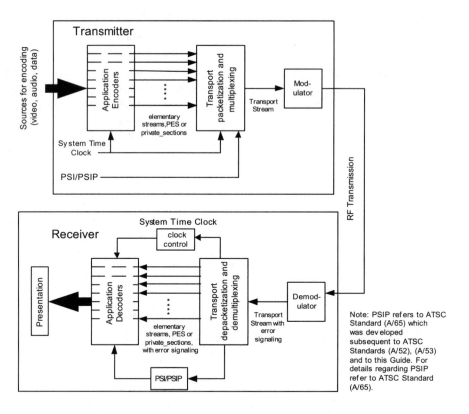

Figure 5.1.3 Sample organization of functionality in a transmitter-receiver pair for a single DTV program. (*From* [1]. *Used with permission.*)

Table 5.1.1 Standardized Video Input Formats

Video Standard	Active Lines	Active Samples/Line
SMPTE 274M, SMPTE 295M (50 Hz)	1080	1920
SMPTE 296M	720	1280
ITU-R Rec. 601-4	483	720
SMPTE 293M (59.94, p)		
SMPTE 294M (59.94, p)		

Noise, interference, and multipath are elements of the terrestrial transmission path, and the receiver circuits are expected to deal with these impairments.

Table 5.1.2 ATSC DTV Compression Format Constraints (*After* [2].)

Vertical Size Value	Horizontal Size Value	Aspect Ratio	Frame-Rate Code	Scanning Sequence
1080[1]	1920	16:9, square pixels	1,2,4,5	Progressive
			4,5	Interlaced
720	1280	16:9, square pixels	1,2,4,5,7,8	Progressive
480	704	4:3, 16:9	1,2,4,5,7,8	Progressive
			4,5	Interlaced
	640	4:3, square pixels	1,2,4,5,7,8	Progressive
			4,5	Interlaced

Frame-rate code: 1 = 23.976 Hz, 2 = 24 Hz, 4 = 29.97 Hz, 5 = 30 Hz, 7 = 59.94 Hz, 8 = 60 Hz

[1] Note that 1088 lines actually are coded in order to satisfy the MPEG-2 requirement that the coded vertical size be a multiple of 16 (progressive scan) or 32 (interlaced scan).

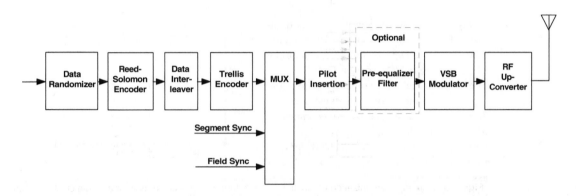

Figure 5.1.4 Simplified block diagram of an example VSB transmitter. (*From* [2]. *Used with permission.*)

5.1.3 DTV Transmission Characteristics

The terrestrial broadcast mode (8-VSB) will support a payload data rate of approximately 19.28 Mbits/s in a 6 MHz channel [2]. (In the interest of simplicity, the bit stream data rate is rounded to two decimal points.) A functional block diagram of a representative 8-VSB terrestrial broadcast transmitter is shown in Figure 5.1.4. The basic parameters of the VSB transmission modes are listed in Table 5.1.3. The input to the transmission subsystem from the transport subsystem is a serial data stream composed of 188-byte MPEG-compatible data packets (including a Sync Byte and 187 bytes of data, which represent a payload data rate of 19.28 Mbits/s).

The incoming data is randomized, then processed for *forward error correction* (FEC) in the form of *Reed-Solomon* (RS) coding (whereby 20 RS parity bytes are added to each packet). This is followed by 1/6-data-field interleaving and 2/3-rate trellis coding. The randomization and FEC processes are not applied to the Sync Byte of the transport packet, which is represented in transmission by a *Data Segment Sync* signal. Following randomization and FEC processing, the data packets are formatted into *Data Frames* for transmission, and Data Segment Sync and *Data Field Sync* are added.

Table 5.1.3 Parameters for VSB Transmission Modes (*After* [1].)

Parameter	Terrestrial Mode	High Data Rate Mode
Channel bandwidth	6 MHz	6 MHz
Guard bandwidth	11.5 percent	11.5 percent
Symbol rate	10.76… Msymbols/s	10.76… Msymbols/s
Bits per symbol	3	4
Trellis FEC	2/3 rate	None
Reed-Solomon FEC	T = 10 (207,187)	T = 10 (207,187)
Segment length	832 symbols	832 symbols
Segment sync	4 symbols per segment	4 symbols per segment
Frame sync	1 per 313 segments	1 per 313 segments
Payload data rate	19.39 Mbps	38.78 Mbps
Analog co-channel rejection	Analog rejection filter in receiver	N/A
Pilot power contribution	0.3 dB	0.3 dB
C/N threshold	~ 14.9 dB	~ 28.3 dB

Figure 5.1.5 shows how the data is organized for transmission. Each Data Frame consists of two *Data Fields*, each containing 313 D*ata Segments*. The first Data Segment of each Data Field is a unique synchronizing signal (Data Field Sync) and includes the *Training Sequence* used by the equalizer in the receiver. The remaining 312 Data Segments each carry the equivalent of the data from one 188-byte transport packet plus its associated FEC overhead. The actual data in each Data Segment comes from several transport packets because of data interleaving. Each Data Segment consists of 832 *symbols*. The first four symbols are transmitted in binary form and provide segment synchronization. This Data Segment Sync signal also represents the *Sync Byte* of the 188-byte MPEG-compatible transport packet. The remaining 828 symbols of each Data Segment carry data equivalent to the remaining 187 bytes of a transport packet and its associated FEC overhead. These 828 symbols are transmitted as 8-level signals and, therefore, carry 3 bits per symbol. Thus, 2484 (828 × 3) bits of data are carried in each Data Segment, which exactly matches the requirement to send a protected transport packet:

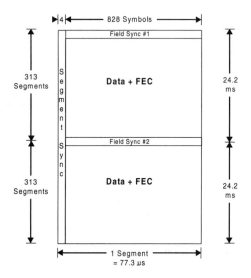

Figure 5.1.5 The basic VSB Data Frame. (*From* [2]. *Used with permission.*)

- 187 data bytes + 20 RS parity bytes = 207 bytes

- 207 bytes × 8 bits/byte = 1656 bits

- 2/3-rate trellis coding requires 3/2 × 1656 bits = 2484 bits.

The exact symbol rate S_r is given by

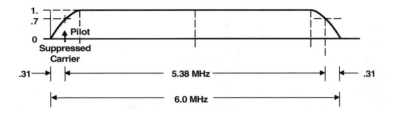

Figure 5.1.6 Nominal VSB channel occupancy. (*From* [2]. *Used with permission.*)

$$S_r = \frac{4.5}{286} \times 684 = 10.76\ldots \text{ MHz} \tag{5.1.4}$$

The frequency of a Data Segment f_{seg} is given by

$$f_{seg} = \frac{S_r}{832} = 12.94\ldots \times 10^3 \text{ Data Segments/s} \tag{5.1.5}$$

The Data Frame rate f_{frame} is given by

$$f_{frame} = \frac{f_{seg}}{626} = 20.66\ldots \text{ frames/s} \tag{5.1.6}$$

The symbol rate S_r and the transport rate T_r are locked to each other in frequency.

The 8-level symbols combined with the binary Data Segment Sync and Data Field Sync signals are used to suppressed-carrier-modulate a single carrier. Before transmission, however, most of the lower sideband is removed. The resulting spectrum is flat, except for the band edges where a nominal square-root raised-cosine response results in 620 kHz transition regions. The nominal VSB transmission spectrum is shown in Figure 5.1.6. It includes a small pilot signal at the suppressed-carrier frequency, 310 kHz from the lower band edge.

5.1.3a Channel Error Protection and Synchronization

All payload data is carried with the same priority [2]. A Data Randomizer is used on all input data to randomize the data payload (except for the Data Field Sync, Data Segment Sync, and RS parity bytes). The Data Randomizer *exclusive-ORs* (XORs) all the incoming data bytes with a 16-bit maximum-length *Pseudorandom Binary Sequence* (PRBS), which is initialized at the beginning of the Data Field. The PRBS is generated in a 16-bit shift register that has nine feedback taps. Eight of the shift register outputs are selected as the fixed randomizing byte, where each bit from this byte is used to individually XOR the corresponding input data bit. The randomizer-generator polynomial and initialization are shown in Figure 5.1.7.

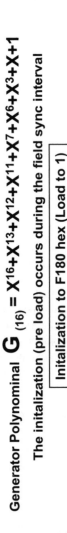

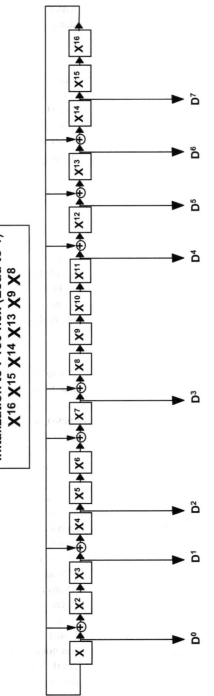

Figure 5.1.7 Randomizer polynomial for the DTV transmission subsystem. (*From* [2]. *Used with permission.*)

Although a thorough knowledge of the channel error protection and synchronization system is not required for typical end-users, a familiarity with the basic principles of operation—as outlined in the following sections—is useful in understanding the important functions performed.

Reed-Solomon Encoder

The Reed-Solomon (RS) code used in the VSB transmission subsystem is a $t = 10$ (207,187) code [2]. The RS data block size is 187 bytes, with 20 RS parity bytes added for error correction. A total RS block size of 207 bytes is transmitted per Data Segment. In creating bytes from the serial bit stream, the most significant bit (MSB) is the first serial bit. The 20 RS parity bytes are sent at the end of the Data Segment. The parity-generator polynomial and the primitive-field-generator polynomial (with the fundamental supporting equations) are shown in Figure 5.1.8.

Reed-Solomon encoding/decoding is expressed as the total number of bytes in a transmitted packet to the actual application payload bytes, where the overhead is the RS bytes used (i.e., 207,187).

Interleaving

The interleaver employed in the VSB transmission system is a 52-data-segment (*intersegment*) convolutional byte interleaver [2]. Interleaving is provided to a depth of about one-sixth of a Data Field (4 ms deep). Only data bytes are interleaved. The interleaver is synchronized to the first data byte of the Data Field. Intrasegment interleaving also is performed for the benefit of the trellis coding process. The convolutional interleave stage is shown in Figure 5.1.9.

Trellis Coding

The 8-VSB transmission subsystem employs a 2/3-rate ($R = 2/3$) trellis code (with one unencoded bit that is precoded). Put another way, one input bit is encoded into two output bits using a 1/2-rate convolutional code while the other input bit is precoded [2]. The signaling waveform used with the trellis code is an 8-level (3-bit), 1-dimensional constellation. Trellis code intrasegment interleaving is used. This requires 12 identical trellis encoders and precoders operating on interleaved data symbols. The code interleaving is accomplished by encoding symbols (0, 12, 24, 36 ...) as one group, symbols (1, 13, 25, 37 ...) as a second group, symbols (2, 14, 26, 38 ...) as a third group, and so on for a total of 12 groups.

In creating serial bits from parallel bytes, the MSB is sent out first: (7, 6, 5, 4, 3, 2, 1, 0). The MSB is precoded (7, 5, 3, 1) and the LSB (least significant bit) is feedback-convolutional encoded (6, 4, 2, 0). Standard 4-state optimal Ungerboeck codes are used for the encoding. The trellis code utilizes the 4-state feedback encoder shown in Figure 5.1.10. Also shown in the figure is the precoder and the symbol mapper. The trellis code and precoder intrasegment interleaver, which feed the mapper shown in Figure 5.1.10, are illustrated in Figure 5.1.11. As shown in the figure, data bytes are fed from the byte interleaver to the trellis coder and precoder; then they are processed as whole bytes by each of the 12 encoders. Each byte produces four symbols from a single encoder.

The output multiplexer shown in Figure 5.1.11 advances by four symbols on each segment boundary. However, the state of the trellis encoder is not advanced. The data coming out of the multiplexer follows normal ordering from encoders 0 through 11 for the first segment of the frame; on the second segment, the order changes, and symbols are read from encoders 4 through 11, and then 0 through 3. The third segment reads from encoder 8 through 11 and then 0 through

$$\prod_{i=0}^{i=2t-1}(X+\alpha^{i}) =$$

$$= X^{20} + X^{19}\alpha^{17} + X^{18}\alpha^{60} + X^{17}\alpha^{79} + X^{16}\alpha^{50} + X^{15}\alpha^{61} + X^{14}\alpha^{163} + X^{13}\alpha^{26} + X^{12}\alpha^{187} + X^{11}\alpha^{202} + X^{10}\alpha^{180} + X^{9}\alpha^{221} + X^{8}\alpha^{225} + X^{7}\alpha^{83} + X^{6}\alpha^{239} + X^{5}\alpha^{156} + X^{4}\alpha^{164} + X^{3}\alpha^{212} + X^{2}\alpha^{212} + X^{1}\alpha^{188} + \alpha^{190}$$

$$= X^{20} + 152X + 185X^{19} + 240X^{18} + 5X^{17} + 111X^{16} + 99X^{15} + 6X^{14} + 220X^{13} + 112X^{12} + 150X^{11} + 69X^{10} + 36X^{9} + 187X^{8} + 22X^{7} + 228X^{6} + 198X^{5} + 121X^{4} + 121X^{3} + 198X^{2} + 165X^{1} + 174$$

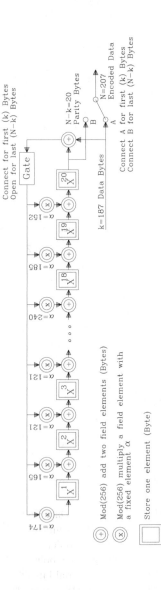

Connect for first (k) Bytes
Open for last (N–k) Bytes

N–k=20 Parity Bytes

k=187 Data Bytes

N=207 Encoded Data

Connect A for first (k) Bytes
Connect B for last (N–k) Bytes

⊕ Mod(256) add two field elements (Bytes)

⊗ Mod(256) multiply a field element with a fixed element α

☐ Store one element (Byte)

Primitive Field Generator Polynomial (Galois Field)

$$G(256) = X^{8} + X^{4} + X^{3} + X^{2} + 1$$

Each shift of the generator produces a field element

Figure 5.1.8 Reed-Solomon (207,187) $t = 10$ parity-generator polynomial. (*From* [2]. *Used with permission.*)

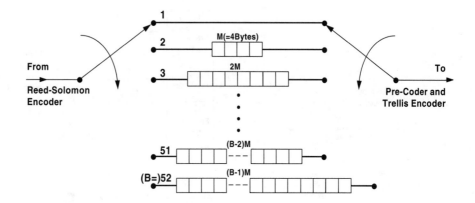

M=4, B=52, N=208, R-S Block =207, BXM=N

Figure 5.1.9 Convolutional interleaving scheme (byte shift register illustration). (*From* [2]. *Used with permission.*)

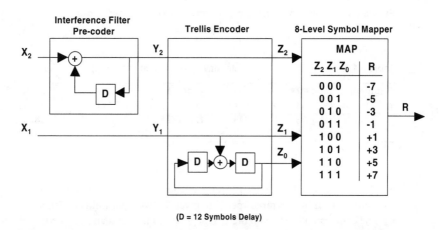

(D = 12 Symbols Delay)

Figure 5.1.10 An 8-VSB trellis encoder, precoder, and symbol mapper. (*From* [2]. *Used with permission.*)

7. This 3-segment pattern repeats through the 312 Data Segments of the frame. Table 5.1.4 shows the interleaving sequence for the first three Data Segments of the frame.

After the Data Segment Sync is inserted, the ordering of the data symbols is such that symbols from each encoder occur at a spacing of 12 symbols.

A complete conversion of parallel bytes to serial bits needs 828 bytes to produce 6624 bits. Data symbols are created from 2 bits sent in MSB order, so a complete conversion operation yields 3312 data symbols, which corresponds to four segments of 828 data symbols. A total of

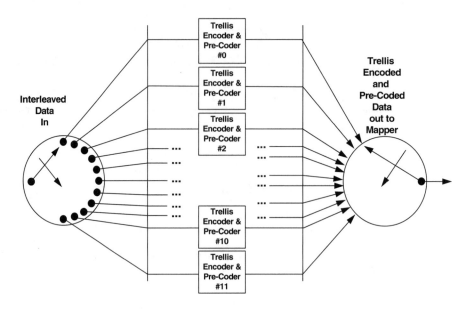

Figure 5.1.11 Trellis code interleaver. (*From* [2]. *Used with permission.*)

Table 5.1.4 Partial Trellis Coding Interleaving Sequence (*After* [2].)

Segment	Block 0				Block 1				...	Block 68			
0	D0	D1	D2	... D11	D0	D1	D2	... D11	...	D0	D1	D2	... D11
1	D4	D5	D6	... D3	D4	D5	D6	... D3	...	D4	D5	D6	... D3
2	D8	D9	D10	... D7	D8	D9	D10	... D7	...	D8	D9	D10	... D7

3312 data symbols divided by 12 trellis encoders gives 276 symbols per trellis encoder, and 276 symbols divided by 4 symbols per byte gives 69 bytes per trellis encoder.

The conversion starts with the first segment of the field and proceeds with groups of four segments until the end of the field. A total of 312 segments per field divided by 4 gives 78 conversion operations per field.

During segment sync, the input to four encoders is skipped, and the encoders cycle with no input. The input is held until the next multiplex cycle, then is fed to the correct encoder.

5.1.3b   Modulation

You will recall that in Figure 5.1.10 the mapping of the outputs of the trellis decoder to the nominal signal levels (–7, –5, –3, –1, 1, 3, 5, 7) was shown. As detailed in Figure 5.1.12, the nominal levels of Data Segment Sync and Data Field Sync are –5 and +5. The value of 1.25 is added to all these nominal levels after the bit-to-symbol mapping function for the purpose of creating a small

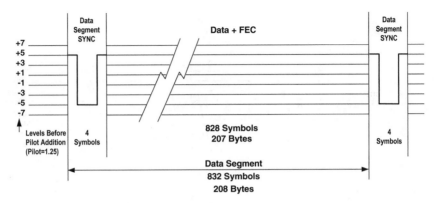

Figure 5.1.12 The 8-VSB Data Segment. (*From* [2]. *Used with permission.*)

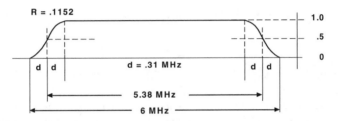

Figure 5.1.13 Nominal VSB system channel response (linear-phase raised-cosine Nyquist filter). (*From* [2]. *Used with permission.*)

pilot carrier [2]. The frequency of the pilot is the same as the suppressed-carrier frequency. The in-phase pilot is 11.3 dB below the average data signal power.

The VSB modulator receives the 10.76 Msymbols/s, 8-level trellis-encoded composite data signal with pilot and sync added. The DTV system performance is based on a *linear-phase raised-cosine* Nyquist filter response in the concatenated transmitter and receiver, as illustrated in Figure 5.1.13. The system filter response is essentially flat across the entire band, except for the transition regions at each end of the band. Nominally, the rolloff in the transmitter has the response of a *linear-phase root raised-cosine* filter.

5.1.3c   High Data-Rate Mode

The high data-rate mode trades off transmission robustness (28.3 dB S/N threshold) for payload data rate (38.57 Mbits/s) [2]. Most parts of the high data-rate mode VSB system are identical, or at least similar, to the terrestrial system. A pilot, Data Segment Sync, and Data Field Sync all are used to provide robust operation. The pilot in the high data-rate mode is 11.3 dB below the data signal power; and the symbol, segment, and field signals and rates all are the same as well, allowing either receiver type to lock up on the other's transmitted signal. Also, the Data Frame

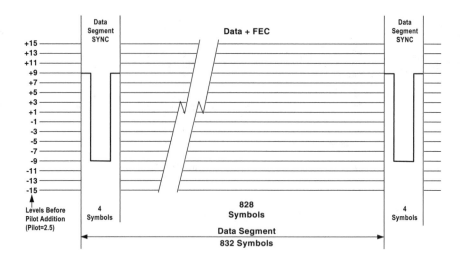

Figure 5.1.14 Typical Data Segment for the 16-VSB mode. (*From* [2]. *Used with permission.*)

definitions are identical. The primary difference is the number of transmitted levels (8 vs.16) and the use of trellis coding and NTSC interference-rejection filtering in the terrestrial system.

The RF spectrum of the high data-rate modem transmitter looks identical to the terrestrial system. Figure 5.1.14 illustrates a typical Data Segment, where the number of data levels is seen to be 16 as a result of the doubled data rate. Each portion of 828 data symbols represents 187 data bytes and 20 Reed-Solomon bytes, followed by a second group of 187 data bytes and 20 Reed-Solomon bytes (before convolutional interleaving).

Figure 5.1.15 shows a functional block diagram of the high data-rate transmitter. It is identical to the terrestrial VSB system, except that the trellis coding is replaced with a mapper that converts data to multilevel symbols. The interleaver is a 26-data-segment intersegment convolutional byte interleaver. Interleaving is provided to a depth of about one-twelfth of a Data Field (2 ms deep). Only data bytes are interleaved. Figure 5.1.16 shows the mapping of the outputs of the interleaver to the nominal signal levels (–15, –13, –11, ..., 11, 13, 15). As shown in Figure 5.1.14, the nominal levels of Data Segment Sync and Data Field Sync are –9 and +9. The value of 2.5 is added to all these nominal levels after the bit-to-symbol mapping for the purpose of creating a small pilot carrier. The frequency of the in-phase pilot is the same as the suppressed-carrier frequency. The modulation method of the high data-rate mode is identical to the terrestrial mode except that the number of transmitted levels is 16 instead of 8.

5.1.3d   Bit Rate Delivered to a Transport Decoder

As noted previously, the symbol rate of the transmission subsystem is given by

$$\frac{4.5}{286} \times 684 \ = \ 10.7 \ ... \ \text{million symbols/second (megabaud)} \tag{5.1.7}$$

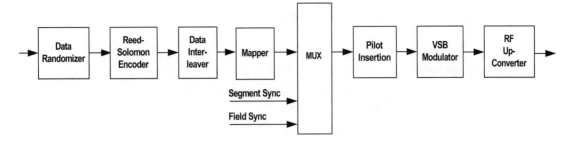

Figure 5.1.15 Functional block diagram of the 16-VSB transmitter. (*From* [2]. *Used with permission.*)

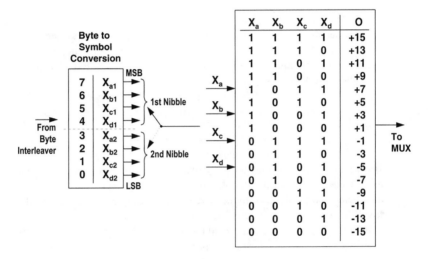

X_a	X_b	X_c	X_d	O
1	1	1	1	+15
1	1	1	0	+13
1	1	0	1	+11
1	1	0	0	+9
1	0	1	1	+7
1	0	1	0	+5
1	0	0	1	+3
1	0	0	0	+1
0	1	1	1	-1
0	1	1	0	-3
0	1	0	1	-5
0	1	0	0	-7
0	0	1	1	-9
0	0	1	0	-11
0	0	0	1	-13
0	0	0	0	-15

Figure 5.1.16 A 16-VSB mapping table. (*From* [2]. *Used with permission.*)

The symbol rate must be locked in frequency to the transport rate.

The numbers in the formula for the ATSC symbol rate in 6 MHz systems are related to NTSC scanning and color frequencies. Because of this relationship, the symbol clock can be used as a basis for generating an NTSC color subcarrier for analog output from a set top box. The repetition rates of Data Segments and Data Frames are deliberately chosen not to have an integer relationship to NTSC or PAL scanning rates, to insure that there will be no discernible pattern in co-channel interference.

The particular numbers used are:

- 4.5 MHz = the center frequency of the audio carrier offset in NTSC. This number was traditionally used in NTSC literature to derive the color subcarrier frequency and scanning rates. In modern equipment, this may start with a precision 10 MHz reference, which is then multiplied by 9/20.

- 4.5 MHz/286 = the horizontal scan rate of NTSC, 15734.2657+...Hz (note that the color sub-carrier is 455/2 times this, or 3579545 +5/11 Hz).

- 684: this multiplier gives a symbol rate for an efficient use of bandwidth in 6 MHz. It requires a filter with Nyquist roll-off that is a fairly sharp cutoff (11 percent excess bandwidth), which is still realizable with a reasonable surface acoustic wave (SAW) filter or digital filter.

In the terrestrial broadcast mode, channel symbols carry three bits/symbol of trellis-coded data. The trellis code rate is 2/3, providing 2 bits/symbol of gross payload. Therefore the gross payload is

$$10.76... \times 2 = 21.52 ... \text{Mbps} \tag{5.1.8}$$

To find the net payload delivered to a decoder it is necessary to adjust Equation (5.1.8) for the overhead of the Data Segment Sync, Data Field Sync, and Reed-Solomon FEC.

To get the net bit rate for an MPEG-2 stream carried by the system (and supplied to an MPEG transport decoder), it is first noted that the MPEG Sync Bytes are removed from the data stream input to the 8-VSB transmitter and replaced with segment sync, and later reconstituted at the receiver. For throughput of MPEG packets (the only allowed transport mechanism) segment sync is simply equivalent to transmitting the MPEG Sync Byte, and does not reduce the net data rate. The net bit rate of an MPEG-2 stream carried by the system and delivered to the transport decoder is accordingly reduced by the Data Field Sync (one segment of every 313) and the Reed-Solomon coding (20 bytes of every 208)

$$21.52... \times \frac{312}{313} \times \frac{188}{208} = 19.39 ... \text{Mbps} \tag{5.1.9}$$

The net bit rate supplied to the transport decoder for the high data rate mode is

$$19.39... \times 2 = 38.78 ... \text{Mbps} \tag{5.1.10}$$

5.1.3e   Performance Characteristics of Terrestrial Broadcast Mode

The terrestrial 8-VSB system can operate in a signal-to-additive-white-Gaussian-noise (S/N) environment of 14.9 dB [1]. The 8-VSB segment error probability curve including 4-state trellis decoding and (207,187) Reed-Solomon decoding in Figure 5.1.17 shows a segment error probability of 1.93×10^{-4}. This is equivalent to 2.5 segment errors/second, which was established by measurement as the *threshold of visibility* (TOV) of errors in the prototype equipment. Particular product designs may achieve somewhat better performance for subjective TOV by means of error masking.

The *cumulative distribution function* (CDF) of the peak-to-average power ratio, as measured on a low power transmitted signal with no non-linearities, is plotted in Figure 5.1.18. The plot shows that 99.9 percent of the time the transient peak power is within 6.3 dB of the average power.

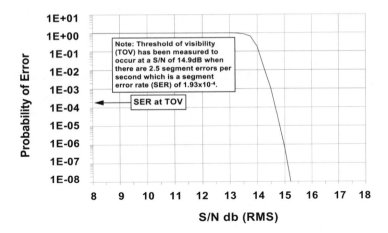

Figure 5.1.17 Segment-error probability for 8-VSB with 4-state trellis coding; RS (207,187). (*From* [1]. *Used with permission.*)

Transmitter Signal Processing

A pre-equalizer filter is recommended for use in over-the-air broadcasts where the high power transmitter may have significant in-band ripple or significant roll off at band edges [1]. Pre-equalization is typically required in order to compensate the high-order filter used to meet a stringent out-of-band emission mask, such as the U.S. FCC required mask [4]. This linear distortion can be measured by an equalizer in a reference demodulator ("ideal" receiver) employed at the transmitter site. A directional coupler, which is recommended to be located at the sending end of the antenna feed transmission line, supplies the reference demodulator a small sample of the antenna signal feed. The equalizer tap weights of the reference demodulator are transferred to the transmitter pre-equalizer for pre-correction of transmitter linear distortion. This is a one-time procedure of measurement and transmitter pre-equalizer adjustment. Alternatively, the transmitter pre-equalizer can be made continuously adaptive. In this arrangement, the reference demodulator is provided with a fixed-coefficient equalizer compensating for its own deficiencies in ideal response.

A pre-equalizer suitable for many applications is an 80 tap, feed-forward transversal filter. The taps are symbol-spaced (93 ns) with the main tap being approximately at the center, giving approximately ± 3.7 µs correction range. The pre-equalizer operates on the I channel data signal (there is no Q channel data signal in the transmitter), and shapes the frequency spectrum of the IF signal so that there is a flat in-band spectrum at the output of the high power transmitter that feeds the antenna for transmission. There is no effect on the out-of-band spectrum of the transmitted signal. If desired, complex equalizers or fractional equalizers (with closer-spaced taps) can provide independent control of the outer portions of the spectrum (beyond the Nyquist slopes).

The transmitter vestigial sideband filtering is sometimes implemented by sideband cancellation, using the phasing method. In this method, the baseband data signal is supplied to digital filtering that generates in-phase and quadrature-phase digital modulation signals for application to respective D/A converters. This filtering process provides the root raised cosine Nyquist filtering

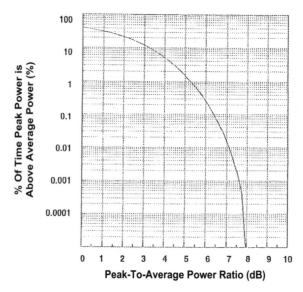

Figure 5.1.18 The cumulative distribution function of 8-VSB peak-to-average power ratio in an ideal linear system. (*From* [1]. *Used with permission.*)

and provides compensation for the $(\sin x) / x$ frequency responses of the D/A converters, as well. The baseband signals are converted to analog form. The in-phase signal modulates the amplitude of the IF carrier at zero degrees phase, while the quadrature signal modulates a 90-degree shifted version of the carrier. The amplitude-modulated quadrature IF carriers are added to create the vestigial sideband IF signal, canceling the unwanted sideband and increasing the desired sideband by 6 dB. The nominal frequency of the IF carrier (and small in-phase pilot) in the prototype hardware was 46.69 MHz, which is equal to the IF center frequency (44.000 MHz) plus the symbol rate divided by 4

$$\frac{10.762}{4} = 2.6905 \text{ MHz} \tag{5.1.11}$$

Additional adjacent-channel suppression (beyond that achieved by sideband cancellation) may be performed by a linear phase, flat amplitude response SAW filter. Other implementations for VSB filtering are possible that may include the prefilter of the previous section.

5.1.3f   Upconverter and RF Carrier Frequency Offsets

Modern analog TV transmitters use a two-step modulation process [1]. The first step usually is modulation of the data onto an IF carrier, which is the same frequency for all channels, followed by translation to the desired RF channel. The digital 8-VSB transmitter applies this same two-step modulation process. The RF upconverter translates the filtered flat IF data signal spectrum to the desired RF channel. For the same approximate coverage as an analog transmitter (at the same frequency), the average power of the DTV signal is on the order of 12 dB less than the analog peak sync power (when operating on the same frequency).

The nominal frequency of the RF upconverter oscillator in DTV terrestrial broadcasts is typically be the same as that used for analog transmitters, (except for offsets required in particular situations).

Note that all examples in this section relate to a 6 MHz DTV system. Values may be modified easily for other channel widths.

Nominal DTV Pilot Carrier Frequency

The nominal DTV pilot carrier frequency is determined by fitting the DTV spectrum symmetrically into the RF channel [1]. This is obtained by taking the bandwidth of the DTV signal—5,381.1189 kHz (the Nyquist frequency difference or one-half the symbol clock frequency of 10,762.2378 kHz)—and centering it in the 6 MHz TV channel. Subtracting 5,381.1189 kHz from 6,000 kHz leaves 618.881119 kHz. Half of that is 309.440559 kHz, precisely the standard pilot offset above the lower channel edge. For example, on channel 45 (656–662 MHz), the nominal pilot frequency is 656.309440559 MHz.

Requirements for Offsets

There are two categories of requirements for pilot frequency offsets [1]:

- Offsets to protect lower adjacent channel analog broadcasts, mandated by FCC rules in the U.S., and which override other offset considerations.

- Recommended offsets for other considerations such as co-channel interference between DTV stations or between DTV and analog stations.

Upper DTV Channel into Lower Analog Channel

This is the overriding case mandated by the FCC rules in the U.S.—precision offset with a lower adjacent analog station, full service or Low Power Television (LPTV) [1]. The FCC Rules, Section 73.622(g)(1), states that: "DTV stations operating on a channel allotment designated with a "c" in paragraph (b) of this section must maintain the pilot carrier frequency of the DTV signal 5.082138 MHz above the visual carrier frequency of any analog TV broadcast station that operates on the lower adjacent channel and is located within 88 kilometers. This frequency difference must be maintained within a tolerance of ± 3 Hz."

This precise offset is necessary to reduce the color beat and high-frequency luminance beat created by the DTV pilot carrier in some receivers tuned to the lower adjacent analog channel. The tight tolerance assures that the beat will be visually cancelled, since it will be out of phase on successive video frames.

Note that the frequency is expressed with respect to the lower adjacent analog video carrier, rather than the nominal channel edge. This is because the beat frequency depends on this relationship, and therefore the DTV pilot frequency must track any offsets in the analog video carrier frequency. The offset in the FCC rules is related to the particular horizontal scanning rate of NTSC, and can easily be modified for PAL. The offset O_f was obtained from

$$O_f = 455 \times \left(\frac{F_h}{2}\right) + 191 \times \left(\frac{F_h}{2}\right) - 29.97 = 5082138 \ \text{Hz} \tag{5.1.12}$$

Where F_h = NTSC horizontal scanning frequency = 15,734.264 Hz.

The equation indicates that the offset with respect to the lower adjacent chroma is an odd multiple (191) of one-half the line rate to eliminate the color beat. However, this choice leaves the possibility of a luma beat. The offset is additionally adjusted by one-half the analog field rate to eliminate the luma beat. While satisfying the exact adjacent channel criteria, this offset is also as close as possible to optimal comb filtering of the analog co-channel in the digital receiver. Note additionally that offsets are to higher frequencies rather than lower, to avoid any possibility of encroaching on the lower adjacent sound. (It also reduces the likelihood of the automatic fine tuning (AFT) in the analog receiver experiencing lock-out because the signal energy including the pilot is moved further from the analog receiver bandpass.

As an example, if a channel 44 NTSC station is operating with a zero offset, the Channel 45 DTV pilot carrier frequency must be 651.250000 MHz plus 5.082138 MHz or 656.332138 MHz; that is, 22.697 kHz above the nominal frequency. If the lower adjacent NTSC channel is offset ±10 kHz, the DTV frequency will have to be adjusted accordingly.

Co-Channel DTV into Analog

In co-channel cases, DTV interference into analog TV appears noise-like [1]. The pilot carrier is low on the Nyquist slope of the IF filter in the analog receiver, so no discernible beat is generated. In this case, offsets to protect the analog channel are not required. Offsets are useful, however to reduce co-channel interference from analog TV into DTV. The performance of the analog rejection filter and clock recovery in the DTV receiver will be improved if the DTV carrier is 911.944 kHz below the NTSC visual carrier. In other words, in the case of a 6 MHz NTSC system, if the analog TV station is not offset, the DTV pilot carrier frequency will be 338.0556 kHz above the lower channel edge instead of the nominal 309.44056 kHz. As before, if the NTSC station is operating with a ±10 kHz offset, the DTV frequency will have to be adjusted in the same direction. The formula for calculating this offset is

$$F_{pilot} = F_{vis(n)} - 70.5 \times F_{seg} = 338.0556 \ \text{Hz (for no NTSC analog offset)} \tag{5.1.13}$$

Where:
F_{pilot} = DTV pilot frequency above lower channel edge
$F_{vis(n)}$ = NTSC visual carrier frequency above lower channel edge
       = 1,250 kHz for no NTSC offset (as shown)
       = 1,240 kHz for minus offset
       = 1,260 kHz for plus offset
F_{seg} = ATSC Data Segment rate; = symbol clock frequency / 832 = 12,935.381971 Hz

The factor of 70.5 is chosen to provide the best overall comb filtering of analog color TV co-channel interference. The use of a value equal to an integer +0.5 results in co-channel analog TV interference being out-of-phase on successive Data Segment Syncs.

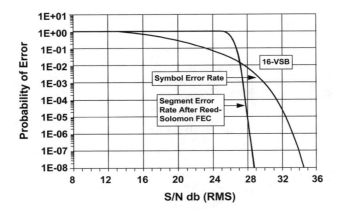

Figure 5.1.19 The error probability of the 16-VSB signal. (*From* [1]. *Used with permission.*)

Note that in this case the frequency tolerance is plus or minus one kHz. More precision is not required. Also note that a different Data Segment rate would be used for calculating offsets for 7 or 8 MHz systems.

Co-channel DTV into DTV

If two DTV stations share the same channel, interference between the two stations can be reduced if the pilot is offset by one and a half times the Data Segment rate [1]. This ensures that the frame and segment syncs of the interfering signal will each alternate polarity and be averaged out in the receiver tuned to the desired signal.

The formula for this offset is

$$F_{offset} = 1.5 \times F_{seg} = 19.4031 \ \text{kHz} \tag{5.1.14}$$

Where:
F_{offset} = offset to be added to one of the two DTV carriers
F_{seg} = 12,935.381971 Hz (as defined previously)

This results in a pilot carrier 328.84363 kHz above the lower band edge, provided neither DTV station has any other offset.

Use of the factor 1.5 results in the best co-channel rejection, as determined experimentally with prototype equipment. The use of an integer +0.5 results in co-channel interference alternating phase on successive segment syncs.

5.1.3g   Performance Characteristics of High Data Rate Mode

The high data rate mode can operate in a signal-to-white-noise environment of 28.3 dB. The error probability curve is shown in Figure 5.1.19 [1].

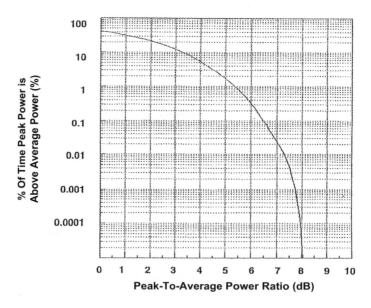

Figure 5.1.20 Cumulative distribution function of the 16-VSB peak-to-average power ratio. (*From* [1]. *Used with permission.*)

The *cumulative distribution function* (CDF) of the peak-to-average power ratio, as measured on a low power transmitted signal with no non-linearities, is plotted in Figure 5.1.20 and is slightly higher than that of the terrestrial mode.

5.1.3h   Enhanced-VSB Transmission Mode

E-VSB is fundamentally a method of adding further error protection coding to part of the 8-VSB signal [2]. It is required to simultaneously provide a performance increase for the E-VSB coded portion, while not degrading the "normal" or "main" portion used by legacy ATSC receivers. Secondarily, but importantly, E-VSB applications require additions to the ATSC transport and PSIP[4] Standards to support functions such as synchronization of separate but related source material in the main and enhanced streams.

The basic technical advantage of the E-VSB stream is an improvement of at least 6 dB in SNR and interference thresholds. This is obtained in exchange for heavier FEC coding, and therefore at the expense of payload data rate for the enhanced part of the transmission. Naturally, designating a portion of the transmitted symbols as enhanced reduces the bandwidth of the main stream, just as in the case of multicasting, where each program uses only part of the 19.39 Mbps ATSC stream.

Applications envisioned for E-VSB include streams unrelated to the main stream, related streams, and synchronized related streams such as fallback audio and/or video.

---

4. Program and System Information Protocol, as defined in ATSC Standard A/65.

Unrelated streams are envisioned for use in carrying secondary channels or data that can be used by portable or PC-based devices with non-optimum antennas, for example a "sub-channel" carrying stock market information, news and weather.

Fallback audio is defined as a duplicate of the main audio that can be switched to in the receiver when the main signal is momentarily lost. The aim is to make this switch as seamless and unnoticeable as possible. This is the most demanding application envisioned, involving all primary and secondary aspects of E-VSB: the physical layer, synchronization of time stamps for the main and fallback, and enhanced PSIP that announces the availability of fallback to the enhanced-capable receiver.

E-VSB Features

E-VSB provides two coding rates designated as "1/2-rate" and "1/4-rate", referring to a choice of two convolutional codes. (The payload is additionally reduced by the ratio 164/188 due to the added RS coding.) This additional coding provides a SNR threshold advantage of 6 dB or 9 dB, respectively, as compared to normal 8-VSB. The arrangement supports applications similar to those of hierarchical transmission, but because the different types of data are actually time multiplexed, they do not interact, and the performance of legacy receivers on the main data is not affected by the presence of enhanced data.

5.1.4   References

1.   ATSC, "Guide to the Use of the Digital Television Standard," Advanced Television Systems Committee, Washington, D.C., Doc. A/54A, December 4, 2003.

2.   ATSC, "ATSC Digital Television Standard," Advanced Television Systems Committee, Washington, D.C., Doc. A/53E, December 27, 2005.

3.   ITU-R Document TG11/3-2, "Outline of Work for Task Group 11/3, Digital Terrestrial Television Broadcasting," June 30, 1992.

4.   Chairman, ITU-R Task Group 11/3, "Report of the Second Meeting of ITU-R Task Group 11/3, Geneva, Oct. 13-19, 1993," p. 40, Jan. 5, 1994.

5.   FCC: "Memorandum Opinion and Order on Reconsideration of the Sixth Report and Order," Federal Communications Commission, Washington, D.C., February 17, 1998.

5.1.5   Bibliography

Rhodes, Charles W.: "Terrestrial High-Definition Television," *The Electronics Handbook*, Jerry C. Whitaker (ed.), CRC Press, Boca Raton, Fla., pp. 1599–1610, 1996.

5.2

Video Compression and Decompression

5.2.1 Introduction[1]

The need for compression in a digital television system is apparent from the fact that the bit rate required to represent an HDTV signal in uncompressed digital form is about 1 Gbits/s and that required to represent a standard-definition television signal is about 200 Mbits/s, while the bit rate that can reliably be transmitted within a standard 6 MHz television channel is about 19 Mbits/s [1]. This implies a need for about a 50:1 or greater compression ratio for HDTV and 10:1 or greater for standard definition.

The ATSC Digital Television Standard specifies video compression using a combination of compression techniques. For reasons of compatibility these compression algorithms have been selected to conform to the specifications of MPEG-2, which is a flexible internationally accepted collection of compression algorithms.

5.2.1a MPEG-2 Levels and Profiles

The MPEG-2 specification is organized into a system of *Profiles* and *Levels*, so that applications can ensure interoperability by using equipment and processing that adhere to a common set of coding tools and parameters [1]. The ATSC Digital Television Standard is based on the MPEG-2 Main Profile. The Main Profile includes three types of frames (*I*-frames, *P*-frames, and *B*-frames), and an organization of luminance and chrominance samples (designated 4:2:0) within the frame. The Main Profile does not include a scalable algorithm, where scalability implies that a subset of the compressed data can be decoded without decoding the entire data stream. The High Level includes formats with up to 1152 active lines and up to 1920 samples per active line, and for the Main Profile is limited to a compressed data rate of no more than 80 Mbits/s. The

---

1. This chapter is based on: ATSC, "Guide to the Use of the Digital Television Standard," Advanced Television Systems Committee, Washington, D.C., Doc. A/54A, December 4, 2003. Readers are encouraged to download the source document from the ATSC Web site (http://www.atsc.org). All ATSC Standards, Recommended Practices, and Information Guides are available at no charge.

parameters specified by the Digital Television Standard represent specific choices within these constraints.

Compatibility with MPEG-2

The video compression system does not include algorithmic elements that fall outside the specifications for MPEG-2 Main Profile [1]. Thus, video decoders that conform to the MPEG-2 MP@HL can be expected to decode bit streams produced in accordance with the ATSC Digital Television Standard. Note that it is not necessarily the case that all video decoders which are based on the Digital Television Standard will be able to properly decode all video bit streams that comply to MPEG-2 MP@HL.

5.2.2   Overview of Video Compression

The video compression system takes in an analog or uncompressed digital video source signal and outputs a compressed digital signal that contains information that can be decoded to produce an approximate version of the original image sequence [1]. The goal is for the reconstructed approximation to be imperceptibly different from the original for most viewers, for most images, for most of the time. In order to approach such fidelity, the algorithms are flexible, allowing for frequent adaptive changes in the algorithm depending on scene content, history of the processing, estimates of image temporal and spatial complexity and perceptibility of distortions introduced by the compression.

Figure 5.2.1 shows the overall flow of signals in the ATSC DTV system. Video signals presented to the system are first digitized (if not already in digital signal form) and sent to the encoder for compression; the compressed data then are transmitted over a communications channel. On being received, the possibly error-corrupted compressed signal is decompressed in the decoder, and reconstructed for display.

5.2.2a   Video Preprocessing

Video preprocessing converts the input signals to digital samples in the form needed for subsequent compression [1]. Analog input signals are typically composite for standard-definition signals or components consisting of luminance (Y) and chrominance (Pb and Pr) for high-definition signals, are first decoded (for composite signals) then digitized as component luminance (Y) and chrominance (Cb and Cr) signals. Digital input signals, both standard-definition and high-definition, are typically serial digital signals carrying Y, Cb, Cr components. The input signals may undergo pre-processing for noise reduction and/or other processing algorithms that improve the efficiency of the compression encoding. Further processing is then carried out for chrominance and luminance filtering and sub-sampling.

5.2.2b   Video Compression Formats

Table 5.2.1 lists the video compression formats allowed in the ATSC Digital Television Standard [1]. In the table, "vertical lines" refers to the number of active lines in the picture. "Pixels" refers to the number of pixels during the active line. "Aspect ratio" refers to the picture aspect ratio.

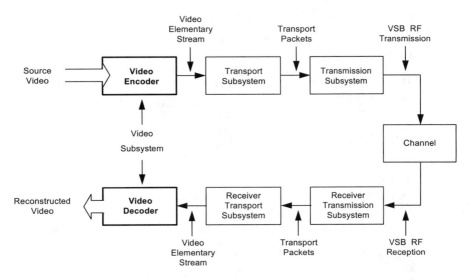

Figure 5.2.1 Video coding in relation to the DTV system. (*From* [1]. *Used with permission.*)

Table 5.2.1 Digital Television Standard Video Formats (*After* [1].)

Vertical Lines	Pixels	Aspect Ratio	Picture Rate
1080	1920	16:9	60I, 30P, 24P
720	1280	16:9	60P, 30P, 24P
480	704	16:9 and 4:3	60P, 60I, 30P, 24P
480	640	4:3	60P, 60I, 30P, 24P

"Picture rate" refers to the number of frames or fields per second. In the values for picture rate, "P" refers to progressive scanning, "I" refers to interlaced scanning. Note that both 60.00 Hz and 59.94 (60 × 1000/1001) Hz picture rates are allowed. Dual rates are allowed also at the picture rates of 30 Hz and 24 Hz.

5.2.2c Possible Video Inputs

While not required by the ATSC Digital Television Standard, there are certain digital television production standards, shown in Table 5.2.2, that define video formats that relate to compression formats specified by the DTV Standard [1]. The compression formats may be derived from one or more appropriate video input formats. It may be anticipated that additional video production standards will be developed in the future that extend the number of possible input formats.

Table 5.2.2 Standardized Video Input Formats (*After* [1].)

Video Standard	Active Lines	Active Samples/ Line	Picture Rate
SMPTE 274M-1998	1080	1920	24P, 30P, 60I
SMPTE 296M-2001	720	1280	24P, 30P, 60P
SMPTE 293M-2003	483	720	60P
ITU-R BT. 601-5	483	720	60I

5.2.2d   Sampling Rates

For the 1080-line format, with 1125 total lines per frame and 2200 total samples per line, the sampling frequency is 74.25 MHz for the 30.00 frames per second (fps) frame rate [1]. For the 720-line format, with 750 total lines per frame and 1650 total samples per line, the sampling frequency is 74.25 MHz for the 60.00 fps frame rate. For the 480-line format using 704 pixels, with 525 total lines per frame and 858 total samples per line, the sampling frequency is 13.5 MHz for the 59.94 Hz field rate. Note that both 59.94 fps and 60.00 fps are acceptable as frame or field rates for the system.

For both the 1080- and 720-line formats, other frame rates, specifically 23.976, 24.00, 29.97, and 30.00 fps rates are acceptable as input to the system. The sample frequency will be either 74.25 MHz (for 24.00 and 30.00 fps) or 74.25/1.001 MHz for the other rates. The number of total samples per line is the same for either of the paired picture rates. See SMPTE 274M [2] and SMPTE 296M [3] for more information.

The six frame rates noted are the only allowed frame rates for the ATSC Digital Television Standard. In this chapter, references to 24 fps include both 23.976 and 24.00 fps, references to 30 fps include both 29.97 and 30.00 fps, and references to 60 fps include both 59.94 and 60.00 fps.

For the 480-line format, there may be 704 or 640 pixels in the active line. The interlaced formats are based on ITU-R BT. 601-5; the progressive formats are based on SMPTE 294M. If the input is based on ITU-R BT. 601-5 or SMPTE 294M, it will have 483 or more active lines with 720 pixels in the active line. Only 480 of these active lines are encoded. The lines to be encoded should be lines 23–262 and 286–525 for 480I and lines 45–524 for 480P, as specified in SMPTE Recommended Practice RP-202, "Video Alignment for MPEG Coding". Only 704 of the 720 pixels are used for encoding; the first eight and the last eight are dropped. The 480-line, 640 pixel picture format is not related to any current video production format. It does correspond to the IBM VGA graphics format and may be used with ITU-R BT. 601-5 sources by using appropriate resampling techniques.

5.2.2e   Colorimetry

For the purposes of the ATSC Digital Television Standard, "colorimetry" means the combination of color primaries, transfer characteristics, and matrix coefficients [1]. Video inputs conforming to SMPTE 274M and SMPTE 296M have the same colorimetry; in this chapter, this will be referred to as SMPTE 274M colorimetry. Note that SMPTE 274M colorimetry is the same as ITU-R BT. 709 Part 2 colorimetry. Video inputs corresponding to ITU-R BT. 601-5 should have SMPTE 170M colorimetry.

ISO/IEC 13818-2 allows the encoder to signal the input colorimetry parameter values to the decoder. If sequence_display_extension()^2 is not present in the bit stream, or if color_description is zero, the color primaries, transfer characteristics, and matrix coefficients are assumed to be implicitly defined by the application. Therefore, the colorimetry should always be explicitly signaled using sequence_display_extension(). If this information is not transmitted, receiver behavior cannot be predicted.

In generating bit streams, broadcasters should understand that some receivers will display 480-line formats according to SMPTE 170M colorimetry (value 0x06) and 720- and 1080-line formats according to SMPTE 274M colorimetry (value 0x01). It is believed that few receivers will display properly the other colorimetry combinations allowed by ISO/IEC 13818-2. Legacy material using SMPTE 240M colorimetry should be treated as if it used ITU-R BT. 709 Part 2 colorimetry.

5.2.2f   Precision of Samples

Samples are typically obtained using analog-to-digital converter circuits with 10-bit precision [1]. After studio processing, the various luminance and chrominance samples will typically be represented using 8 or 10 bits per sample for luminance and 8 bits per sample for each chrominance component. The limit of precision of the MPEG-2 Main Profile is 8 bits per sample for each of the luminance and chrominance components.

5.2.2g   Source-Adaptive Processing

The image sequences that constitute the source signal can vary in spatial resolution (480 lines, 720 lines, or 1080 lines), in temporal resolution (60 fps, 30 fps, or 24 fps), and in scanning format (2:1 interlaced or progressive scan) [1]. The ATSC DTV video compression system accommodates the differences in source material to maximize the efficiency of compression.

5.2.2h   Film Mode

Material originated at 24 frames per second, such as that shot on film, is typically converted to 30 or 60 frame-per-second video for broadcast [1]. In the case of 30 fps interlaced television, this means that each four frames of film are converted to ten fields, or five frames of video. In the case of 60 fps progressive-scan television, each four frames of film are converted into ten frames of video. This conversion is done using the so-called 3:2 pulldown sequence; prior to the introduction of 24P video equipment it was an inherent part of the telecine process.

In the 3:2 pulldown, the first frame of film is converted to two pictures (frames or fields, depending on whether the output format is 60P or 30I respectively). The second frame is converted to three pictures, the third to two pictures and the fourth to three pictures.

When describing the sequence, the film frames are conventionally labelled A, B, C and D; the video fields or frames 1–5 (interlaced) or 1–10 (progressive). In the interlaced case, the third

2. ATSC Digital Television Standard documents contains symbolic references to syntactic elements used in the audio, video, and transport coding subsystems. These references are typographically distinguished by the use of a different font (e.g., restricted), may contain the underscore character (e.g., sequence_end_code) and may consist of character strings that are not English words (e.g., dynrng). Note also that certain syntax element names are capitalized; e.g., Transport Stream.

field generated from film frame A is field 1 of Frame 3; the third field generated from film frame C is field 2 of Frame 5. Note that in the interlaced case, Frame 3 will contain video from film Frames B and C and Frame 4 will contain video from film Frames C and D.

It is inefficient to code these sequences directly; not only is there a great deal of repeated information, but in interlace, Frames 3 and 4 each contain fields from two different film frames, so there may be motion differences between the two fields. MPEG therefore provides tools specifically for coding these sequences; these are top_field_first and repeat_first_field.

It is relatively straightforward for the encoder to detect the repeated frames in progressive-scan video derived from 24 fps material. It is less straightforward to detect the repeated fields in interlaced video. Particularly with interlaced material, it is important that the 3:2 pulldown sequence be maintained; if it is not, encoder efficiency and picture quality may suffer.

5.2.2i   Color Component Separation and Processing

The input video source to the ATSC DTV video compression system is in the form of RGB components matrixed into luminance (Y) and chrominance (Cb and Cr) components using a linear transformation (3-by-3 matrix) [1]. The luminance component represents the intensity, or black-and-white picture, while the chrominance components contain color information. While the original RGB components are highly correlated with each other; the resulting Y, Cb, and Cr signals have less correlation and are thus easier to code efficiently. The luminance and chrominance components correspond to functioning of the biological vision system; that is, the human visual system responds differently to the luminance and chrominance components.

The coding process may take advantage also of the differences in the ways that humans perceive luminance and chrominance. In the Y, Cb, Cr color space, most of the high frequencies are concentrated in the Y component; the human visual system is less sensitive to high frequencies in the chrominance components than to high frequencies in the luminance component. To exploit these characteristics the chrominance components are low-passed filtered and sub-sampled by a factor of two along both the horizontal and vertical dimensions, producing chrominance components that are one-fourth the spatial resolution of the luminance component.

It must be noted that the luminance component Y is not true luminance as this term is used in color science; this is because the RGB-to-YC matrixing operation is performed after the opto-electronic transfer characteristic (gamma) is applied. For this reason, some experts refer to the Y component as luma rather than luminance. While the preponderance of the luminance information is present in the luma, some of it ends up in the chroma, and it can be lost when the chroma components are sub-sampled.

Anti-Alias Filtering

The Y, Cb, and Cr components are applied to appropriate low-pass filters that shape the frequency response of each of the three components [1]. Prior to horizontal and vertical sub-sampling of the two chrominance components, they may be processed by half-band filters in order to prevent aliasing.

5.2.2j   Number of Lines Encoded

The video coding system requires that the coded picture area has a number of lines that is a multiple of 32 for an interlaced format, and a multiple of 16 for a non-interlaced format [1]. This means that for encoding the 1080-line format, a coder must actually deal with 1088 lines (1088 = 32 × 34). The extra eight lines are in effect "dummy" lines having no content, and the coder designers will choose dummy data that simplifies the implementation. The extra eight lines are always the last eight lines of the encoded image. These dummy lines do not carry useful information, but add little to the data required for transmission.

5.2.2k Concatenated Sequences

The MPEG-2 video standard that underlies the ATSC Digital Television Standard clearly specifies the behavior of a compliant video decoder when processing a single video sequence [1]. A coded video sequence commences with a *Sequence Header*, typically contains repeated Sequence Headers and one or more coded pictures, and is terminated by an end-of-sequence code. A number of parameters are specified in the Sequence Header that are required to remain constant throughout the duration of the sequence. The sequence level parameters include, but are not limited to:

- Horizontal and vertical resolution

- Frame rate

- Aspect ratio

- Chroma format

- Profile and Level

- All-progressive indicator

- Video buffering verifier (VBV) size

- Maximum bit rate

It is assumed that it will be common for coded bit streams to be spliced for editing, insertion of commercial advertisements, and other purposes in the video production and distribution chain. If one or more of the sequence level parameters differ between the two bit streams to be spliced, then an end-of-sequence code must be inserted to terminate the first bit stream and a new Sequence Header must exist at the start of the second bit stream (unless the insertion equipment is capable of scaling those parameters in real time). Thus, the situation of concatenated video sequences arises.

Regarding concatenated sequences, the MPEG-2 Video Standard (13818-2) states: "The behavior of the decoding process and display process for concatenated sequences is not within the scope of this Recommendation | International Standard. An application that needs to use concatenated sequences must ensure by private arrangement that the decoder will be able to decode and play concatenated sequences."

While it is recommended, the Digital Television Standard does not require the production of well-constrained concatenated sequences. Well-constrained concatenated sequences are defined as having the following characteristics:

- The extended decoder buffer never overflows, and may only underflow in the case of low-delay bit streams. Here, "extended decoder buffer" refers to the natural extension of the MPEG-2 decoder buffer model to the case of continuous decoding of concatenated sequences.

- When field parity is specified in two coded sequences that are concatenated, the parity of the first field in the second sequence is opposite that of the last field in the first sequence.

- Whenever a progressive sequence is inserted between two interlaced sequences, the exact number of progressive frames should be such that the parity of the interlaced sequences is preserved as if no concatenation had occurred.

5.2.2I   Guidelines for Refreshing

While the ATSC Digital Television Standard does not require refreshing at less than the intra-macroblock refresh rate as defined in IEC/ISO 13818-2, the following is recommended [1]:

- Sequence layer information is very helpful and it is important that it be sent before every I-frame, independent of the interval between I-frames. Use of intra-macroblock refresh in the decoder can improve receiver channel-change performance.

- Some receivers rely on periodic transmission of I-frames for refreshing. The frequency of occurrence of I-frames may determine the channel-change time performance of the receiver. It is recommended that I-frames be sent at least once every 0.5 second in order to have acceptable channel-change performance in such receivers.

- In order to spatially localize errors due to transmission, intra-coded slices should contain fewer macroblocks than the maximum number allowed by the DTV standard. It is recommended that there be four to eight slices in a horizontal row of intra-coded macroblocks for the intra-coded slices in the I-frame refresh case as well as for the intraframe coded regions in the progressive refresh case. The size of non-intra-coded slices can be larger than that of intra-coded slices.

5.2.3   Active Format Description (AFD)

Active Format Description (AFD) solves a troublesome problem in the transition from conventional 4:3 display devices to widescreen 16:9 displays, and also addresses the variety of aspect ratios that have been used over the years by the motion picture industry to produce feature films [1].

There are, of course, a number of different types of video displays in common usage—ranging from 4:3 CRTs to widescreen projection devices and flat-panel displays of various design. Each of these devices may have varying abilities to process incoming video. In terms of input interfaces, these displays may likewise support a range of input signal formats—from composite analog video to IEEE 1394.

Possible video source devices include cable, satellite, or terrestrial broadcast set-top (or integrated receiver-decoder) boxes, media players (such as DVDs), analog or digital tape players, and personal video recorders.

Although choice is good, this wide range of consumer options presented two problems to be solved:

- First, no standard method had been agreed upon to communicate to the display device the "active area" of the video signal. Such a method would be able, for example to signal that the 4:3 signal contains within it a letterboxed 16:9 video image.

- Second, no standard method had been agreed upon to communicate to the display device, for all interface types, that a given image is intended for 16:9 display.

The AFD solves these problems and, in the process, provides the following additional benefits:

- Active area signaling allows the display device to process the incoming signal to make the highest-resolution and most accurate picture possible. Furthermore, the display can take advantage of the knowledge that certain areas of video are currently unused and can implement algorithms that reduce the effects of uneven screen aging.

- Aspect ratio signaling allows the display device to produce the best image possible. In some scenarios, lack of a signaling method translates to restrictions in the ability of the source device to deliver certain otherwise desirable output formats.

5.2.3a Active Area Signaling

A consumer device such as a cable or satellite set-top box cannot reliably determine the active area of video on its own [1]. Even though certain lines at the top and bottom of the screen may be black for periods of time, the situation could change without warning. The only sure way to know active area is for the service provider to include this data at the time of video compression and to embed it into the video stream.

Figure 5.2.2 shows 4:3- and 16:9-coded images with various possible active areas. The group on the left is either coded explicitly in the MPEG-2 video syntax as 4:3, or the uncompressed signal provided in NTSC timing and aspect ratio information (if present) indicates 4:3. The group on the right are coded explicitly in the MPEG-2 video syntax as 16:9, provided with NTSC timing and an aspect ratio signal indicating 16:9, or provided uncompressed with 16:9 timing across the interface.

As can be seen in the figure, a pillar-boxed display results when a 4:3 active area is displayed within a 16:9 area, and a letterboxed display results when a 16:9 active area is displayed within a 4:3 area. It is also apparent that double-boxing can also occur, for example when 4:3 material is delivered within a 16:9 letterbox to a 4:3 display. Or, when 16:9 material is delivered within a 4:3 pillar-box to a 16:9 display.

For the straight letter- or pillar-box cases, if the display is aware of the active area it may take steps to mitigate the effects of uneven screen aging. Such steps could, for example, involve using gray instead of black bars. Some amount of linear or nonlinear stretching and/or zooming may be done as well using the knowledge that video outside the active area can safely be discarded.

The two double-boxed cases can occur as a result of poor or uninformed production choices made by the service provider, in some cases in concert with the content provider. Whenever 4:3 material is coded as 16:9, double boxing occurs when the 4:3 display places the 16:9 coded frame on screen. Whenever 16:9 material is coded as 4:3, double boxing occurs when the 16:9 display pillar-boxes the 4:3 coded frame.

A common situation that will cause double-boxing on a 16:9 digital TV display occurs when a 4:3 NTSC signal is encoded as 480I MPEG video, but the NTSC image is a letterboxed widescreen movie. Regardless of the cause, two aspects of the problem are of prime importance:

- The display device should not be expected to process the double-boxed image to fill the screen to make up for incorrectly coded content.

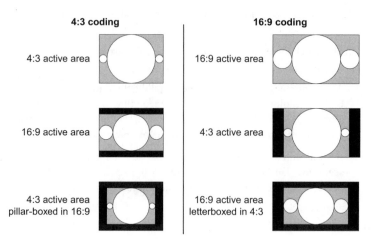

4:3 coding

4:3 active area

16:9 active area

4:3 active area
pillar-boxed in 16:9

16:9 coding

16:9 active area

4:3 active area

16:9 active area
letterboxed in 4:3

Figure 5.2.2 Coding and active area. (*From* [1]. *Used with permission.*)

- Content and service providers should be expected to deliver properly coded content. Native 4:3 content must be delivered coded as 4:3. Native 16:9 content must be delivered coded as 16:9. Letterboxed widescreen video in NTSC should not be coded as 4:3, but should be coded into a 16:9 coded frame.[3]

5.2.3b   Existing Standards

Several industry standards include some form of active area information, including EIA-608-C and DVB Active Format Description [1]. The EIA-608-B standard is applicable only to NTSC analog video. The DVB description data applies only to compressed MPEG-2 video. Also working in this area, MPEG adopted an amendment to the MPEG-2 Video Standard to include active area data.

Letterboxed movies can be seen on cable, satellite, and terrestrial channels today. If one observes closely, considerable variability in the size of the black bar areas can be seen. In fact, variations can sometimes be seen even over the course of one movie.

As mentioned previously, a display device may wish to mitigate the effects of uneven screen aging by substituting gray video for the black areas. It is problematic for the display to be required to actively track a varying letterbox area, and real-time tracking of variations from frame to frame would be difficult.

Clearly, two approaches are possible. First, include—on a frame-by-frame basis—a video parameter identifying the number of black lines (for letterbox) or number of black pixels (for pillar-box). Second, standardize on just two standard aspect ratios: 16:9 and 4:3.

---

3. Letterboxed content inside a 4:3 frame results in vertical resolution less than standard-definition television.

5.2.3c   Active Areas Greater than 16:9

Any wide-screen source material can be coded into a 16:9-coded frame [1]. No aspect ratio for coded frames exceeding 16:9 is standardized for cable, terrestrial broadcast, or satellite transmission in the U.S. If the aspect ratio of given content exceeds 16:9, the coded image will be letterboxed inside the 16:9 frame, as shown in Figure 5.2.3, where 2.35:1 material is letterboxed inside the 16:9 frame.

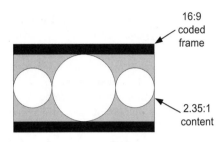

It can be helpful for a display to know the actual aspect ratio of the active portion of the 16:9 coded frame for a variety of reasons, including:

Figure 5.2.3 Example of active video area greater than 16:9 aspect ratio. (*From* [1]. *Used with permission.*)

- Reduction in the effects of uneven screen aging. The display device controller may wish to use gray instead of black for the bars.

- The display may offer the user a "zoom" option to make better use of available display area, and knowledge of the aspect ratio can automate the selection of this display option. The zoom feature can be vertical scaling only, or a combination of horizontal and vertical where the leftmost and rightmost portions of the image are sacrificed to fill the screen area vertically.

Several standards include aspect ratio data. The MPEG-2 video syntax includes horizontal and vertical size data and aspect ratio indication of the coded image. An NTSC signal is normally thought to be intended for 4:3 display, but this is not always the case. EIA-608-C includes a "squeezed" bit, and IEC 61880 defines a method for NTSC VBI line 20. The line-20 method is currently used for playback of anamorphically coded DVDs, when the DVD player supports it and is properly set up by the user.

5.2.3d   Active Format Description (AFD) and Bar Data

In recognition of these issues, the ATSC undertook a study of available options and decided to endorse the basic signaling structure developed by the DVB consortium [1]. The benefits of common active format description signaling across many different markets is easily understood.

Some common active video formats represented by the 4-bit AFD field include:

- The aspect ratio of the active video area is 16:9; when associated with a 4:3 coded frame, the active video is top-justified.

- Active video area is 16:9; when associated with a 4:3 coded frame, the active video is centered vertically.

- Active video area is 4:3; when associated with a 16:9 coded frame, the active video is centered horizontally.

- Active video area exceeds 16:9 aspect ratio; active video is centered vertically (in whatever coded frame is used).

- Active video area is 14:9; when associated with a 4:3 coded frame the active video is centered vertically; when associated with a 16:9 coded frame, it is centered horizontally.

It should be noted that certain active formats signal to the receiver that active video may be safely cropped in the receiver display.

In addition to AFD, ATSC defined another data structure, bar_data(), also for use in the video Elementary Stream. The bar_data() structure, like AFD, appears in the picture user_data() area of the video syntax. While the AFD gives a general view of the relationship between the coded frame and the geometry of the active video within it, bar_data() is able to indicate precisely the number of lines of black video at the top and bottom of a letterboxed image, or the number of black pixels at the left and right side of a pillar-boxed image.

For the ATSC system, AFD and/or bar_data() are included in video user data whenever the rectangular picture area containing useful information does not extend to the full height or width of the coded frame. Such data may optionally also be included in user data when the rectangular picture area containing useful information extends to the full height and width of the coded frame.

The AFD and bar_data() are carried in the user data of the video Elementary Stream. After each sequence start (and repeat sequence start) the default aspect ratio of the area of interest is signalled by the Sequence Header and sequence display extension parameters. After introduction, each type of active format data remain in effect until the next sequence start or until another instance is introduced. Receivers are expected to interpret the absence of AFD and bar_data() in a sequence start to mean the active format is the same as the coded frame. Since it is not able to represent non-standard video aspect ratios, AFD may be only an approximation of the actual active video area. However when bar_data() is present, it should be assumed to be exact. If the bar_data() and the AFD are in conflict, the bar_data() should take precedence.

5.2.4   References

1.   ATSC, "Guide to the Use of the Digital Television Standard," Advanced Television Systems Committee, Washington, D.C., Doc. A/54A, December 4, 2003.

2.   "SMPTE Standard for Television—1920 × 1080 Scanning and Analog and Parallel Digital Interfaces for Multiple-Picture Rates," SMPTE 274-1998, SMPTE, White Plains, N.Y., 1998.

3.   "SMPTE Standard for Television—1280 × 720 Scanning, Analog and Digital Representation and Analog Interface," SMPTE 296M-2001, SMPTE, White Plains, N.Y., 1997.

5.3

DTV Audio Encoding and Decoding

5.3.1   Introduction[1]

Monophonic sound is the simplest form of aural communication. A wide range of acceptable listening positions are practical, although it is obvious from most positions that the sound is originating from one source rather than occurring in the presence of the listener. Consumers have accepted this limitation without much thought in the past because it was all that was available. However, monophonic sound creates a poor illusion of the sound field that the program producer might want to create.

Two channel stereo improves the illusion that the sound is originating in the immediate area of the reproducing system. Still, there is a smaller acceptable listening area. It is difficult to keep the sound image centered between the left and right speakers, so that the sound and the action stay together as the listener moves in the room.

The AC-3 surround sound system is said to have 5.1 channels because there is a left, right, center, left surround, and right surround, which make up the 5 channels. A sixth channel is reserved for the lower frequencies and consumes only 120 Hz of the bandwidth; it is referred to as the 0.1 or *low-frequency enhancement* (LFE) channel. The center channel restores the variety of listening positions possible with monophonic sound.

The AC-3 system is effective in providing either an enveloping (ambient) sound field or allowing precise placement and movement of special effects because of the channel separation afforded by the multiple speakers in the system.

For efficient and reliable interconnection of audio devices, standardization of the interface parameters is of critical importance. The primary interconnection scheme for professional digital audio systems is AES Audio.

---

1. This chapter is based on: ATSC, "Digital Audio Compression Standard (AC-3)," Advanced Television Systems Committee, Washington, D.C., Doc. A/52B, June 14, 2005; and ATSC, "Guide to the Use of the Digital Television Standard," Advanced Television Systems Committee, Washington, D.C., Doc. A/54A, December 4, 2003. Readers are encouraged to download the source documents from the ATSC Web site (http://www.atsc.org). All ATSC Standards, Recommended Practices, and Information Guides are available at no charge.

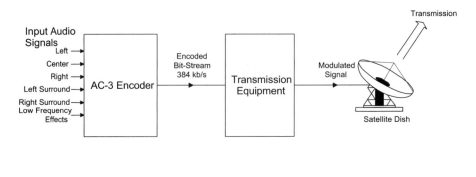

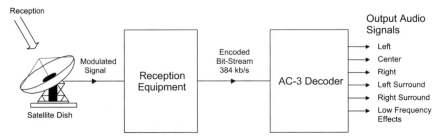

Figure 5.3.1 Example application of the AC-3 audio subsystem for satellite audio transmission. (*From* [1]. *Used with permission.*)

5.3.2 Audio Compression

Efficient recording and/or transmission of digital audio signals demands a reduction in the amount of information required to represent the aural signal [1]. The amount of digital information needed to accurately reproduce the original PCM samples taken of an analog input may be reduced by applying a digital compression algorithm, resulting in a digitally compressed representation of the original signal. (In this context, the term *compression* applies to the digital information that must be stored or recorded, not to the dynamic range of the audio signal.) The goal of any digital compression algorithm is to produce a digital representation of an audio signal which, when decoded and reproduced, sounds the same as the original signal, while using a minimum amount of digital information (bit rate) for the compressed (or encoded) representation. The AC-3 digital compression algorithm specified in the ATSC DTV system can encode from 1 to 5.1 channels of source audio from a PCM representation into a serial bit stream at data rates ranging from 32 to 640 kbits/s.

A typical application of the bit-reduction algorithm is shown in Figure 5.3.1. In this example, a 5.1 channel Audio Program is converted from a PCM representation requiring more than 5 Mbits/s (6 channels × 48 kHz × 18 bits = 5.184 Mbits/s) into a 384 kbits/s serial bit stream by the AC-3 encoder. Radio frequency (RF) transmission equipment converts this bit stream into a modulated waveform that is applied to a satellite transponder. The amount of bandwidth and power thus required by the transmission has been reduced by more than a factor of 13 by the AC-3 digital compression system. The received signal is demodulated back into the 384 kbits/s serial bit stream, and decoded by the AC-3 decoder. The result is the original 5.1 channel Audio Pro-

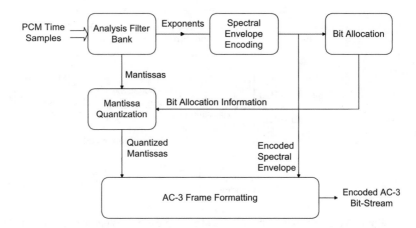

Figure 5.3.2 Overview of the AC-3 audio-compression system encoder. (*From* [1]. *Used with permission.*)

gram. Note that in 2001, the ATSC modified its A/53 DTV Standard to increase the maximum allowable audio bit rate from 384 kbits/s to 448 kbits/s to permit greater flexibility in implementing new services.

Digital compression of audio is useful wherever there is an economic benefit to be obtained by reducing the amount of digital information required to represent the audio signal. Typical applications include the following:

- Terrestrial audio broadcasting

- Delivery of audio over metallic or optical cables, or over RF links

- Storage of audio on magnetic, optical, semiconductor, or other storage media

5.3.2a Encoding

The AC-3 encoder accepts PCM audio and produces the encoded bit stream for the ATSC DTV standard [1]. The AC-3 algorithm achieves high *coding gain* (the ratio of the input bit rate to the output bit rate) by coarsely quantizing a frequency-domain representation of the audio signal. A block diagram of this process is given in Figure 5.3.2. The first step in the encoding chain is to transform the representation of audio from a sequence of PCM time samples into a sequence of blocks of frequency coefficients. This is done in the *analysis filterbank*. Overlapping blocks of 512 time samples are multiplied by a time window and transformed into the frequency domain. Because of the overlapping blocks, each PCM input sample is represented in two sequential transformed blocks. The frequency-domain representation then may be decimated by a factor of 2, so that each block contains 256 frequency coefficients. The individual frequency coefficients are represented in binary exponential notation as a *binary exponent* and a *mantissa*. The set of exponents is encoded into a coarse representation of the signal spectrum, referred to as the *spectral envelope*. This spectral envelope is used by the core bit-allocation routine, which determines

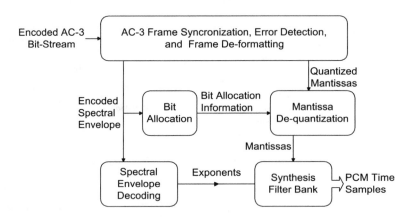

Figure 5.3.3 Overview of the AC-3 audio-compression system decoder. (*From* [1]. *Used with permission.*)

how many bits should be used to encode each individual mantissa. The spectral envelope and the coarsely quantized mantissas for six audio blocks (1536 audio samples) are formatted into an AC-3 *frame*. The AC-3 bit stream is a sequence of AC-3 frames.

The actual AC-3 encoder is more complex than shown in the simplified system of Figure 5.3.2. The following functions also are included:

- A frame header is attached, containing information (bit rate, sample rate, number of encoded channels, and other data) required to synchronize to and decode the encoded bit stream.

- Error-detection codes are inserted to allow the decoder to verify that a received frame of data is error-free.

- The analysis filterbank spectral resolution may be dynamically altered to better match the time/frequency characteristic of each audio block.

- The spectral envelope may be encoded with variable time/frequency resolution.

- A more complex bit-allocation may be performed, and parameters of the core bit-allocation routine may be modified to produce a more optimum bit allocation.

- The channels may be coupled at high frequencies to achieve higher coding gain for operation at lower bit rates.

- In the 2-channel mode, a rematrixing process may be selectively performed to provide additional coding gain, and to allow improved results to be obtained in the event that the 2-channel signal is decoded with a matrix surround decoder.

5.3.2b   Decoding

The decoding process is, essentially, the inverse of the encoding process [1]. The basic decoder, shown in Figure 5.3.3, must synchronize to the encoded bit stream, check for errors, and deformat the various types of data (i.e., the encoded spectral envelope and the quantized mantissas).

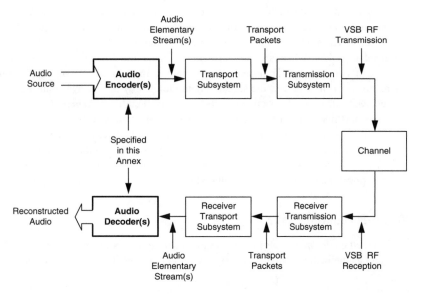

Figure 5.3.4 The audio subsystem in the DTV standard. (*From* [2]. *Used with permission.*)

The bit-allocation routine is run, and the results are used to unpack and dequantize the mantissas. The spectral envelope is decoded to produce the exponents. The exponents and mantissas are transformed back into the time domain to produce the decoded PCM time samples. Additional steps in the audio decoding process include the following:

- Error concealment or muting may be applied in the event a data error is detected.

- Channels that have had their high-frequency content coupled must be decoupled.

- Dematrixing must be applied (in the 2-channel mode) whenever the channels have been *rematrixed*.

- The synthesis filterbank resolution must be dynamically altered in the same manner as the encoder analysis filterbank was altered during the encoding process.

5.3.3   Implementation of the AC-3 System

As illustrated in Figure 5.3.4, the audio subsystem of the ATSC DTV standard comprises the audio-encoding/decoding function and resides between the audio inputs/outputs and the transport subsystem [2]. The audio encoder is responsible for generating the *Audio Elementary Stream*, which is an encoded representation of the baseband audio input signals. (Note that more than one audio encoder may be used in a system.) The flexibility of the transport system allows multiple Audio Elementary Streams to be delivered to the receiver. At the receiver, the transport subsystem is responsible for selecting which audio streams to deliver to the audio subsystem. The audio subsystem is then responsible for decoding the Audio Elementary Stream back into baseband audio.

An Audio Program source is encoded by a *Digital Television Audio Encoder*. The output of the audio encoder is a string of bits that represent the audio source (the Audio Elementary Stream). The transport subsystem packetizes the audio data into PES (*Packetized Elementary Stream*) packets, which are then further packetized into *Transport Packets*. The transmission subsystem converts the Transport Packets into a modulated RF signal for transmission to the receiver. At the receiver, the signal is demodulated by the receiver transmission subsystem. The receiver transport subsystem converts the received audio packets back into an Audio Elementary Stream, which is decoded by the digital television audio decoder.

The partitioning shown in Figure 5.3.4 is conceptual, and practical implementations may differ. For example, the transport processing may be broken into two blocks; the first would perform PES packetization, and the second would perform transport packetization. Or, some of the transport functionality may be included in either the audio coder or the transmission subsystem.

5.3.3a   Audio-Encoder Interface

The audio system accepts baseband inputs with up to six channels per Audio Program bit stream in a channelization scheme consistent with ITU-R Rec. BS-775 [3]. The six audio channels are:

- Left

- Center

- Right

- Left surround

- Right surround

- Low-frequency enhancement (LFE)

Multiple audio elementary bit streams may be conveyed by the transport system.

The bandwidth of the LFE channel is limited to 120 Hz. The bandwidth of the other (main) channels is limited to 20 kHz. Low-frequency response may extend to dc, but it is more typically limited to approximately 3 Hz (–3 dB) by a dc-blocking high-pass filter. Audio-coding efficiency (and thus audio quality) is improved by removing dc offset from audio signals before they are encoded. The input audio signals may be in analog or digital form.

For analog input signals, the input connector and signal level are not specified [2]. Conventional broadcast practice may be followed. One commonly used input connector is the 3-pin XLR female (the incoming audio cable uses the male connector) with pin 1 ground, pin 2 hot or positive, and pin 3 neutral or negative.

Likewise, for digital input signals, the input connector and signal format are not specified. Commonly used formats such as the AES3 2-channel interface are suggested. When multiple 2-channel inputs are used, the preferred channel assignment is:

- Pair 1: Left, Right

- Pair 2: Center, LFE

- Pair 3: Left surround, Right surround

Sampling Parameters

The AC-3 system conveys digital audio sampled at a frequency of 48 kHz, locked to the 27 MHz system clock [2]. If analog signal inputs are employed, the A/D converters should sample at 48 kHz. If digital inputs are employed, the input sampling rate should be 48 kHz, or the audio encoder should contain sampling rate converters that translate the sampling rate to 48 kHz. The sampling rate at the input to the audio encoder must be locked to the video clock for proper operation of the audio subsystem.

In general, input signals should be quantized to at least 16-bit resolution. The audio-compression system can convey audio signals with up to 24-bit resolution.

5.3.3b   Output Signal Specification

Conceptually, the output of the audio encoder is an Elementary Stream that is formed into PES packets within the transport subsystem [2]. It is possible that digital television systems will be implemented wherein the formation of audio PES packets takes place within the audio encoder. In this case, the output of the audio encoder would be PES packets. Physical interfaces for these outputs (Elementary Streams and/or PES packets) may be defined as voluntary industry standards by SMPTE or other organizations; they are not, however, specified in the core ATSC standard.

5.3.4   Operational Details of the AC-3 Standard

The AC-3 audio-compression system consists of three basic operations, as illustrated in Figure 5.3.5 [4]. In the first stage, the representation of the audio signal is changed from the time domain to the frequency domain, which is a more efficient domain in which to perform psychoacoustically based audio compression. The resulting frequency-domain coefficients are then encoded. The frequency-domain coefficients may be coarsely quantized because the resulting quantizing noise will be at the same frequency as the audio signal, and relatively low S/N ratios are acceptable because of the phenomenon of psychoacoustic masking. Based on a psychoacoustic model of human hearing, a bit-allocation operation determines the actual S/N acceptable for each individual frequency coefficient. Finally, the frequency coefficients are coarsely quantized to the necessary precision and formatted into the Audio Elementary Stream.

The basic unit of encoded audio is the AC-3 *Sync Frame*, which represents 1536 audio samples. Each Sync Frame of audio is a completely independent encoded entity. The elementary bit stream contains the information necessary to allow the audio decoder to perform the identical (to the encoder) bit allocation. This permits the decoder to unpack and dequantize the elementary bit-stream frequency coefficients, resulting in the reconstructed frequency coefficients. The synthesis filterbank is the inverse of the analysis filterbank, and it converts the reconstructed frequency coefficients back into a time-domain signal.

5.3.4a   Transform Filterbank

The process of converting the audio from the time domain to the frequency domain requires that the audio be blocked into overlapping blocks of 512 samples [4]. For every 256 new audio sam-

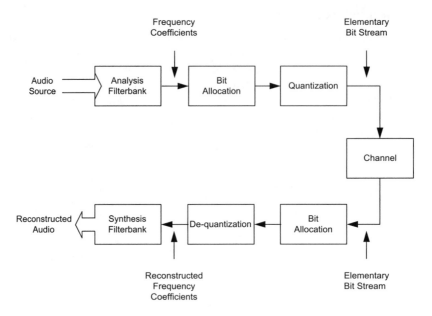

Figure 5.3.5 Overview of the AC-3 audio-compression system. (*From* [3]. *Used with permission.*)

ples, a 512-sample block is formed from the 256 new samples and the 256 previous samples. Each audio sample is represented in two audio blocks, so the number of samples to be processed initially is doubled. The overlapping of blocks is necessary to prevent audible blocking artifacts. New audio blocks are formed every 5.33 ms. A group of six blocks is coded into one AC-3 Sync Frame.

Window Function

Prior to being transformed into the frequency domain, the block of 512 time samples is *windowed* [4]. The windowing operation involves a vector multiplication of the 512-point block with a 512-point window function. The window function has a value of 1.0 in its center, tapering down to almost zero at the ends. The shape of the window function is such that the overlap/add processing at the decoder will result in a reconstruction free of blocking artifacts. The window function shape also determines the shape of each individual filterbank filter.

Time-Division Aliasing Cancellation Transform

The analysis filterbank is based on the fast Fourier transform [4]. The particular transformation employed is the oddly stacked *time-domain aliasing cancellation* (TDAC) transform. This particular transformation is advantageous because it allows removal of the 100 percent redundancy that was introduced in the blocking process. The input to the TDAC transform is 512 windowed time-domain points, and the output is 256 frequency-domain coefficients.

Transient Handling

When extreme time-domain transients exist (an impulse, such as a castanets click), there is a possibility that quantization error—incurred by coarsely quantizing the frequency coefficients of the transient—will become audible as a result of *time smearing* [4]. The quantization error within a coded audio block is reproduced throughout the block. It is possible for the portion of the quantization error that is reproduced prior to the impulse to be audible. Time smearing of quantization noise may be reduced by altering the length of the transform that is performed. Instead of a single 512-point transform, a pair of 256-point transforms may be performed—one on the first 256 windowed samples, and one on the last 256 windowed samples. A transient detector in the encoder determines when to alter the transform length. The reduction in transform length prevents quantization error from spreading more than a few milliseconds in time, which is adequate to prevent audibility.

5.3.4b   Coded Audio Representation

The frequency coefficients that result from the transformation are converted to a binary floating point notation [4]. The scaling of the transform is such that all values are smaller than 1.0. An example value in binary notation (base 2) with 16-bit precision would be

0.0000 0000 1010 11002

The number of leading zeros in the coefficient, 8 in this example, becomes the *raw exponent*. The value is left-shifted by the exponent, and the value to the right of the decimal point (1010 1100) becomes the *normalized mantissa* to be coarsely quantized. The exponents and the coarsely quantized mantissas are encoded into the bit stream.

Exponent Coding

A certain amount of processing is applied to the raw exponents to reduce the amount of data required to encode them [4]. First, the raw exponents of the six blocks to be included in a single AC-3 Sync Frame are examined for block-to-block differences. If the differences are small, a single exponent set is generated that is usable by all six blocks, thus reducing the amount of data to be encoded by a factor of 6. If the exponents undergo significant changes within the frame, exponent sets are formed over blocks where the changes are not significant. Because of the frequency response of the individual filters in the analysis filterbank, exponents for adjacent frequencies rarely differ by more than ± 2. To take advantage of this fact, exponents are encoded differentially in frequency. The first exponent is encoded as an absolute, and the difference between the current exponent and the following exponent then is encoded. This reduces the exponent data rate by a factor of 2. Finally, where the spectrum is relatively flat, or an exponent set only covers 1 or 2 blocks, differential exponents may be shared across 2 or 4 frequency coefficients, for an additional savings of a factor of 2 or 4.

The final coding efficiency for AC-3 exponents is typically 0.39 bits/exponent (or 0.39 bits/ sample, because there is an exponent for each audio sample). Exponents are coded only up to the frequency needed for the perception of full frequency response. Typically, the highest audio frequency component in the signal that is audible is at a frequency lower than 20 kHz. In the case that signal components above 15 kHz are inaudible, only the first 75 percent of the exponent values are encoded, reducing the exponent data rate to less than 0.3 bits/sample.

The exponent processing changes the exponent values from their original values. The encoder generates a local representation of the exponents that is identical to the decoded representation that will be used by the decoder. The decoded representation then is used to shift the original frequency coefficients to generate the normalized mantissas that are subsequently quantized.

Mantissas

The frequency coefficients produced by the analysis filterbank have a useful precision that is dependent upon the word length of the input PCM audio samples as well as the precision of the transform computation [4]. Typically, this precision is on the order of 16 to 18 bits, but may be as high as 24 bits. Each normalized mantissa is quantized to a precision from 0 to 16 bits. Because the goal of audio compression is to maximize the audio quality at a given bit rate, an optimum (or near-optimum) allocation of the available bits to the individual mantissas is required.

5.3.4c   Bit Allocation

The number of bits allocated to each individual mantissa value is determined by the bit-allocation routine [4]. The identical core routine is run in both the encoder and the decoder, so that each generates an identical bit allocation.

The core bit-allocation algorithm is considered *backward adaptive*, in that some of the encoded audio information within the bit stream (fed back into the encoder) is used to compute the final bit allocation. The primary input to the core allocation routine is the decoded exponent values, which give a general picture of the signal spectrum. From this version of the signal spectrum, a *masking curve* is calculated. The calculation of the masking model is based on a model of the human auditory system. The masking curve indicates, as a function of frequency, the level of quantizing error that may be tolerated. Subtraction (in the log power domain) of the masking curve from the signal spectrum yields the required S/N as a function of frequency. The required S/N values are mapped into a set of *Bit-Allocation Pointers* (BAPs) that indicate which quantizer to apply to each mantissa.

Forward Adaptive

The AC-3 encoder may employ a more sophisticated psychoacoustic model than that used by the decoder [4]. The core allocation routine used by both the encoder and the decoder makes use of a number of adjustable parameters. If the encoder employs a more sophisticated psychoacoustic model than that of the core routine, the encoder may adjust these parameters so that the core routine produces a better result. The parameters are subsequently inserted into the bit stream by the encoder and fed forward to the decoder.

In the event that the available bit-allocation parameters do not allow the ideal allocation to be generated, the encoder can insert explicit codes into the bit stream to alter the computed masking curve, hence the final bit allocation. The inserted codes indicate changes to the base allocation and are referred to as *delta bit-allocation codes*.

5.3.4d   Rematrixing

When the AC-3 encoder is operating in a 2-channel stereo mode, an additional processing step is inserted to enhance interoperability with Dolby Surround 4-2-4 matrix encoded programs [4]. This extra step is referred to as *rematrixing*.

The signal spectrum is broken into four distinct rematrixing frequency bands. Within each band, the energy of the left, right, sum, and difference signals are determined. If the largest signal energy is in the left and right channels, the band is encoded normally. If the dominant signal energy is in the sum and difference channels, then those channels are encoded instead of the left and right channels. The decision as to whether to encode left and right or sum and difference is made on a band-by-band basis and is signaled to the decoder in the encoded bit stream.

5.3.4e   Coupling

In the event that the number of bits required to transparently encode the audio signals exceeds the number of bits that are available, the encoder may invoke *coupling* [4]. Coupling involves combining the high-frequency content of individual channels and sending the individual channel signal envelopes along with the combined coupling channel. The psychoacoustic basis for coupling is that within narrow frequency bands, the human ear detects high-frequency localization based on the signal envelope rather than on the detailed signal waveform.

The frequency above which coupling is invoked, and the channels that participate in the process, are determined by the AC-3 encoder. The encoder also determines the frequency banding structure used by the coupling process. For each coupled channel and each coupling band, the encoder creates a sequence of *coupling coordinates*. The coupling coordinates for a particular channel indicate what fraction of the common coupling channel should be reproduced out of that particular channel output. The coupling coordinates represent the individual signal envelopes for the channels. The encoder determines the frequency with which coupling coordinates are transmitted. If the signal envelope is steady, the coupling coordinates do not need to be sent every block, but can be reused by the decoder until new coordinates are sent. The encoder determines how often to send new coordinates, and it can send them as often as each block (every 5.3 ms).

5.3.4f   Bit Stream Elements and Syntax

An AC-3 serial-coded audio bit stream is made up of a sequence of *Synchronization Frames*, as illustrated in Figure 5.3.6 [4]. Each Synchronization Frame contains six coded audio blocks, each of which represent 256 new audio samples. A *Synchronization Information* (SI) header at the beginning of each frame contains information needed to acquire and maintain synchronization. A *Bit-Stream Information* (BSI) header follows each SI, containing parameters describing the coded audio service. The coded audio blocks may be followed by an auxiliary data (Aux) field. At the end of each frame is an error-check field that includes a CRC word for error detection. An additional CRC word, the use of which is optional, is located in the SI header.

A number of bit-stream elements have values that may be transmitted, but whose meaning has been reserved. If a decoder receives a bit stream that contains reserved values, the decoder may or may not be able to decode and produce audio.

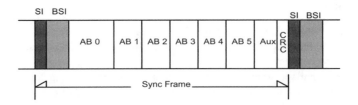

Figure 5.3.6 The AC-3 Synchronization Frame. (*From* [4]. *Used with permission.*)

Splicing and Insertion

The ideal place to splice encoded audio bit streams is at the boundary of a Sync Frame [4]. If a bit stream splice is performed at the boundary of the Sync Frame, the audio decoding will proceed without interruption. If a bit stream splice is performed randomly, there will be an audio interruption. The frame that is incomplete will not pass the decoder's error-detection test, and this will cause the decoder to mute. The decoder will not find sync in its proper place in the next frame, and it will enter a sync search mode. After the sync code of the new bit stream is found, synchronization will be achieved, and audio reproduction will resume. This type of outage will be on the order of two frames, or about 64 ms. Because of the windowing process of the filterbank, when the audio goes to mute, there will be a gentle fadedown over a period of 2.6 ms. When the audio is recovered, it will fade up over a period of 2.6 ms. Except for the approximately 64 ms of time during which the audio is muted, the effect of a random splice of an AC-3 Elementary Stream is relatively benign.

Error-Detection Codes

Each AC-3 Sync Frame ends with a 16-bit CRC error-check code [4]. The decoder may use this code to determine whether a frame of audio has been damaged or is incomplete. Additionally, the decoder may make use of error flags provided by the transport system. In the case of detected errors, the decoder may try to perform error concealment, or it may simply mute.

5.3.4g   Loudness and Dynamic Range

It is important for the digital television system to provide uniform subjective loudness for all Audio Programs [4]. Consumers often find it annoying when audio levels fluctuate between broadcast channels (observed when channel hopping) or between program segments on a particular channel (such as commercials being much louder than entertainment programs). One element found in most Audio Programming is the human voice. Achieving an approximate level match for dialogue (spoken in a normal voice, without shouting or whispering) in all Audio Programming is a desirable goal. The AC-3 audio system provides syntactical elements that make this goal achievable.

Because the digital audio-coding system can provide more than 100 dB of dynamic range, there is no technical reason for dialogue to be encoded anywhere near 100 percent, as it commonly is in NTSC television. However, there is no assurance that all program channels, or all programs or program segments on a given channel, will have dialogue encoded at the same (or even a similar) level. Without a uniform coding level for dialogue (which would imply a uniform headroom available for all programs), there would be inevitable audio-level fluctuations between program channels or even between program segments.

Dynamic Range Compression

It is common practice for high-quality programming to be produced with wide dynamic range audio, suitable for the highest-quality audio reproduction environment [4]. Because they serve audiences with a wide range of receiver capabilities, however, broadcasters typically process audio to reduce its dynamic range. This processed audio is more suitable for most of the audience, which does not have an audio reproduction environment that matches the original audio production studio. In the case of NTSC, all viewers receive the same audio with the same dynamic range; it is impossible for any viewer to enjoy the original wide dynamic range of the audio production.

For DTV, the audio-coding system provides an embedded Dynamic Range Control scheme that allows a common encoded bit stream to deliver programming with a dynamic range appropriate for each individual listener. A *Dynamic Range Control value* (DynRng²) is provided in each audio block (every 5 ms). These values are used by the audio decoder to alter the level of the reproduced sound for each audio block. Level variations of up to ± 24 dB can be indicated.

5.3.4h   Encoding the AC-3 Bit Stream

Because the ATSC DTV standard AC-3 audio system is specified by the syntax and decoder processing, the encoder itself is not precisely specified [1]. The only normative requirement on the encoder is that the output elementary bit stream follow the AC-3 syntax. Therefore, encoders of varying levels of sophistication may be produced. More sophisticated encoders may offer superior audio performance, and they may make operation at lower bit rates acceptable. Encoders are expected to improve over time, and all decoders will benefit from encoder improvements. The encoder described in this section, although basic in operation, provides good performance and offers a starting point for encoder designs. A flow chart diagram of the encoding process is given in Figure 5.3.7.

Input Word Length/Sample Rate

The AC-3 encoder accepts audio in the form of PCM words [1]. The internal dynamic range of AC-3 allows input word lengths of up to 24 bits to be useful.

The input sample rate must be locked to the output bit rate so that each AC-3 Sync Frame contains 1536 samples of audio. If the input audio is available in a PCM format at a different sample rate than that required, sample rate conversion must be performed to conform the sample rate.

Individual input channels may be high-pass filtered. Removal of dc components of the input signals can allow more efficient coding because the available data rate then is not used to encode

---

2. The ATSC AC-3 Standard contains symbolic references to syntactic elements used in the various subsystems. These references are typographically distinguished by the use of a different font (e.g., restricted), may contain the underscore character (e.g., sequence_end_code) and may consist of character strings that are not English words (e.g., dynrng). Note also that certain syntax element names are capitalized; e.g., Transport Stream.

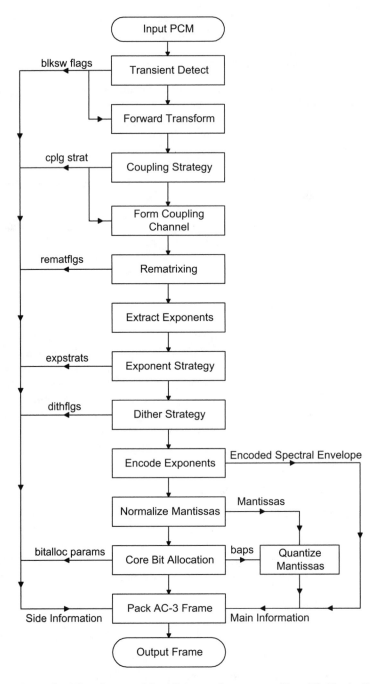

Figure 5.3.7 Generalized flow diagram of the AC-3 encoding process. (*From* [1]. *Used with permission.*)

dc. However, there is the risk that signals that do not reach 100 percent PCM level before high-pass filtering will exceed the 100 percent level after filtering, and thus be clipped. A typical encoder would high-pass filter the input signals with a single pole filter at 3 Hz.

The LFE channel normally is low-pass-filtered at 120 Hz. A typical encoder would filter the LFE channel with an 8th-order elliptic filter whose cutoff frequency is 120 Hz.

Transients are detected in the full-bandwidth channels to decide when to switch to short-length audio blocks to improve pre-echo performance. High-pass filtered versions of the signals are examined for an increase in energy from one subblock time segment to the next. Subblocks are examined at different time scales. If a transient is detected in the second half of an audio block in a channel, that channel switches to a short block.

The transient detector is used to determine when to switch from a *long transform block* (length 512) to a *short transform block* (length 256). It operates on 512 samples for every audio block. This is done in two passes, with each pass processing 256 samples. Transient detection is broken down into four steps:

- High-pass filtering

- Segmentation of the block into submultiples

- Peak amplitude detection within each subblock segment

- Threshold comparison

5.3.4i   AC-3/MPEG Bit Stream

The AC-3 elementary bit stream is included in an MPEG-2 multiplex bit stream in much the same way an MPEG-1 audio stream would be included, with the AC-3 bit stream packetized into PES packets [1]. An MPEG-2 multiplex bit stream containing AC-3 Elementary Streams must meet all audio constraints described in the MPEG model. It is necessary to unambiguously indicate that an AC-3 stream is, in fact, an AC-3 stream, and not an MPEG audio stream. The MPEG-2 standard does not explicitly state codes to be used to indicate an AC-3 stream. Also, the MPEG-2 standard does not have an Audio Descriptor adequate to describe the contents of the AC-3 bit stream in its internal tables. The solution to this problem is beyond the scope of this chapter; interested readers should consult [1] for additional information on the subject.

5.3.4j   Decoding the AC-3 Bit Stream

An overview of AC-3 decoding is diagrammed in Figure 5.3.8, where the decoding process flow is shown as a sequence of blocks down the center of the illustration, and some of the key information flow is indicated by arrowed lines at the sides [1]. This decoder should be considered only as an example; other methods certainly exist to implement decoders, and those other methods may have advantages in certain areas (such as instruction count, memory requirements, number of transforms required, and other parameters). The input bit stream typically will come from a transmission or storage system. The interface between the source of AC-3 data and the AC-3 decoder is not specified in the ATSC DTV standard.

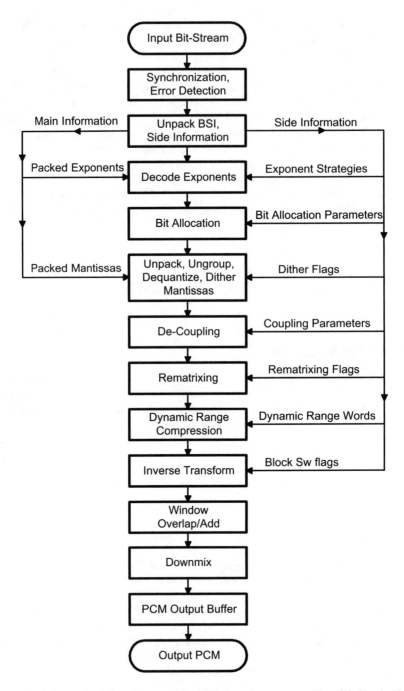

Figure 5.3.8 Generalized flow diagram of the AC-3 decoding process. (*From* [1]. *Used with permission.*)

Continuous or Burst Input

The encoded AC-3 data may be input to the decoder as a continuous data stream at the nominal bit rate, or chunks of data may be burst into the decoder at a high rate with a low duty cycle [1]. For burst-mode operation, either the data source or the decoder may be the master controlling the burst timing. The AC-3 decoder input buffer may be smaller if the decoder can request bursts of data on an as-needed basis, but the external buffer memory may need to be larger.

Most applications of the AC-3 Standard will convey the elementary AC-3 bit stream with byte or (16-bit) word alignment. The Sync Frame is always an integral number of words in length. The decoder may receive data as a continuous serial stream of bits without any alignment, or the data may be input to the decoder with either byte or word alignment. Byte or word alignment of the input data may allow some simplification of the decoder. Alignment does reduce the probability of false detection of the Sync Word.

Synchronization and Error Detection

The AC-3 bit steam format allows for rapid synchronization [1]. The 16-bit Sync Word has a low probability of false detection. With no input stream alignment, the probability of false detection of the Sync Word is 0.0015 percent per input stream bit position. For a bit rate of 384 kbits/s, the probability of false Sync Word detection is 19 percent per frame. Byte alignment of the input stream drops this probability to 2.5 percent, and word alignment drops it to 1.2 percent.

When a sync pattern is detected, the decoder may be estimated to be in sync, and one of the CRC words (CRC1 or CRC2) may be checked. Because CRC1 comes first and covers the first five-eighths of the frame, the result of a CRC1 check may be available after only five-eighths of the frame has been received. Or, the entire frame size can be received and CRC2 checked. If either CRC word checks, the decoder may safely be presumed to be in sync, and decoding and reproduction of audio may proceed. The chance of false sync in this case would be the concatenation of the probabilities of a false Sync Word detection and a CRC misdetection of error. The CRC check is reliable to 0.0015 percent. This probability, concatenated with the probability of a false sync detection in a byte-aligned input bit stream, yields a probability of false synchronization of 0.000035 percent (or about once in 3 million synchronization attempts).

If this small probability of false sync is too large for a specific application, several methods may be used to reduce it. The decoder may only presume correct sync in the case that both CRC words check properly. The decoder also may require multiple Sync Words to be received with the proper alignment. If the data transmission or storage system is aware that data is in error, this information may be made known to the decoder.

Inherent to the decoding process is the *unpacking* (demultiplexing) of the various types of information included in the bit stream. Among the options for distribution of this bit stream information are:

- Selected data may be copied from the input buffer to dedicated registers.

- Data from the input buffer may be copied to specific working memory locations.

- The data may simply be located in the input buffer, with pointers to the data saved to another location for use when the information is required.

Decoding Components

The AC-3 audio-compression system exponents are delivered in the bit stream in an encoded form [1]. To unpack and decode the exponents, two types of "side information" are required:

- The number of exponents must be known.

- The exponent "strategy" in use by each channel must be known.

The *bit-allocation computation* reveals how many bits are used for each mantissa. The inputs to the bit-allocation computation are the decoded exponents and the bit-allocation side information. The outputs of the bit-allocation computation are a set of Bit-Allocation Pointers, one BAP for each coded mantissa. The BAP indicates the quantizer used for the mantissa, and how many bits in the bit stream were used for each mantissa.

The coarsely quantized mantissas make up the bulk of the AC-3 data stream. Each mantissa is quantized to a level of precision indicated by the corresponding BAP. To pack the mantissa data more efficiently, some mantissas are grouped together into a single transmitted value. For instance, two 11-level quantized values are conveyed in a single 7-bit code (3.5 bits/value) in the bit stream.

The mantissa data is unpacked by peeling off groups of bits as indicated by the BAPs. Grouped mantissas must be ungrouped. The individual coded mantissa values are converted into a dequantized value. Mantissas that are indicated as having zero bits may be reproduced as either zero or by a random dither value (under control of a dither flag).

Other steps in the decoding process include the following:

- *Decoupling.* When *coupling* is in use, the channels that are coupled must be decoupled. Decoupling involves reconstructing the high-frequency section (exponents and mantissas) of each coupled channel, from the common coupling channel and the coupling coordinates for the individual channel. Within each coupling band, the coupling-channel coefficients (exponent and mantissa) are multiplied by the individual channel coupling coordinates.

- *Rematrixing.* In the 2/0 audio-coding mode, rematrixing may be employed as indicated by a R*ematrix Flag*. When the flag indicates that a band is rematrixed, the coefficients encoded in the bit stream are sum and difference values, instead of left and right values.

- *Dynamic range compression.* For each block of audio, a Dynamic Range Control value may be included in the bit stream. The decoder, by default, will use this value to alter the magnitude of the coefficient (exponent and mantissa) as required to properly process the data.

- *Inverse transform.* The decoding steps described in this section will result in a set of frequency coefficients for each encoded channel. The inverse transform converts these blocks of frequency coefficients into blocks of time samples.

- *Window, overlap/add.* The individual blocks of time samples must be windowed, and adjacent blocks are overlapped and added together to reconstruct the final continuous-time-output PCM audio signal.

- *Downmixing.* If the number of channels required at the decoder output is smaller than the number of channels that are encoded in the bit stream, then downmixing is required. Downmixing in the time domain is shown in the example decoder of Figure 5.3.8. Because the inverse transform is a linear operation, it also is possible to downmix in the frequency domain prior to transformation.

- *PCM output buffer*. Typical decoders will provide PCM output samples at the PCM sampling rate. Because blocks of samples result from the decoding process, an output buffer typically is required.

- *Output PCM*. The output PCM samples are delivered in a form suitable for interconnection to a digital-to-analog converter (D/A), or in some other form required by the receiver.

5.3.4k   Algorithmic Details

The actual audio information conveyed by the AC-3 bit stream consists of the quantized frequency coefficients [1]. The coefficients, delivered in floating point form, are 5-bit values that indicate the number of leading zeros in the binary representation of a frequency coefficient. The exponent acts as a scale factor for each mantissa, equal to 2^{-exp}. Exponent values are allowed to range from 0 (for the largest-value coefficients with no leading zeros) to 24. Exponents for coefficients that have more than 24 leading zeros are fixed at 24, and the corresponding mantissas are allowed to have leading zeros. Exponents require 5 bits to represent all allowed values.

AC-3 bit streams contain coded exponents for all independent channels, all coupled channels, and for the coupling and low-frequency effects channels (when they are enabled). Because audio information is not shared across frames, block 0 of every frame will include new exponents for every channel. Exponent information may be shared across blocks within a frame, so blocks 1 through 5 may reuse exponents from previous blocks.

AC-3 exponent transmission employs *differential coding*, in which the exponents for a channel are differentially coded across frequency. These differential exponents are combined into groups in the audio block. This grouping is done by one of three methods, which are referred to as *exponent strategies*. The number of grouped differential exponents placed in the audio block for a particular channel depends on the exponent strategy and on the frequency bandwidth information for that channel. The number of exponents in each group depends only on the exponent strategy.

An AC-3 audio block contains two types of fields with exponent information. The first type defines the exponent coding strategy for each channel, and the second type contains the actual coded exponents for channels requiring new exponents. For independent channels, frequency bandwidth information is included along with the exponent strategy fields. For coupled channels, and the coupling channel, the frequency information is found in the coupling strategy fields.

5.3.4l   Bit Allocation

The bit allocation routine analyzes the spectral envelope of the audio signal being coded with respect to masking effects to determine the number of bits to assign to each transform coefficient mantissa [1]. In the encoder, the bit allocation is performed globally on the ensemble of channels as an entity, from a common bit pool. Because there are no preassigned exponent or mantissa bits, the routine is allowed to flexibly allocate bits across channels, frequencies, and audio blocks in accordance with signal demand.

The bit allocation contains a parametric model of human hearing for estimating a noise-level threshold, expressed as a function of frequency, which separates audible from inaudible spectral components. Various parameters of the hearing model can be adjusted by the encoder depending upon signal characteristics. For example, a prototype masking curve is defined in terms of two piecewise continuous line segments, each with its own slope and y-axis intercept. One of several possible slopes and intercepts is selected by the encoder for each line segment. The encoder may iterate on one or more such parameters until an optimal result is obtained. When all parameters used to estimate the noise-level threshold have been selected by the encoder, the final bit allocation is computed. The model parameters are conveyed to the decoder with other side information. The decoder then executes the routine in a single pass.

The estimated noise-level threshold is computed over 50 bands of nonuniform bandwidth (an approximate 1/6-octave scale). The defined banding structure is independent of sampling frequency. The required bit allocation for each mantissa is established by performing a table lookup based upon the difference between the input signal *power spectral density* (PSD), evaluated on a fine-grain uniform frequency scale, and the estimated noise-level threshold, evaluated on the coarse-grain (*banded*) frequency scale. Therefore, the bit allocation result for a particular channel has spectral granularity corresponding to the exponent strategy employed.

5.3.5   Audio System Level Control

The AC-3 system provides elements that allow the encoded bit stream to satisfy listeners in many different situations. Two principal techniques are used to control the subjective loudness of the reproduced audio signals:

- Dialogue normalization

- Dynamic range compression

5.3.5a   Dialogue Normalization

The *dialogue normalization* (DialNorm) element permits uniform reproduction of spoken dialogue when decoding any AC-3 bit stream [1]. When audio from different sources is reproduced, the apparent loudness often varies from source to source. Examples include the following:

- Audio elements from different program segments during a broadcast (for example, a movie vs. a commercial message)

- Different broadcast channels

- Different types of media (for example, disc vs. tape)

The AC-3 coding technology solves this problem by explicitly coding an indication of loudness into the AC-3 bit stream.

The subjective level of normal spoken dialogue is used as a reference. The 5-bit dialogue normalization word that is contained in the bit stream, DialNorm, is an indication of the subjective loudness of normal spoken dialogue compared with digital 100 percent. The 5-bit value is interpreted as an unsigned integer (most significant bit transmitted first) with a range of possible values from 1 to 31. The unsigned integer indicates the headroom in decibels above the subjective dialogue level. This value also may be interpreted as an indication of how many decibels the subjective dialogue level is below digital 100 percent.

The DialNorm value is not directly used by the AC-3 decoder. Rather, the value is used by the section of the sound reproduction system responsible for setting the reproduction volume, such as the system volume control. The system volume control generally is set based on listener input as to the desired loudness, or *sound-pressure level* (SPL). The listener adjusts a volume control that directly adjusts the reproduction system gain. With AC-3 and the DialNorm value, the reproduction system gain becomes a function of both the listener's desired reproduction sound-pressure level for dialogue, and the DialNorm value that indicates the level of dialogue in the audio signal. In this way, the listener is able to reliably set the volume level of dialogue, and the subjective level of dialogue will remain uniform no matter which AC-3 program is decoded.

Example Situation

An example will help to illustrate the DialNorm concept [1]. The listener adjusts the volume control to 67 dB. (With AC-3 dialogue normalization, it is possible to calibrate a system volume control directly in sound-pressure level, and the indication will be accurate for any AC-3 encoded audio source). A high quality entertainment program is being received, and the AC-3 bit stream indicates that the dialogue level is 25 dB below the 100 percent digital level. The reproduction system automatically sets the reproduction system gain so that full-scale digital signals reproduce at a sound-pressure level of 92 dB. Therefore, the spoken dialogue (down 25 dB) will reproduce at 67 dB SPL.

The broadcast program cuts to a commercial message, which has dialogue level at −15 dB with respect to 100 percent digital level. The system level gain automatically drops, so that digital 100 percent is now reproduced at 82 dB SPL. The dialogue of the commercial (down 15 dB) reproduces at a 67 dB SPL, as desired.

For the dialogue normalization system to work, the DialNorm value must be communicated from the AC-3 decoder to the system gain controller so that DialNorm can interact with the listener-adjusted volume control. If the volume-control function for a system is performed as a digital multiplier inside the AC-3 decoder, then the listener-selected volume setting must be communicated into the AC-3 decoder. The listener-selected volume setting and the DialNorm value must be combined to adjust the final reproduction system gain.

Adjustment of the system volume control is not an AC-3 function. The AC-3 bit stream simply conveys useful information that allows the system volume control to be implemented in a way that automatically removes undesirable level variations between program sources.

5.3.5b   Dynamic Range Compression

The *dynamic range compression* (DynRng) element allows the program provider to implement subjectively pleasing dynamic range reduction for most of the intended audience, while allowing individual members of the audience the option to experience more (or all) of the original dynamic range [1].

A consistent problem in the delivery of audio programming is that members of the audience may prefer differing amounts of dynamic range. Original high-quality programs (such as feature films) typically are mixed with quite a wide dynamic range. Using dialogue as a reference, loud sounds, such as explosions, often are at least 20 dB louder; faint sounds, such as rustling leaves, may be 50 dB quieter. In many listening situations, it is objectionable to allow the sound to become very loud, so the loudest sounds must be compressed downward in level. Similarly, in many listening situations, the very quiet sounds would be inaudible, and they must be brought upward in level to be heard. Because most of the television audience will benefit from a limited program dynamic range, motion picture soundtracks that have been mixed with a wide dynamic range generally are compressed. The dynamic range is reduced by bringing down the level of the loud sounds and bringing up the level of the quiet sounds. Although this satisfies the needs of much of the audience, some audience members may prefer to experience the original sound program in its intended form. The AC-3 audio-coding technology solves this conflict by allowing Dynamic Range Control values to be placed into the AC-3 bit stream.

The Dynamic Range Control values, DynRng, indicate a gain change to be applied in the decoder to implement dynamic range compression. Each DynRng value can indicate a gain change of ± 24 dB. The sequence of DynRng values constitute a compression control signal. An AC-3 encoder (or a bit stream processor) will generate the sequence of DynRng values. Each value is used by the AC-3 decoder to alter the gain of one or more audio blocks. The DynRng values typically indicate gain reductions during the loudest signal passages and gain increases during the quiet passages. For the listener, it is often desirable to bring the loudest sounds down in level, toward dialogue level, and bring the quiet sounds up in level, again toward dialogue level. Sounds that are at the same loudness as normal spoken dialogue typically will not have their gain changed.

The compression actually is applied to the audio in the AC-3 decoder. The encoded audio has full dynamic range. It is permissible for the AC-3 decoder to (optionally, under listener control) ignore the DynRng values in the bit stream. This will result in reproduction of the full dynamic range of the audio. It also is permissible (again under listener control) for the decoder to use some fraction of the DynRng control value and to use a different fraction of positive or negative values. Therefore, the AC-3 decoder can reproduce sounds according to one of the following parameters:

- Fully compressed audio (as intended by the compression control circuit in the AC-3 encoder)

- Full dynamic range audio

- Audio with partially compressed dynamic range, with different amounts of compression for high-level and low-level signals.

Example Situation

A feature film soundtrack is encoded into AC-3 [1]. The original program mix has dialogue level at −25 dB. Explosions reach a full-scale peak level of 0 dB. Some quiet sounds that are intended to be heard by all listeners are 50 dB below dialogue level (−75 dB). A compression control signal (a sequence of DynRng values) is generated by the AC-3 encoder. During those portions of the Audio Program when the audio level is higher than dialogue level, the DynRng values indicate negative gain, or gain reduction. For full-scale 0 dB signals (the loudest explosions), a gain reduction of −15 dB is encoded into DynRng. For very quiet signals, a gain increase of 20 dB is encoded into DynRng.

A listener wishes to reproduce this soundtrack quietly so as not to disturb anyone, but wishes to hear all of the intended program content. The AC-3 decoder is allowed to reproduce the default, which is full compression. The listener adjusts dialogue level to 60 dB SPL. The explosions will go only as loud as 70 dB (they are 25 dB louder than dialogue but receive −15 dB applied gain), and the quiet sounds will reproduce at 30 dB SPL (20 dB of gain is applied to their original level of 50 dB below dialogue level). The reproduced dynamic range, therefore, will be 70 dB − 30 dB = 40 dB.

The listening situation changes, and the listener now wishes to raise the reproduction level of dialogue to 70 dB SPL, but still wishes to limit the loudness of the program. Quiet sounds may be allowed to play as quietly as before. The listener instructs the AC-3 decoder to continue to use the DynRng values that indicate gain reduction, but to attenuate the values that indicate gain increases by a factor of 1/2. The explosions still will reproduce 10 dB above dialogue level, which is now 80 dB SPL. The quiet sounds now are increased in level by 20 dB/2 = 10 dB. They now will be reproduced 40 dB below dialogue level, at 30 dB SPL. The reproduced dynamic range is now 80 dB – 30 dB = 50 dB.

Another listener prefers the full original dynamic range of the audio. This listener adjusts the reproduced dialogue level to 75 dB SPL and instructs the AC-3 decoder to ignore the Dynamic Range Control signal. For this listener, the quiet sounds reproduce at 25 dB SPL, and the explosions hit 100 dB SPL. The reproduced dynamic range is 100 dB – 25 dB = 75 dB. This reproduction is exactly as intended by the original program producer.

For this Dynamic Range Control method to be effective, it must be used by all program providers. Because all broadcasters wish to supply programming in the form that is most usable by their audiences, nearly all will apply dynamic range compression to any Audio Program that has a wide dynamic range. This compression is not reversible unless it is implemented by the technique embedded in AC-3. If broadcasters make use of the embedded AC-3 Dynamic Range Control system, listeners can have significant control over the reproduced dynamic range at their receivers. Broadcasters must be confident that the compression characteristic that they introduce into AC-3 will, by default, be heard by the listeners. Therefore, the AC-3 decoder must, by default, implement the compression characteristic indicated by the DynRng values in the data stream. AC-3 decoders may optionally allow listener control over the use of the DynRng values, so that the listener may select full or partial dynamic range reproduction.

5.3.5c   Heavy Compression

The *compression* (COMPR) element allows the program provider (or broadcaster) to implement a large dynamic range reduction (heavy compression) in a way that ensures that a monophonic downmix will not exceed a certain peak level [1]. The heavily compressed Audio Program may be desirable for certain listening situations, such as movie delivery to a hotel room or to an airline seat. The peak level limitation is useful when, for example, a monophonic downmix will feed an RF modulator, and overmodulation must be avoided.

Some products that decode the AC-3 bit stream will need to deliver the resulting audio via a link with very restricted dynamic range. One example is the case of a television signal decoder that must modulate the received picture and sound onto an RF channel to deliver a signal usable by a low-cost television receiver. In this situation, it is necessary to restrict the maximum peak output level to a known value—with respect to dialogue level—to prevent overmodulation. Most of the time, the Dynamic Range Control signal, DynRng, will produce adequate gain reduction so that the absolute peak level will be constrained. However, because the Dynamic Range Control system is intended to implement a subjectively pleasing reduction in the range of perceived loudness, there is no assurance that it will control instantaneous signal peaks adequately to prevent overmodulation.

To allow the decoded AC-3 signal to be constrained in peak level, a second control signal, COMPR, (COMPR2 for channel 2 in 1+1 mode) may be included in the AC-3 data stream. This control signal should be present in all bit streams that are intended to be received by, for example, a television set-top decoder. The COMPR control signal is similar to the DynRng control signal in that it is used by the decoder to alter the reproduced audio level. The COMPR control signal has twice the control range as DynRng (± 48 dB compared with ± 24 dB) with half the resolution (0.5 vs. 0.25 dB).

5.3.6   Audio System Features

The DTV audio subsystem offers a host of services and features to meet varied applications and audiences [2]. An AC-3 Elementary Stream contains the encoded representation of a single audio service. Multiple audio services are provided by multiple Elementary Streams. Each Elementary Stream is conveyed by the transport multiplex with a unique *Program ID* (PID). A number of audio service types may be coded (individually) into each Elementary Stream; each AC-3 Elementary Stream is tagged as to its service type. There are two types of *Main Service* and six types of *Associated Service*. Each Associated Service may be tagged (in the AC-3 Audio Descriptor) as being associated with one or more main audio services. Each AC-3 Elementary Stream also may be tagged with a language code.

Associated Services may contain complete program mixes or only a single program element. Associated Services that are complete mixes may be decoded and used "as is." Associated Services that contain only a single program element are intended to be combined with the program elements from a main audio service.

In general, a complete Audio Program (what is presented to the listener over the set of loudspeakers) may consist of a main audio service, an associated audio service that is a complete mix, or a main audio service combined with an associated audio service. The capability to simultaneously decode one Main Service and one Associated Service is required in order to form a complete Audio Program in certain service combinations. This capability may not exist in some receivers.

5.3.6a   Complete Main Audio Service (CM)

The CM type of main audio service contains a complete Audio Program (complete with dialogue, music, and effects) [2]. This is the type of audio service normally provided. The CM service may contain from 1 to 5.1 audio channels, and it may be further enhanced by means of the VI, HI, C, E, or VO Associated Services described in the following sections. Audio in multiple languages may be provided by supplying multiple CM services, each in a different language.

5.3.6b   Main Audio Service, Music and Effects (ME)

The ME type of main audio service contains the music and effects of an Audio Program, but not the dialogue for the program [2]. The ME service may contain from 1 to 5.1 audio channels. The primary program dialogue is missing and (if any exists) is supplied by simultaneously encoding a D Associated Service. Multiple D Associated Services in different languages may be associated with a single ME service.

5.3.6c   Visually Impaired (VI)

The VI Associated Service typically contains a narrative description of the visual program content [2]. In this case, the VI service is a single audio channel. The simultaneous reproduction of both the VI Associated Service and the CM main audio service allows the visually impaired user to enjoy the main multichannel Audio Program, as well as to follow (by ear) the on-screen activity.

The Dynamic Range Control signal in this type of service is intended to be used by the audio decoder to modify the level of the main Audio Program. Thus, the level of the main audio service will be under the control of the VI service provider, and the provider may signal the decoder (by altering the Dynamic Range Control words embedded in the VI Audio Elementary Stream) to reduce the level of the main audio service by up to 24 dB to ensure that the narrative description is intelligible.

Besides being provided as a single narrative channel, the VI service may be provided as a complete program mix containing music, effects, dialogue, and the narration. In this case, the service may be coded using any number of channels (up to 5.1), and the Dynamic Range Control signal would apply only to this service.

5.3.6d   Hearing Impaired (HI)

The HI Associated Service typically contains only dialogue that is intended to be reproduced simultaneously with the CM service [2]. In this case, the HI service is a single audio channel. This dialogue may have been processed for improved intelligibility by hearing-impaired users. Simultaneous reproduction of both the CM and HI services allows the hearing-impaired users to hear a mix of the CM and HI services in order to emphasize the dialogue while still providing some music and effects.

Besides being available as a single dialogue channel, the HI service may be provided as a complete program mix containing music, effects, and dialogue with enhanced intelligibility. In this case, the service may be coded using any number of channels (up to 5.1).

5.3.6e   Dialogue (D)

The D Associated Service contains program dialogue intended for use with an ME main audio service [2]. The language of the D service is indicated in the AC-3 bit stream and in the Audio Descriptor. A complete Audio Program is formed by simultaneously decoding the D service and the ME service, then mixing the D service into the center channel of the ME Main Service (with which it is associated).

If the ME main audio service contains more than two audio channels, the D service is mono-phonic (1/0 mode). If the main audio service contains two channels, the D service may also contain two channels (2/0 mode). In this case, a complete Audio Program is formed by simultaneously decoding the D service and the ME service, mixing the left channel of the ME service with the left channel of the D service, and mixing the right channel of the ME service with the right channel of the D service. The result will be a 2-channel stereo signal containing music, effects, and dialogue.

Audio in multiple languages may be provided by supplying multiple D services (each in a different language) along with a single ME service. This is more efficient than providing multiple CM services, but, in the case of more than two audio channels in the ME service, requires that dialogue be restricted to the center channel.

Some receivers may not have the capability to simultaneously decode an ME and a D service.

5.3.6f   Commentary (C)

The commentary Associated Service is similar to the D service, except that instead of conveying essential program dialogue, the C service conveys optional Program Commentary [2]. The C service may be a single audio channel containing only the commentary content. In this case, simultaneous reproduction of a C service and a CM service will allow the listener to hear the added Program Commentary.

The Dynamic Range Control signal in the single-channel C service is intended to be used by the audio decoder to modify the level of the main Audio Program. Thus, the level of the main audio service will be under the control of the C service provider; the provider may signal the decoder (by altering the Dynamic Range Control words embedded in the C Audio Elementary Stream) to reduce the level of the main audio service by up to 24 dB to ensure that the commentary is intelligible.

Besides providing the C service as a single commentary channel, the C service may be provided as a complete program mix containing music, effects, dialogue, and the commentary. In this case, the service may be provided using any number of channels (up to 5.1).

5.3.6g   Emergency (E)

The E Associated Service is intended to allow the insertion of emergency or high priority announcements [2]. The E service is always a single audio channel. An E service is given priority in transport and in audio decoding. Whenever the E service is present, it will be delivered to the audio decoder. Whenever the audio decoder receives an E-type Associated Service, it will stop reproducing any Main Service being received and reproduce only the E service out of the center channel (or left and right channels if a center loudspeaker does not exist). The E service also may be used for nonemergency applications. It may be used whenever the broadcaster wishes to force all decoders to quit reproducing the main Audio Program and reproduce a higher priority single audio channel.

5.3.6h   Voice-Over (VO)

The VO Associated Service is a single-channel service intended to be reproduced along with the main audio service in the receiver [2]. It allows typical voice-overs to be added to an already encoded Audio Elementary Stream without requiring the audio to be decoded back to baseband and then reencoded. The VO service is always a single audio channel and has second priority; only the E service has higher priority. It is intended to be simultaneously decoded and mixed into the center channel of the main audio service. The Dynamic Range Control signal in the VO service is intended to be used by the audio decoder to modify the level of the main Audio Program. Thus, the level of the main audio service may be controlled by the broadcaster, and the broadcaster may signal the decoder (by altering the Dynamic Range Control words embedded in the VO Audio Elementary Stream) to reduce the level of the main audio service by up to 24 dB during the voice-over.

Some receivers may not have the capability to simultaneously decode and reproduce a voice-over service along with a program audio service.

Table 5.3.1 Typical Bit Rates for Various Services (*After* [2].)

Type of Service	Number of Channels	Typical Bit Rates
CM, ME, or associated audio service containing all necessary program elements	5	384–448 kbits/s
CM, ME, or associated audio service containing all necessary program elements	4	320–384 kbits/s
CM, ME, or associated audio service containing all necessary program elements	3	192–320 kbits/s
CM, ME, or associated audio service containing all necessary program elements	2	128–256 kbits/s
VI, narrative only	1	64–128 kbits/s
HI, narrative only	1	64–96 kbits/s
D	1	64–128 kbits/s
D	2	96–192 kbits/s
C, commentary only	1	64–128 kbits/s
E	1	64–128 kbits/s
VO	1	64–128 kbits/s

5.3.6i   Typical Audio Bit Rates

Table 5.3.1 provides a general guideline as to the audio bit rates for various services [4]. For Main Services, the use of the LFE channel is optional and will not affect the indicated data rates.

Audio Bit Rate Limitations

The audio decoder input buffer size is determined by the maximum bit rate which must be decoded [4]. The syntax of the AC-3 standard supports bit rates ranging from a minimum of 32 kbits/s up to a maximum of 640 kbits/s per individual elementary bit stream. The bit rate utilized in the digital television system is restricted in order to reduce the size of the input buffer in the audio decoder, and thus the receiver cost. Receivers can be expected to support the decoding of a main audio service, or an associated audio service which is a complete service (containing all necessary program elements), at a bit rate up to and including 448 kbits/s. Transmissions may contain main audio services, or associated audio services which are complete services (containing all necessary program elements), encoded at a bit rate up to and including 448 kbits/s. Transmissions may contain single-channel associated audio services intended to be simultaneously decoded along with a Main Service encoded at a bit rate up to and including 128 kbits/s. Transmissions may contain dual-channel dialogue Associated Services intended to be simultaneously decoded along with a Main Service encoded at a bit rate up to and including 192 kbits/s. Transmissions have a further limitation that the combined bit rate of a main and an Associated Service which are intended to be simultaneously reproduced is less than or equal to 576 kbits/s.

5.3.7   Enhanced AC-3

Enhanced AC-3 (E-AC-3), added to the A/52 Standard in 2005, was developed to fundamentally improve overall system performance and provide new features that allow operation over a wide range of bit-rates and channel configurations. All E-AC-3 decoders will also decode AC-3 bit streams. In addition, although the new enhanced audio format is not directly compatible with current AC-3 decoders, it is feasible to perform a modest-complexity conversion into a compliant AC-3 bit stream syntax, thus enabling backwards compatibility to legacy decoders that have S/PDIF bit stream inputs.

Important technical capabilities of Enhanced AC-3 that relate directly to ATSC broadcast applications include the following:

- **Expanded data rate flexibility**. E-AC-3 allows the number of blocks per Sync Frame and the number of compressed data bits per frame to be adjusted to achieve significantly more data rate flexibility than standard AC-3, including a greater maximum theoretical data rate and finer data rate granularity.

- **Spectral extension**. Enhanced AC-3 decoders support a new coding technique called *spectral extension*. Like channel coupling, spectral extension codes the highest frequency content of the signal more efficiently. Spectral extension recreates a signal's high frequency spectrum from side data transmitted in the bit stream that characterizes the original signal, as well as from actual signal content from the lower frequency portion of the signal. Because it may be desirable, in some circumstances, to use channel coupling for a mid-range portion of the frequency spectrum and spectral extension for the higher-range portion of the frequency spectrum, spectral extension is fully compatible with channel coupling.

- **Transient pre-noise processing**. This is an optional decoder tool that improves audible performance through the substitution of audio segments just before transients to reduce the duration of pre-noise distortions. This technique is called *time scaling synthesis*, where synthesized PCM audio segments are used to eliminate the transient pre-noise, thereby improving the perceived quality of low-bit rate audio coded transient material. To enable the decoder to efficiently perform transient pre-noise processing with no impact on decoding latency, transient location detection and time scaling synthesis analysis is performed by the encoder and the information transmitted to the decoder.

- **Adaptive hybrid transform processing**. In 1995, the transform employed in A/52 AC-3—based on a modified discrete cosine transform (MDCT) of length 256 frequency samples—provided a reasonable tradeoff between audio coding gain and decoder implementation cost. With continuing advances in silicon manufacturing processes over the years, the integrated circuit complexity that constitutes a reasonable level has now increased. This increase in chip performance provides an opportunity to improve the coding gain of AC-3, and hence perceptual audio quality at a given bit-rate, by increasing the length of the transform. This is accomplished through use of the Adaptive Hybrid Transform (AHT), which adds a second transform in cascade in order to generate a single transform with 1536 frequency samples.

- **Enhanced coupling**. This is a new tool that improves the imaging properties of coupled signals by adding phase compensation to the amplitude-based processing of conventional coupling. Prior to down-mixing the coupled channels to a single composite signal, the encoder derives both amplitude and additionally interchannel phase information on a subband basis for each channel. The phase information includes a decorrelation scale factor as a measure of the variation of the phase within a frame. This side chain information is transmitted to the decoder once per frame. The decoder uses the information to recover the multiple output channels from the composite signal using a combination of both amplitude scaling and phase rotation. The result is an improvement in soundstage imaging over conventional coupling.

Additional features of E-AC-3 of particular interest to applications outside of DTV include:

- **Channel and program extensions.** The Enhanced AC-3 bit stream syntax allows for time-multiplexed *substreams* to be present in a single bit stream. With this capability, the Enhanced AC-3 bit stream syntax enables a single program with greater than 5.1 channels, multiple programs of up to 5.1 channels, or a mixture of programs with up to 5.1 channels and programs with greater than 5.1 channels, to be carried in a single bit stream. These extra channels do not affect a two or 5.1 channel decoder in ATSC broadcast applications.

- **Sample rate processing.** Additional metadata is reserved for applications that involve source material sampled at 2x the nominal rate, such as 96 kHz and 88.2 kHz.

- **Mixing control processing.** Additional metadata is reserved for applications that involve the mixing of two program streams. These applications require control of the mixing process and resultant Dynamic Range Control metadata; this feature reserves data capacity to accomplish this task.

5.3.8 References

1. ATSC: "Digital Audio Compression Standard (AC-3)," Advanced Television Systems Committee, Washington, D.C., Doc. A/52B, June 14, 2005.

2. ATSC: "Digital Television Standard," Advanced Television Systems Committee, Washington, D.C., Doc. A/53E, 2006.

3. ITU-R Recommendation BS-775, "Multi-channel Stereophonic Sound System with and Without Accompanying Picture."

4. ATSC: "Guide to the Use of the Digital Television Standard," Advanced Television Systems Committee, Washington, D.C., Doc. A/54A, December 4, 2003.

5.3.9 Bibliography

Ehmer, R. H.: "Masking of Tones Vs. Noise Bands," *J. Acoust. Soc. Am.*, vol. 31, pp. 1253–1256, September 1959.

Ehmer, R. H.: "Masking Patterns of Tones," *J. Acoust. Soc. Am.*, vol. 31, pp. 1115–1120, August 1959.

Moore, B. C. J., and B. R. Glasberg: "Formulae Describing Frequency Selectivity as a Function of Frequency and Level, and Their Use in Calculating Excitation Patterns," *Hearing Research*, vol. 28, pp. 209–225, 1987.

Todd, C., et. al.: "AC-3: Flexible Perceptual Coding for Audio Transmission and Storage," AES 96th Convention, Preprint 3796, Audio Engineering Society, New York, February 1994.

Zwicker, E.: "Subdivision of the Audible Frequency Range Into Critical Bands (Frequenzgruppen)," *J. Acoust. Soc. of Am.*, vol. 33, pg. 248, February 1961.

5.4

DTV Transport System

5.4.1 Introduction[1]

The ATSC DTV system described in core documents A/52 and A/53 provides the framework for conveying information to consumers [1]. Built into this framework is a toolkit of features that can be used to extend the capabilities of the DTV system far beyond what the initial designers might have envisioned. This *extensibility* is, perhaps, the greatest benefit of digital technology.

5.4.1a Transport System

The ATSC DTV transport system employs the fixed-length Transport Stream *packetization* approach defined in ISO/IEC13818-1, which is usually referred to as the MPEG-2 Systems Standard [2]. This approach is well-suited to the needs of terrestrial broadcast and cable television transmission of digital television [1]. The use of relatively short, fixed-length packets matches well with the needs and techniques for error protection in both terrestrial broadcast and cable television distribution environments.

The DTV transport may carry a number of television programs. The MPEG-2 term "Program" corresponds to an individual digital TV channel or data service, where each Program is composed of a number of MPEG-2 Program Elements (i.e., related video, audio, and data streams). The MPEG-2 Systems Standard support for multiple channels or services within a single, multiplexed bit stream enables the deployment of practical, bandwidth-efficient digital broadcasting systems. It also provides great flexibility to accommodate the initial needs of the service to multiplex video, audio, and data while providing a well-defined path to add additional services in the future in a fully backward-compatible manner. By basing the transport subsystem on MPEG-2, maximum interoperability with other media and standards is maintained.

---

1. This chapter is based on: ATSC, "Guide to the Use of the Digital Television Standard," Advanced Television Systems Committee, Washington, D.C., Doc. A/54A, December 4, 2003. Readers are encouraged to download the source document from the ATSC Web site (http://www.atsc.org). All ATSC Standards, Recommended Practices, and Information Guides are available at no charge.

The transport system resides between the application (e.g., audio or video) encoding and decoding functions and the transmission system. At its lowest layer, the encoder transport system is responsible for formatting the encoded bits and multiplexing the different components of the Program for transmission. At the receiver, it is responsible for recovering the bit streams for the individual application decoders and for the corresponding error signaling. The transport system also incorporates other higher-level functionality related to identification of applications and synchronization of the receiver.

5.4.2   MPEG-2 Basics

The MPEG-2 Standards are built upon the foundations of the MPEG-1 Standards. While the MPEG-1 Standards were developed primarily to address the then upcoming video CD market-place need for an interoperable solution for compressed digital video storage and real-time play-back at rates of about 1.5 Mbits/s, MPEG-2 was developed to primarily address the broadcast digital television and DVD markets and includes [1]:

- Improved video and audio compression technologies

- Encoding support for both 4:2:0 and 4:2:2 video

- Support for the transmission of the coded bit streams in error-prone environments

- Support for multiple Programs ("channels") in a single, multiplexed stream. This includes improved synchronization with the capability for each Program to have a unique time-base, and the ability to describe and identify a network consisting of multiple multiplexed streams, each containing multiple Programs

- Conditional access support

- Stream buffer management including buffer initialization

- Private data transport support

In contrast to previously developed standards, the MPEG-2 standards were designed to support full ITU-R 601 standard-definition resolutions, high-definition resolutions, and interlaced sequences. The MPEG-2 Standards were also designed to support multi-channel networks carried in error-prone environments (such as terrestrial broadcasting), and the basic constructs used to encapsulate private data and a multitude of data essence formats. MPEG-2 Standards are the foundation of several digital television technologies including digital set top boxes (STB), high-definition television (HDTV), and data broadcasting.

The MPEG-2 Systems Standard defines the bit stream syntax and the methods necessary for (de)multiplexing, transporting, and synchronizing coded video, coded audio, and other data (including data essence not defined by the MPEG Standards, referred to as "private data"). The standard includes the definition of packet formats, the synchronization and timing model, the mechanism for identifying content carried in the bit stream, and the buffer models used to enable a receiving device to properly decode and reconstruct the video, audio, and/or data presentation. The MPEG-2 Systems Standard as constrained and extended by the ATSC is the basis for the remainder of this chapter.

5.4.2a   Standards Layering

The MPEG-2 Systems Standard (ISO/IEC 13818-1) provides a toolkit that can be used to create the DTV transport bit stream [1]. This toolkit can be thought of as providing general purpose functionality. Users of the MPEG-2 Standards (such as the ATSC) choose tools from the toolkit and specify how they may be used (i.e., specify constraints on the syntax and semantics of the MPEG-2 Standards). A/53 describes which portions of the MPEG-2 Systems Standard are to be used in creating the ATSC bit stream and also describes the constraints imposed.

In addition to constraining the MPEG-2 Systems Standard, the ATSC has also created compatible extensions to the standard. Some syntactical fields in the MPEG-2 Systems Standard are user defined—other fields have user private ranges. The ATSC is considered a "user" of the MPEG-2 standards and has used the user private areas to create ATSC standardized extensions to the MPEG-2 standards.

5.4.2b   MPEG-2 Transport Stream Packet

An MPEG-2 Transport Stream is a continuous series of MPEG-2 Transport Stream packets. An MPEG-2 Transport Stream packet is 188 bytes in length and always begins with the synchronization byte 0x47.

MPEG-2 TS Packet Structure

The first four bytes of the MPEG-2 Transport Stream packet are the Transport Stream *Packet Header* [1]. The remaining 184 bytes of an MPEG-2 Transport Stream packet may contain an optional *Adaptation Field* and up to 184 bytes of Transport Stream packet *payload*. If the Adaptation Field is present, it immediately follows the last byte of the Transport Stream Packet Header. The Adaptation Field is not part of the Transport Stream Packet Header nor the Transport Stream packet payload. When the Adaptation Field is present, the MPEG-2 Transport Stream packet payload size is 184 bytes minus the length of the Adaptation Field.

The definition of the contents of an MPEG-2 Transport Stream packet payload may differ depending upon the MPEG-2 stream_type[2] and the encapsulation method.

MPEG-2 Transport Stream Packet Syntax

In the Packet Header, the *Packet Identifier* (PID) is a 13-bit value used to identify multiplexed packets within the MPEG-2 Transport Stream [1]. Assigning a unique PID value to each bit stream allows Transport Stream packets from up to 8192 (2^{13}) separate bit streams to be simultaneously carried within the MPEG-2 Transport Stream. Note that not all bit streams are MPEG-2 Program Elements (e.g., PSI), but all Program Elements are bit streams. The PID provides a unique bit stream (and, therefore, Program Element) association for each Transport Stream packet.

2. ATSC Digital Television Standard documents contains symbolic references to syntactic elements used in the audio, video, and transport coding subsystems. These references usually contain the underscore character (e.g., sequence_end_code) and may consist of character strings that are not English words (e.g., dynrng). In this book they are identified by a distinctive font. Note also that MPEG element names are capitalized; e.g., Transport Stream.

The payload_unit_start_indicator is used to signal to the decoder (by being set to '1') that the first byte of something "interesting" can be found within the payload of the current MPEG-2 Transport Stream packet (an MPEG-2 PES packet or MPEG-2 Section. This form of signaling, combined with hardware filtering in the decoder, allows for considerable efficiencies in decoding the contents of the stream. A PES packet must always commence as the first byte of the Transport Stream packet payload and only a single PES packet may begin in a Transport Stream packet. Thus, two PES packets (or portions thereof) are not permissible in a single Transport Stream packet.

For MPEG-2 Sections (PSI and Private Sections) carried as payload, when the payload_unit_start_indicator field is set to '1', then the first byte of the MPEG-2 Transport Stream packet payload carries the pointer_field, which indicates the byte offset from the start of the Transport Stream packet payload to the beginning of the next PSI or Private Section. If the payload_unit_start_indicator field is set to '0', then the first byte of the Transport Stream packet payload is not a pointer_field. Instead, the Transport Stream packet payload contains the continuation of a previously started PSI or Private Section along with any necessary Stuffing Bytes.

The transport_scrambling_control field indicates if the MPEG-2 Transport Stream packet payload has been scrambled. The MPEG-2 Transport Stream Packet Header, the optional Adaptation Field, and the payload of a null MPEG-2 Transport Stream packet are never scrambled.

The adaptation_field_control field signals the inclusion of the optional Adaptation Field. The most significant bit of the two-bit field always indicates the presence of the Adaptation Field. The least significant bit indicates the presence of payload.

The continuity_counter field is a 4-bit rolling counter associated with MPEG-2 Transport Stream packets carrying the same PID. The counter is incremented by one for each consecutive Transport Stream packet for a given PID except when the adaptation_field_control field is set to indicate that the Transport Stream packet contains an Adaptation Field only (no payload) or if it is set to the 'reserved' value, or if the Transport Stream packet is a duplicate[3] (these exception cases are known as "non-incrementing conditions"). The continuity_counter is considered "continuous" if it has incremented by one from the continuity_counter value in the previous Transport Stream packet of the same PID or when any of the non-incrementing conditions have been met. The continuity counter is considered "discontinuous" if it has not incremented by one from the continuity counter value in the previous Transport Stream packet having the same PID and a non-incrementing condition has not been met. Except in the case when the discontinuity_indicator flag[4] has been set to '1' to signal a discontinuous continuity_counter, if a receiver encounters a situation where the continuity_counter is discontinuous, then it should assume that some number of MPEG-2 Transport Stream packets have been lost.

Two other fields, the transport_error_indicator and the transport_priority, which are not typically used in ATSC Transport Streams, are also carried in the Packet Header. The

---

3. The MPEG-2 Systems Standard defines a duplicate Transport Stream packet to be the second of two-and only two-consecutive Transport Stream packets having the same PID that are carrying payload and contain identical byte-by-byte contents (except for the Program clock reference, if present). Duplicate Transport Stream packets may be used for additional error resilience purposes.
4. The MPEG-2 Systems Standard defines the discontinuity_indicator as a flag in the Adaptation Field syntax. Among other uses, it may be set to indicate a discontinuous continuity_counter value.

transport_error_indicator may be used to indicate that at least one uncorrectable bit error exists in the Transport Stream packet. The transport_priority field may be used to indicate that a Transport Stream packet with the field set to '1' is of higher priority than other Transport Stream packets having the same PID which do not have the field set to '1'.

The payload field carries the data content. The data content can be one of many types; for example, an MPEG-2 PES packet (which itself may contain an elementary stream) or one or more PSI or Private Sections.

The MPEG-2 Transport Stream Null Packet

The MPEG-2 Transport Stream *Null Packet* is a special Transport Stream packet designed to pad an MPEG-2 Transport Stream [1]. While individual MPEG-2 Programs (services) within a multiplexed bit stream may have variable bit-rate characteristics, the overall MPEG-2 Transport Stream must have a constant bit rate. MPEG-2 Transport Stream Null Packets are transmitted when there are no other packets ready to be transmitted. This is necessary, since the MPEG-2 equipment creating the Transport Stream must maintain a constant bit rate output. Note that Null Packets may be added and/or removed by any re-multiplexing process within the data path.

MPEG-2 Transport Stream Null Packets are always identified by a PID with value 0x1FFF. The Transport Stream Null Packet payload may contain any data values. The continuity_counter of a null Transport Stream packet is undefined, carries no information, and should be ignored.

5.4.2c   MPEG-2 Transport Stream Data Structures

MPEG-2 Systems defines two fundamental bit stream data structures. The first, generically called a "Section," is used to encapsulate either descriptive information about the data essence streams (coded video, coded audio, or data) within the Transport Stream Service Multiplex (e.g., stream type, information needed to extract the streams, Program guide information) or a "private data" essence stream itself [1]. The second, called a *Packetized Elementary Stream* (PES) packet is used to encapsulate elementary stream data essence (e.g., coded video, coded audio, or data).

Tables and Sections

The MPEG-2 Systems Standard defines tables that provide information necessary to act on or to further describe the data essence streams within the Transport Stream Service Multiplex [1]. The logical tables are constructed by using one or more *Sections*. For example, the *Program Map Table* (PMT) contains information about what elementary streams are parts of which MPEG-2 Programs. The PMT is composed of one or more TS_Program_map_section Sections. A *table* is the aggregation of the *Sections* that comprise it. A Section is divided as necessary to be packetized into the payload of one or more MPEG-2 Transport Stream packets so that it may be incorporated into the Transport Stream Service Multiplex along with other bit streams.

The MPEG-2 Systems Standard defines several different tables, collectively called *Program Specific Information* (PSI). Using the private_section, which is the MPEG-2 Systems-defined generic Section data structure, the ATSC standards define many other tables.

MPEG-2 Private Section

The term "Section" is a generic term referring to any data structure that is based on the MPEG-2 private_section syntax [1]. The MPEG-2 private_section defines a data encapsulation method used to place private data (that is, data that the MPEG-2 standards do not define, including ATSC-defined Sections) into an MPEG-2 Transport Stream packet with a minimum amount of structure.

A Section, or more specifically the MPEG-2 private_section, always begins with an 8-bit table_id, which uniquely identifies the table of which the Section is part. Another field, the section_syntax_indicator, determines whether the "short" or "long" form of the private_section syntax is used. The short form Section includes a minimal amount of header information and is limited to carrying a payload of at most 4093 bytes. The long form Section incorporates additional header fields, which allow the segmentation of large data structures into multiple parts. A collection of long form Sections may accommodate 256×4084 bytes of payload (maximum size of 1,045,504 bytes).

In practice, most receivers incorporate hardware Section filtering allowing the receiver to specify filtering criteria for the first eight bytes of a Section. This length equates to the byte count necessary to filter the long form private_section header. Hardware assisted filtering offloads the processing burden from the host processor and enables the receiver to specify exact Section identification syntax for the Section it is interested in acquiring.

The long form Section header contains a version_number field, which identifies the revision of the contents of the Section. Any time the Section's payload bytes are modified, the version_number must be incremented so that a receiver will be able to determine that the Section's contents have changed.

The long form Section contains a CRC_32 field as the first byte following the last payload byte, which is used for error detection purposes. The receiver's 32-bit CRC decoder (the CRC decoder model is described in MPEG-2 Systems, Annex A) calculates the CRC result over all the bytes that comprise a Section beginning with the table_id through the last byte of the CRC_32 field itself. A CRC accumulator result of zero indicates that the Section was received without error.

One or more Sections may be placed into an MPEG-2 Transport Stream packet depending on the Section's size. If the Section length is smaller than a Transport Stream packet's payload, then there may be multiple Sections contained within the single MPEG-2 Transport Stream packet. Sections that are larger than a single MPEG-2 Transport Stream packet are segmented across multiple MPEG-2 Transport Stream packets. Once the process of packetizing a Section commences, a new Section will not be packetized into Transport Stream packets having the same PID until the previous Section's packetization has completed. When a Section does not completely fill an MPEG-2 Transport Stream packet's payload area and there is no new Section ready to begin filling the remainder of the payload area, the remaining bytes of the MPEG-2 Transport Stream packet are stuffed, or filled, with the value 0xFF. To prevent Stuffing Byte emulation, the MPEG-2 Systems Standard forbids the use of 0xFF as a table_id value.

MPEG-2 PSI

MPEG-2 *Program Specific Information* (PSI) provides data necessary to identify an MPEG-2 *Program* (i.e., the desired service) and to demultiplex (i.e., separate and extract) the Program and its *Program Elements* from the MPEG-2 single or multi-Program Transport Stream Service Multiplex. The MPEG-2 Systems Standard defines five PSI tables:

- Program Association Table (PAT)

- Program Map Table (PMT)

- Conditional Access Table (CAT)

- Network Information Table (NIT)

- Transport Stream Description Table (TSDT).

The Program Association Table provides a complete list of all the MPEG-2 Programs (services) within the Transport Stream. The PAT establishes a relationship between each MPEG-2 Program, via the program_number, and its corresponding Program Map Section (properly defined as TS_Program_map_section), via the PID value assigned to the corresponding Program Map Section. Transport Stream packets that contain the PAT are assigned to PID 0x0000.

Each Program Map Section contains the mapping between an MPEG-2 Program and the Program Elements that define the Program (this mapping is called a *Program Definition*). Specifically, a Program Definition establishes a mapping (establishing the relationship) between an MPEG-2 Program number and the list of the PIDs that identify the individual Program Elements comprising the MPEG-2 Program. The PMT is defined as the complete collection of individual Program Definitions within the Transport Stream, with one TS_Program_map_section per MPEG-2 Program. The PMT is unique among the PSI tables in that its contents may be carried as part of different bit streams (i.e., within Transport Stream packets that have different PIDs). This simplifies the addition, deletion, or modification of the PSI for individual MPEG-2 Programs, as each can be altered independently. This also simplifies the demultiplexing process as only relevant portions of the Transport Stream need to be parsed by the receiver. In comparison, the other PSI tables are each required to be in its own unique bit stream (within Transport Stream packets of a single, unique PID).

However, even though an MPEG-2 Program is announced in a TS_Program_map_section, there is no requirement in MPEG-2 that the individual Program Elements are currently present in the Transport Stream. Furthermore, there is no MPEG-2 requirement that all PIDs currently in use are described by any PSI table.

Whenever an MPEG-2 Program's bit stream is scrambled (i.e., the contents are only decodable with the use of a conditional access system process), a CAT must be present in the Transport Stream. The CAT associates aspects of the conditional access system (CA system or CAS), such as access rights sent in *Entitlement Management Messages* (EMMs), with the scrambled streams. Transport Stream packets which contain the CAT are assigned to PID 0x0001. CA systems provide scrambling of MPEG-2 Programs or individual Program Elements along with end-user authorization. While MPEG-2 Programs or individual Program Elements may be scrambled, all of the tables that comprise the PSI are never scrambled. The MPEG Standards do not define the contents of the CAT payload.

The function of the Network Information Table is to carry information that applies network-wide (i.e., to all Transport Stream Service Multiplexes in the delivery/emission network). ATSC standards do not specify the use of the NIT.

The function of the Transport Stream Description Table is to carry descriptors that apply to an entire MPEG-2 Transport Stream Service Multiplex. A/53 neither constrains nor specifies the use of the TSDT [3].

MPEG-2 Packetized Elementary Stream (PES) Packet

The MPEG-2 Systems Standard includes a mechanism for efficiently and reliably conveying continuous streams of data (bit streams of compressed audio, compressed video, and/or data) in real-time over a variety of network environments, including terrestrial broadcasting [1]. Each bit stream (Program Element) is segmented into variable-length packets, called Packetized Elementary Stream (PES) packets, which are conveyed in the MPEG-2 Transport Stream and then reassembled at the receiver. MPEG-2 PES packets are used to segment and encapsulate elementary streams such as coded video, coded audio, and private data streams, along with stream synchronization information. Elementary streams are each independently carried in separate PES packets; thus, a PES packet contains data from one and only one elementary stream. A PES packet is further segmented into fixed-length packets, called MPEG-2 Transport Stream packets. The set of TS packets so created all share a single, common Packet Identifier (PID).

The MPEG-2 PES packet consists of a PES Packet Header followed by the PES packet payload. Each PES packet may have a variable length. A length field allows explicitly signaling the size of the PES packet (up to 65,536 Bytes) or, in the case of Video Elementary Streams, the size may be indicated as unbounded by setting the packet length field to zero. When encapsulating data into a PES packet, the elementary stream is first segmented into variable byte-sized segments and these segments are encapsulated using the MPEG-2 PES packet syntax. ATSC Standard A/53 has placed constraints on PES packets that encapsulate Video Elementary Streams: an MPEG-2 PES packet may only contain one coded video frame and must be signaled as being unbounded in size by defining the length field as 0x0000.

MPEG-2 PES packets carry stream synchronization information in the PES Packet Header using Presentation Time Stamps (PTS) and Decoding Time Stamps (DTS) fields. The timestamps enable decoding the access units and presenting the access units respectively. The PTS and the DTS are each 33-bits long with units in 90 kHz clock periods.

MPEG-2 PES Packet Segmentation, Encapsulation, and Packetization

In order to transport an MPEG-2 PES packet, it is first segmented into the payload of one or more MPEG-2 Transport Stream packets [1]. The first byte of a PES packet must always be the first byte of a Transport Stream packet payload field. When the first byte of a PES packet appears in an MPEG-2 Transport Stream packet, the MPEG-2 Transport Stream Packet Header's payload_unit_start_indicator flag must be set to '1'. The payload_unit_start_indicator is set to '0' in all subsequent MPEG-2 Transport Stream packets carrying the remaining portion of the PES packet. PES packets are typically much larger than an MPEG-2 Transport Stream packet; however, they can be smaller than an MPEG-2 Transport Stream packet. Only a single PES packet may be packetized into an MPEG-2 Transport Stream packet.

Stuffing and the MPEG-2 PES Packet

Since the MPEG-2 Transport Stream is composed of autonomous units of Transport Stream packets, "stuffing" is needed when there is insufficient PES packet data to completely fill a Transport Stream packet payload [1]. "Stuffing" is the process of filling out the remainder of a Transport Stream packet with data bytes that carry no useful information, but only take up the remaining available Transport Stream packet payload bytes. For Transport Stream packets carrying PES packets, stuffing is accomplished by defining an Adaptation Field longer than the sum of the lengths of the data elements in the Adaptation Field, so that the payload bytes remaining after the Adaptation Field exactly accommodate the available PES packet data. This extra space in the Adaptation Field is filled with Stuffing Bytes.

5.4.2d   Multiplex Concepts

The MPEG-2 term "Program" corresponds to an individual digital TV channel or data service [1]. The MPEG-2 Systems Standard's support for multiple channels or services within a single, multiplexed bit stream (known as a multi-Program Transport Stream or Service Multiplex) enables the deployment of practical, bandwidth-efficient digital broadcasting systems. This approach enables the delivery of services at various bit rates in one defined construct.

The Packet Identifier (PID), contained in each Transport Stream packet, is the key to sorting out the components or elements in the Transport Stream. The PID is used to reassemble higher level constructs that make up different bit stream elements within the multiplex and can change from Transport Stream packet to packet. This identification mechanism enables the time-based interleaving or multiplexing of services at differing bit rates. For example, video essence typically requires a much higher bit rate than audio essence. A series of Transport Stream packets identified by the same PID contain either a Program Element or descriptive information about one or more Program Elements (a series of Transport Stream packets having the same PID is often referred to as a bit stream).

The MPEG-2 Systems Standard has set aside a few special PIDs to directly identify Transport Stream packets that contain constructs that assist in locating the individual MPEG-2 Programs and their associated Program Elements. These constructs are collectively called Program Specific Information (PSI).

A related set of one or more Program Elements is called an MPEG-2 Program. Figure 5.4.1 illustrates how two MPEG-2 Programs, each consisting of a video and audio Program Element (in these cases each Program Element is also an elementary stream), might be multiplexed into an MPEG-2 Transport Stream. The Transport Stream packet payload contents are reassembled into a higher level construct (with different packet sizes and structure). For coded audio and video, this higher layer of packetization is called a Packetized Elementary Stream (PES) packet.

In Figure 5.4.1, Program P1's video stream is illustrated to consist of three MPEG-2 Transport Stream packets identified by PID 0x1024. Each MPEG-2 Transport Stream packet has a continuity_counter associated with the specific PID that enables a receiver to determine if a loss has occurred. In this example, the continuity_counter values begin at 0x3 and end with 0x5 for Program P1's video stream. The individual MPEG-2 Transport Stream packets that contain this PID are extracted from the multiplexed bit stream and reassembled, in this case making up part of an MPEG-2 PES packet carrying a Video Elementary Stream.

Program P1 also has an associated audio stream of packets identified by PID 0x1025. Two MPEG-2 Transport Stream packets from Program P1's audio stream are shown with the continuity_counter values of 0x2 and 0x3 respectively. Similarly, in Figure 5.4.1, Program P2's packet composition is illustrated. In Program P2's video stream identified by PID 0x0377, the second to last MPEG-2 Transport Stream packet's continuity_counter is 0xB rather than the expected value of 0x9. This condition may indicate an error and the loss of possibly three MPEG-2 Transport Stream packets having this PID. The next expected and received continuity_counter value is 0xC as illustrated in the diagram.

As discussed later in this chapter, the mechanism for recreating the original System Time Clock (STC) in the decoder uses the actual arrival time of the packets carrying the individual Program clock references (PCR) as compared to the value carried in the PCR field.

Because of this, MPEG-2 Transport Stream packets with a given PID value cannot casually be rearranged in the MPEG-2 Transport Stream. This limitation exists because shifting the relative location of a Transport Stream packet carrying the PCR introduces jitter into the data stream,

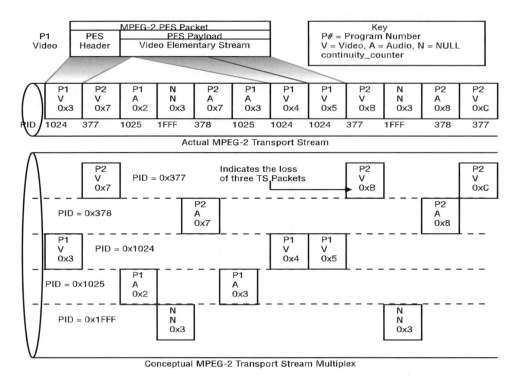

Figure 5.4.1 MPEG-2 Transport Stream Program multiplex. (*From* [1]. *Used with permission.*)

which may cause the decoder's System Time Clock (STC) to vary. The temporal location of the individual MPEG-2 Transport Stream packet payload delivery conforms to the buffer model associated with the encapsulation type. Shifting or rearranging the MPEG-2 Transport Stream packets potentially causes buffer model violations by either overflowing or underflowing the buffer, unless such is done without violation of these constraints.

Also notice the null MPEG-2 Transport Stream packets that were interleaved. These MPEG-2 Transport Stream packets (identified by PID 0x1FFF) may appear anywhere in the stream and are often used to set the Transport Stream Service Multiplex at a known, fixed overall bit rate, regardless of the total bit rate of all the MPEG-2 Programs it contains. For illustrative purposes, a value of 0x03 is shown in the figure for the continuity_counter for the Null Packets. In practice any value may be used, as the continuity_counter for Null Packets is ignored.

5.4.2e   MPEG-2 Timing and Buffer Model

Key elements of the MPEG-2 Systems Standard include a model for system timing and another for buffering [1]. The timing model allows the synchronization of the components making up MPEG-2 Programs. The buffer model ensures interoperability between encoders and decoders for information delivery (i.e., ensuring that the necessary information is always available when needed for decoding).

MPEG-2 System Timing

One of the basic concepts of the MPEG-2 Standards revolves around the system timing model [1]. The timing model was developed to enable the synchronization of video and audio Program Elements that are delivered as separate streams, with differing delivery rates and different sized presentation units. As will be discussed below, elements that enable the synchronization are clock references, which allow the decoder to recreate a clock that very closely tracks that used in the encoder, and time stamps, which are used to temporally coordinate the presentation of video and audio presentation units. This basic timing model is applicable to other forms of Program Elements, including data.

Timing Model

The MPEG-2 timing model requires that the clock used to encode the content be regenerated (within specified tolerances) at the receiver and used to decode the content [1]. Video and audio consist of discrete presentation units, which must be delivered from the decoder at the same rate as they entered the encoder in order to achieve correct reproduction. For video, the presentation unit is a picture (a frame or field of video). For audio, the presentation unit is a block of audio samples (also known as an *audio frame*). The presentation unit for data is dependant upon the form of the data, but the basic concept is similar. The output rate at the decoder must match the input rate at the encoder.

In developing the timing model, the MPEG-2 Systems Standard adopted two basic concepts: a constant end-to-end delay and an instantaneous decoding process (see Figure 5.4.2). The MPEG-2 Systems Standard does not specify how the encoders or decoders operate; rather, it specifies the format of the bit stream (the syntax and semantics) and a theoretical decoding buffer model. With these concepts applied to the bit stream, it is possible to develop implementations of both encoders and decoders that consider real-world constraints and will interoperate. In real systems, the delay through the encoding and decoding buffers is variable [2] and the decoding process takes a finite, non-zero and possibly variable, amount of time.

The MPEG-2 Systems Standard's timing and buffer models solve the issues of synchronization of individual elements by use of a common time reference shared by all individual Program Elements of an MPEG-2 Program. This common time clock is referred to as the System Time Clock (STC).

System Time Clock (STC)

The System Time Clock (STC) is the master clock reference for all encoding and decoding processes [1]. Each encoder samples the STC as needed to create time stamps associated with the data's Presentation Units. A time stamp associated with a Presentation Unit is referred to as the Presentation Time Stamp (PTS). A time stamp associated with the decoding start time, known as Decoding Time Stamp (DTS), may also appear in the bit stream.

The STC is not a normative element in the MPEG-2 Systems standard; however, it is required for synchronized services (including video and audio), meaning that all practical implementations require its use. The STC is represented by a 42-bit counter in units of 27 MHz (27 MHz equals approximately 37 ns per clock period).

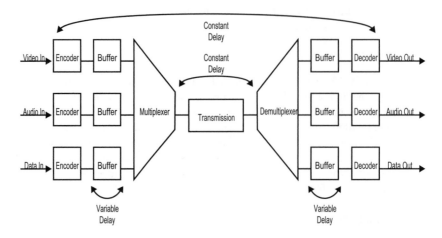

Figure 5.4.2 MPEG-2 constant delay buffer model. (*From* [1]. *Used with permission.*)

The STC must be recreated in the decoder in such a way that it very closely matches (within specified tolerances) the STC at the encoder for both buffer management and synchronization reasons. In order for a decoder to reconstruct this clock, the STC is periodically sampled and transmitted in the MPEG-2 Transport Stream packet's adaptation_field, as clock references known as Program Clock References (PCRs). Figure 5.4.3 illustrates a general decoder circuit used to recreate the STC.

Each MPEG-2 Program may have its own STC or multiple MPEG-2 Programs may share a common STC (by referring to the same Program Element that carries the PCR values). There may be situations where an MPEG-2 Program does not require any form of synchronization and will not need an STC. Also, Program Elements may or may not reference a Program's STC.

The STC increases linearly in time and monotonically. The exception is when there are discontinuities, which are discussed below. Since the STC value is contained within a finite size field, it wraps back to zero when the maximum bit count is achieved, approximately every 26.5 hours.

System Clock Frequency

The System Time Clock is derived from the system_clock_frequency specified as 27,000,000 Hz ±810 Hz [1]. The STC period is 1/27 MHz or approximately 37 ns per clock period.

Program Clock Reference

The Program Clock Reference (PCR) is a 42-bit value used to lock the decoder's 27 MHz clock to the encoder's 27 MHz clock, thereby matching the decoder's STC to the encoder's STC [1]. The PCR is carried in the MPEG-2 Transport Stream packet's adaptation_field using the program_clock_reference_base and the program_clock_reference_extension fields. The MPEG-2 Systems standard mandates that the PCR be sent at least every 100 ms or 10 times a second. The

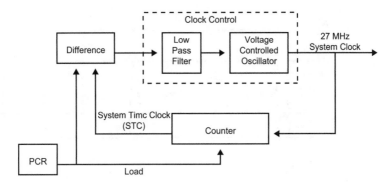

Figure 5.4.3 MPEG-2 System Time Clock. (*From* [1]. *Used with permission.*)

PCR may be sent more frequently if desired. In addition, the standard limits the amount of PCR jitter for a compliant stream to no more than ±500 ns.

The decoder uses the arrival time of the MPEG-2 Transport Stream packet carrying a PCR value, and the PCR value itself, in comparison to the current value of the STC to adjust the clock control component. Figure 5.4.3 illustrates an example of how the PCR is used to exactly recreate the STC.

The program_clock_reference_base is constructed by dividing the value of the 27 MHz clock reference count by 300. This operation creates a 33-bit value in units of 90 kHz clock periods. The program_clock_reference_extension contains the remainder of the previous division (i.e., the 27 MHz clock modulo 300).

The location of the Program Element carrying the PCR for an MPEG-2 Program is signaled in the TS_Program_map_section PCR_PID field. The PCR may be carried on the same PID as a video, audio, or data Program Element as the PCR field is independent of the encapsulated data payload. Different MPEG-2 Programs may share the same STC, by referring to the same PCR_PID.

MPEG-2 Programs not requiring synchronized decoding and presentation to an STC set the PCR_PID field to the value 0x1FFF indicating that there is not a Program Element carrying a PCR.

Presentation Time Stamp (PTS)

The Presentation Time Stamp (PTS) is a 33-bit quantity measured in units of 90 kHz clock periods (approximately 11.1 microsecond ticks) carried in the MPEG-2 PES Packet Header's PTS or DTS fields [1]. The PTS, when compared against the System Time Clock (STC), indicates when the associated Presentation Unit should be "presented" to the viewer. In the case of video, a picture is displayed and in the case of audio the next audio frame is emitted by the receiver. The PTS must be contained in the MPEG-2 Transport Stream at intervals no longer than 700 ms and the ATSC requires that the PTS be inserted at the beginning of every access unit (i.e., coded picture or audio frame).

The PTS, when included, is divided into two fifteen-bit quantities and a 3-bit quantity spread across 36 bits. There are also three "marker bits", always set to '1', interspersed among

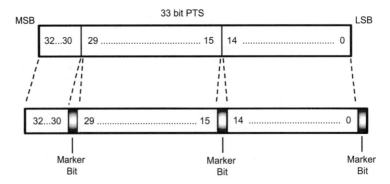

Figure 5.4.4 The MPEG-2 PTS and marker_bits.

the three groups. This division into three parts, along with the inclusion of the marker_bits, avoids start_code emulation in the MPEG-2 PES Packet Header. Avoiding the emulation of the start_code prevents decoders from incorrectly identifying the start of an elementary stream. Figure 5.4.4 illustrates the PTS and marker_bits.

Decoding Time Stamp (DTS)

The Decoding Time Stamp (DTS) is a 33-bit quantity, measured in units of 90 kHz clock periods (approximately 11.1 microseconds) that may be carried in the MPEG-2 PES Packet Header's DTS field [1]. The MPEG-2 Systems Standard only defines a normative meaning for the DTS field for video. Generally speaking, a video stream is the only stream type that may need the DTS due to picture re-ordering (bi-directionally interpolated pictures are decoded after the "future" frame it references has been decoded. The DTS value, compared to the System Time Clock (STC), indicates when the access unit should be removed from the buffer and decoded.

The DTS must always be accompanied by a PTS. If the DTS contains the same value as the PTS, then the DTS is omitted and the decoder assumes that the DTS is equal to the PTS. The DTS, if present, must be contained in the MPEG-2 Transport Stream at intervals no longer than 700 ms. ATSC mandates that the DTS must be inserted at the beginning of every access unit (i.e., coded picture or audio frame), except when the DTS value matches the PTS value.

The DTS is encoded in the same manner as the PTS—splitting the 33-bit quantity into three portions and incorporating the marker bits.

Discontinuities

MPEG-2 Program and MPEG-2 Transport Stream discontinuities are a reality in digital television [1]. Planned discontinuities, where the interruption is not the result of an error, can occur in any number of situations. As an example, the splicing of a commercial into the video and audio streams is a typical planned discontinuity scenario. Other planned discontinuity scenarios include switching between content sources or a new MPEG-2 Program commencing. In each of these cases, the System Time Clock (STC) may be interrupted and set to some new random value from which the count then continues, thus creating a discontinuity in the timeline.

The decoder, in all of these instances, should be notified of the upcoming interruption by the MPEG-2 PES Packet Header's discontinuity_indicator. The discontinuity_indicator is used to indicate a discontinuity in the STC or a disruption in the continuity_counter. The signaling of continuity_counter disruptions via the discontinuity_indicator is limited in its practical usefulness. The discontinuity_indicator can be used by the multiplexing process to indicate a known and expected discontinuity in the Program time line. MPEG-2 Transport Stream packet loss is not signaled via this indicator bit.

An STC interruption is the result of a receiver receiving a new PCR value associated with the MPEG-2 Program that is out of range of a reasonable variance from the expected value, regardless of whether or not the discontinuity_indicator had been set. Receivers receiving the next PCR after either an explicit discontinuity or a PCR out of range must adjust themselves accordingly. In cases where a discontinuity has been signaled explicitly, a receiver typically will simply use the next PCR value received to reset its internal clock phase circuitry without making any frequency adjustments. In cases where a discontinuity has not be signalled explicitly, a receiver typically will begin a clock-error recovery process. This may include tracking PCR values during a predefined time window to make an "intelligent" determination of what adjustments need to be made to the STC, if any. Besides the STC reference changing, another discontinuity that may be encountered as part of the stream changeover or Program interruption involves the MPEG-2 Transport Stream Packet Header's continuity_counter value. The continuity_counter may skip to a new value when the newly encoded stream is inserted. Thus, the decoder upon seeing the discontinuity_indicator is made aware of an upcoming continuity_counter change and this change should not be treated as an error or indicative of lost packets.

Buffer Model

The MPEG-2 standards define the bit stream syntax itself and the meaning (or semantics) of the bit stream syntax. In order to ensure interoperability of equipment designed to the specifications, the MPEG-2 Systems Standard also precisely specifies the exact definitions of byte arrival and decoding events—and the times at which these occur—through the use of a hypothetical decoder called the Transport Stream System Target Decoder (T-STD) [1]. The T-STD is a conceptual model that is used solely for the purpose of defining terms precisely and to model the decoding process during construction or verification of Transport Streams; neither the architecture of the T-STD nor the timing described is meant to preclude the design of practical solutions that implement the buffer and timing model in other ways. The buffers are therefore "virtual" since they may or may not exist within real physical decoders.

The T-STD model is structured as follows:

- Defines several "virtual buffers"

- Rules for when bytes enter and leave each buffer

- Rules that constrain buffer fullness

- The number of "virtual buffers" in the model varies depending upon the number of streams in the Transport Stream

For Video Elementary Streams, the T-STD consists of three buffers: the Transport Buffer, the Multiplex Buffer, and the Elementary Stream Buffer. For Audio Elementary Streams and system-related streams (e.g., PSI tables), the T-STD consists of two buffers, the Transport Buffer and the Main Buffer. The rules define when bytes enter and leave the buffers in terms of where they occur within the bit stream and either the transport rate or the specified maximum bit rate for the type of elementary stream, depending on the buffer type and the elementary stream type. The transport rate computation is determined by a mathematical formula based on the Program clock reference fields encoded in the bit stream, and the maximum bit rate is determined from the profile and level or similar inherent definition. Buffer sizes are defined by specific mathematical formulas based on the buffer type and the elementary stream type. The decoding time is specified in terms of embedded or inferred decoding or Presentation Time Stamps (DTS or PTS, respectively) and may be delayed due to any re-ordering of pictures that is needed (in the case of Video Elementary Streams only).

Buffer management is needed to ensure that none of the buffers overflow or, in some cases, underflow. Constraints on buffer fullness say whether a particular buffer is allowed to underflow, whether a buffer must empty occasionally, etc. These rules are all clearly defined in the T-STD model.

A suitable analysis tool can verify whether or not a bit stream conforms to the T-STD. It is more difficult to verify that a decoder conforms to the T-STD because true conformance may only be determined by demonstrating that the decoder is capable of decoding all conformant bit streams properly.

5.4.2f   MPEG-2 Descriptors

The MPEG-2 descriptor is a generic structure used to carry information within other MPEG-2 data structures, typically Sections (PSI or private) [1]. The use of standardized descriptors is often optional. Descriptors can be viewed as a mechanism to extend the information conveyed within another MPEG-2 structure.

A descriptor cannot stand alone in the MPEG-2 Transport Stream; rather, it must be contained within a larger syntactic structure, typically within a descriptor loop (an area set aside to carry an arbitrary number of descriptors). The basic descriptor format is a tag byte, followed by a length byte followed by data [2]. The tag byte uniquely identifies the descriptor and the length byte specifies the number of data bytes that immediately follow the length field. The form of the data varies for each specific descriptor.

In future versions of a given ATSC standard, additional information may be added to any defined descriptor by simply adding new semantic fields at the end. Receiver designers should always process the length field of all descriptors to ensure that if a receiver finds any information beyond the known fields, then it discards such information but continues parsing the stream at the first byte beyond that indicated by the length field. Receivers may also encounter descriptors that they do not recognize (such as could be added to a new version of the standard created after the receiver was built). To ensure that newly defined descriptors do not cause operational problems in existing equipment, all descriptors defined will adhere to the existing structure. This provides an inherent escape mechanism to allow receivers that don't understand a particular descriptor to easily skip over it. By jumping the number of bytes listed in the length field, the receiver can proceed to the next item in the loop.

Because a receiver that does not recognize a descriptor of a certain type is expected to simply ignore it, the addition of new features via new descriptor definitions is a powerful way to add new features to the protocol while maintaining backward compatibility.

5.4.2g   ATSC Descriptors

ATSC-defined descriptors follow the same behavior as described previously for MPEG-2 descriptors and may be used for similar purposes [1]. ATSC standards have also described the usage of some MPEG descriptors. The following descriptors are defined by ATSC Standards A/52 [4] and A/53 [3].

AC-3 Audio Descriptor

An AC-3 Audio Descriptor [4] describes an audio service present in an ATSC Transport Stream. In addition to describing a possible audio service(s) that a broadcaster might send, this descriptor(s) provides the receiver with audio set up information such as whether the Program is in stereo or surround sound [1]. This descriptor is optionally present in the Program Element loop of the TS_Program_map_section that describes the AC-3 Audio Elementary Stream.

ATSC Private Information Descriptor

The ATSC Private Information Descriptor [3] provides a method for carrying private information within this descriptor and for unambiguous identification of its registered owner [1]. Since both the identification and the private information are self-contained within a single descriptor, more than one ATSC Private Information Descriptor may appear within a single descriptor loop.

The format identifier field appears in both the ATSC Private Information Descriptor and the MPEG Registration Descriptor. Its purpose is the same in both: it identifies the company or organization that has supplied the associated private data. Only values of the format identifier field registered by the ISO-assigned Registration Authority, the Society of Motion Picture Engineers (SMPTE), may be used.

5.4.2h   MPEG Descriptors Constrained by ATSC

The following descriptors have been defined in MPEG-2 Systems (13818-1) [2], but their usage has been constrained by A/53 [3].

Data Stream Alignment Descriptor

The ATSC requires this descriptor to be present in the Program Element loop of the TS_Program_map_section that describes the Video Elementary Stream [1]. In this context, the descriptor specifies the alignment of video stream syntax with respect to the start of the PES packet payload. The ATSC has constrained the alignment to be the first byte of the start code for a video access unit (alignment_type 0x02). Because a video access unit follows immediately after a GOP or Sequence header, this does not preclude alignment from the beginning of a GOP or Sequence. It does, however, prevent alignment from being at the start of a slice.

The use of this descriptor for other stream types is not defined.

ISO 639 Language Descriptor

In the ATSC digital television system, if the ISO_639_language_descriptor (defined in ISO/IEC 13818-1 Section 2.6.18 [2]) is present then it is used to indicate the language of Audio Elementary Stream components.

If present, then the ISO_639_language_descriptor is included in the descriptor loop immediately following the ES_info_length field in the TS_Program_map_section for each Elementary Stream of stream_type 0x81 (AC-3 audio). This descriptor will be present when the number of Audio Elementary Streams in the TS_Program_map_section having the same value of bit stream mode (bsmod in the AC-3 Audio Descriptor) is two or more.

As an example, consider an MPEG-2 Program that includes two audio ES components: a Complete Main (CM) audio track (bsmod = 0) and a Visually Impaired (VI) audio track (bsmod = 2). Inclusion of the ISO_639_language_descriptor is optional for this Program. If a second CM track were to be added, however, it would then be necessary to include ISO_639_language_descriptors in the TS_Program_map_section.

The audio_type field in any ISO_639_language_descriptor used ATSC standards is set to 0x00 (meaning "undefined"). An ISO_639_language_descriptor may be present in the TS_Program_map_section in other positions as well, for example to indicate the language or languages of a textual data service Program Element.

MPEG-2 Registration Descriptor

Under certain circumstances, the MPEG-2 Registration Descriptor (MRD) is used to provide unambiguous identification of privately defined fields or private data bytes in associated syntactical structures [1]. The detailed rules for the use of the MRD are found in A/53 [3]. Note that no more than one MRD should appear in any given descriptor loop, since the semantics of this situation are unspecified. This usage restriction does not apply to the ATSC Private Information Descriptor discussed previously. The MRD does not contain the private data itself, while the ATSC Private Information Descriptor is designed to carry the actual private data.

Smoothing Buffer Descriptor

A/53 requires the TS_Program_map_section that describes each Program to have a Smoothing Buffer Descriptor pertaining to that Program [3]. This descriptor signals the required size and leak rate of the smoothing buffer (SBn) to avoid errors in decoding that could be caused by over- or under-flow. During the continuous existence of a Program, the value of the elements of the Smoothing Buffer Descriptor are not allowed to change.

Code Point Conflict Avoidance

MPEG standards have numerous syntactical fields set aside for private use [1]. When fields, tags, and table identifier fields are assigned a value by MPEG or by an MPEG user, such as ATSC, the values are then known as "code points." In addition, many other fields have ranges defined as user private. The user (not a standards body) may define one or more of these private fields, tags, and table values. Without some type of coordination mechanism, use of ATSC user private fields and ranges may lead to conflicts between privately defined services. Furthermore, without some form of scoping and registration, different organizations may inadvertently choose to use the same values for these fields, but with different meanings for the semantics of the information carried. The ATSC Digital Television Standard A/53 [3] has placed constraints on the use of private fields and ranges to avoid code point conflicts, through the use of the MPEG-2 Registration Descriptor mechanism.

If an organization uses user private fields and/or ranges, to comply with the ATSC standards, one or more MRDs are used as described in A/53.

5.4.2i   The MPEG-2 Registration Descriptor

MPEG-2 Systems defines a registration descriptor: "The registration_descriptor provides a method to uniquely and unambiguously identify formats of private data." The Society of Motion Picture and Television Engineers (SMPTE) is the ISO-designated registration authority for the 32 bit format_identifier field carried within this descriptor, guaranteeing that every assigned value will be unique [1].

The following Sections discuss the use of the MRD to avoid collisions. There are some circumstances where an MRD cannot be used for scoping (for example, the MRD has no significance for other descriptors in the same descriptor loop). The ATSC Private Information Descriptor (described previously) has been defined to allow the carriage of private information in a descriptor.

Private Information in an MPEG-2 Program

The scoping of the use of private structures within an MPEG-2 Program may be done by placing an MRD in the Program loop in the PMT (otherwise known as the "outer" loop—the descriptor loop following the Program_info_length field) [1]. When used in this location, the scope of the MRD is the entire MPEG-2 Program, meaning all of the Program Elements defined in this instance of the Program Map Table. When the MRD is used to identify the owner of private data, then the identification applies to all Program Elements comprising the MPEG-2 Program.

Private Information in an MPEG-2 Program Element

MRDs may be placed in the Program Element loop in the PMT (otherwise known as the "inner" loop—the descriptor loop following the es_info_length field) [1]. When used in this location, the scope of the MRD is the individual Program Element to which the MRD is bound. When the MRD is used to identify the owner of private data, then the identification applies to the single Program Element. The scope of the MRD also covers the stream_type used for this Program Element, in the case that a privately defined stream_type is used.

Multiple MRDs

At most, one MRD for any entity at any level will appear; in other words, no more than one MRD will appear in the PMT Program loop; no more than one MRD will appear in the PMT Program Element loop for a particular Program Element [1]. There is no guarantee of how remultiplexing equipment will behave in the presence of multiple MRDs in a single loop especially in regards to retaining the original ordering of descriptors in the loop. Multiple MRD's at the same level would be ambiguous to a receiver.

MRDs used at different levels are intended to be complimentary, with a deeper level MRD refining the meaning of a higher level MRD. However, certain combinations of MRDs at different levels may result in streams that may cause problems for standard receivers if the combinations are not expected. As an example, a combination of MRDs that identify the Program as defined by company X and a particular Program Element as defined by company Y would lead to contradictions in interpreting semantic elements. The behavior of a receiver upon receiving a non-conformant stream of this type cannot be specified and construction of streams of this type should be avoided.

5.4.3   References

1.   ATSC: A/54A, "Guide to the Use of the Digital Television Standard," Advanced Television Systems Committee, Washington, D.C., December 4, 2003.

2.   ISO/IEC IS 13818-1:2000 (E), International Standard, Information technology – Generic coding of moving pictures and associated audio information: Systems.

3.   ATSC: A/53E, "ATSC Digital Television Standard," Advanced Television Systems Committee, Washington, D.C., December 27, 2006.

4.   ATSC: A/52B, "Digital Audio Compression (AC-3)," Advanced Television Systems Committee, Washington, D.C., June 14, 2005.

5.   ATSC: A/65C, "Program and System Information Protocol," Advanced Television Systems Committee, Washington, D.C., January 2, 2006.

Understanding ATSC/MPEG Syntax Tables

5.5.1 Introduction[1]

ATSC and MPEG-2 standards use a common convention for specifying how to construct the data structures defined in the standards. This convention consists of a table specifying the syntax (the in-order concatenation of the fields), following by a section specifying the semantics (the detailed definitions of the syntax fields). The syntax is specified using C-language "like" ("C-like") constructs, meaning statements that take the form of the computer language, but would not necessarily be expected to produce reasonable results if run through a compiler.

5.5.2 About the Tables

The tables typically have three columns, as shown in the fragment in Table 5.6.1:

- **Syntax**: The name of the field or a "C-like" construct

- **Number of Bits**: The size of the field in bits.

- **Format**: Either how to order the bits in the field; an acronym or mnemonic is used, which is defined earlier in the standard (for example, uimsbf means *unsigned integer, most significant bit first*)—or, when the field has a pre-defined value, the value itself (typically in either binary or hex notation).

It should be noted that when the data structures are constructed, the fields are concatenated using big-endian byte-ordering. This means that for a multi-byte field, the most significant byte is encountered first. A common practice for implementation is to step through the syntax struc-

---

1. This chapter is based on: ATSC, "Guide to the Use of the Digital Television Standard," Advanced Television Systems Committee, Washington, D.C., Doc. A/54A, December 4, 2003. Readers are encouraged to download the source document from the ATSC Web site (http://www.atsc.org). All ATSC Standards, Recommended Practices, and Information Guides are available at no charge.

ture and copy the values for the fields to a memory buffer. The end result of this type of operation may vary for multi-byte fields, depending upon the computer architecture.

5.5.2a   Formatting

The curly-bracket characters ('{' and '}') are used to group a series of fields together. In the sample shown in Table 5.6.1, the curly-bracket characters are used to indicate that all of the fields between the paired curly-brackets belong to the "typical_PSI_table()"[2]. For the conditional and loop statements that follow (shown in Table 5.6.2), curly-bracket pairs are used to indicate the fields affected by either the conditional or loop statements.

The syntax column uses indentation as an aid to the reader (in a similar fashion to a common convention when writing C-code). When a series of fields is grouped, then the convention is to indent them.

Table 5.6.1 Basic Table Format

Syntax	No. of Bits	Format
typical_PSI_table() {		
table_id	8	uimsbf
section_syntax_indicator	1	'1'
....	...	...
}		

5.5.2b   Conditional Statements

In many of the constructs, a series of fields is included only if certain conditions are met. This situation is indicated using an "if (condition) { }" statement as shown in Table 5.6.2a. When this type of statement is encountered, the fields grouped by brackets are included only if the condition is true.

As with C-code, an alternate path may be indicated by an "else" statement, as illustrated in Table 5.6.2b: If the condition is true, then field_2 is used; otherwise, field_3 is used.

Table 5.6.2a IF Statement

Syntax	No. of Bits	Format
typical_PSI_table() {		
field_1	8	uimsbf
if (condition) {		
field_2	8	uimsbf
field_3	8	uimsbf
}		
....	...	...
}		

Table 5.6.2b IF Else Statement

Syntax	No. of Bits	Format
typical_PSI_table() {		
field_1	8	uimsbf
if (condition) {		
field_2	8	uimsbf
} else {		
field_3	8	uimsbf
}		
....	...	...
}		

---

2. ATSC Digital Television Standard documents contains symbolic references to syntactic elements used in the audio, video, and transport coding subsystems. These references usually contain the underscore character (e.g., sequence_end_code) and may consist of character strings that are not English words (e.g., dynrng). In this book they are identified by a distinctive font. Note also that certain syntax names are capitalized; e.g., Transport Stream.

5.5.2c   Loop Statements

For-loop statements are commonly used in the syntax tables and have the widest variation in style and interpretation[3]. The for-loop takes the following form:

```
for ( i=0; i<N; i++ ) {
...fields
}
```

This type of statement indicates that the fields between curly-brackets should be included a number of times, but can't necessarily be interpreted the way a C compiler would, due to variations in the meaning of the end-point (N in the example above) and nesting of for-loops with reuse of the counter variables. The syntax tables always provide enough information to understand the meaning of the for-loop. Unfortunately, in many cases, common-sense and insight must be used. The following example fragments illustrate how to interpret the for-loop for different types of usage:

Example 1: In the example shown in Table 5.6.3, the interpretation is quite straightforward: the end-point variable, private_data_length, which is given a value in the field immediately above the for-loop. It simply indicates the number of private_data_bytes that follow.

Table 5.6.3 For-Loop Example 1

Syntax	No. of Bits	Format
...		
field_1	8	uimsbf
private_data_length	8	uimsbf
for (I=0; I<private_data_length; I++) {		
private_data_bytes	8	uimsbf
}		
....	...	...

Example 2: Upon examination of Table 5.6.4, one quickly notices that there are two nested for-loops, both using the same counter variable (i). As opposed to the interpretation in a real C-program where a change in the value held by the variable in the inner loop will be reflected in the outer, these counter variables do not affect each other. In practice, the actual variable name chosen should be ignored; one should simply view the for-loop construct as a loop that should be traversed some number of times.

Furthermore, the inner and outer loops of this example represent different ways of interpreting how many times to traverse the loop. The outer loop represents a fairly conventional interpretation—the loop is to be traversed "num_channels_in_section" times; each traversal includes all of the fields between this level of paired curly-brackets. As in the example in Table 5.6.3, the value for "num_channels_in_section" is set by the field preceding the loop.

---

3. The for-loop statement represents the biggest divergence from actual C-code usage.

Table 5.6.4 For-Loop Example 2

Syntax	No. of Bits	Format
...	8	uimsbf
num_channels_in_section	8	uimsbf
for(i=0; i<num_channels_in_section; i++) {		
field_1	8	uimsbf
...		
descriptors_length	10	uimsbf
for (i=0;i<N;i++) {		
descriptor()		
}		
}		
...	6	'111111'

Example 3: The inner loop has a different interpretation. In this case, contents of the loop are descriptors. Different descriptors have different lengths, but as shown in Table 5.6.5, the second byte of each descriptor always specifies the length (in bytes) of the remaining descriptor. The end-point variable "N" is not explicitly set, but may be inferred from a knowledge of how descriptors are constructed. For the inner loop example, the "descriptors_length" field specifies how many bytes make up the fields included in all traversals of this particular loop. In practice, one would follow the steps listed below when parsing this portion of the syntax:

- Read descriptors_length field

- Read the descriptor_tag and descriptor_length fields that follow

- If the descriptor_tag is understood, interpret the following descriptor_length bytes according to the syntax of the particular descriptor, otherwise skip over these bytes

- Increment the number of bytes traversed by the descriptor_length field + 2 bytes (to include the descriptor_tag and descriptor_length fields)

- If the number of bytes traversed is less than the value in the descriptors_length field, go to step 2; otherwise, the loop has been fully traversed.

Table 5.6.5 General Descriptor Format, Example 3

Syntax	No. of Bits	Format
descriptor () {		
descriptor_tag	8	uimsbf
descriptor_length	8	uimsbf
fields		
...		
}		

Example 4: Upon examining Table 5.6.6 one encounters a noteworthy usage of the for-loop. In this case, the end-point variable is listed simply as N, with no indication of what N might be[4]. In some cases, the for-loop is immediately preceded by a length field. For this type of situation, N would take the value of the length field and be interpreted as in one of the cases above.

The particular case illustrated in Table 5.6.6 (taken from the A/65 System Time Table) requires a little more insight to understand. The field immediately above the for-loop provides no information as to what value to use for N. Examining the entire table, one finds that the only portion of the syntax with variable length is the for-loop. In addition, one of the earlier fields in the table is the section_length field, which specifies the size of the overall table. These two observations provide enough information to allow the calculation of how many bytes would be included in the for-loop carrying the descriptors.

Table 5.6.6 For-Loop Example 4

Syntax	No. of Bits	Format
system_time_table_section () {		
table_id	8	0xCD
…		
section_length	12	uimsbf
…		
daylight_savings	16	uimsbf
for (i= 0;i< N;i++) {		
descriptor()		
}		
CRC_32	32	rpchof
}		

5.5.2d   Length Fields

Many of the MPEG-2 syntactical structures include length fields, which indicate the number of bytes remaining in the structure; for example, the general descriptor illustrated in Table 5.6.5. Some of these structures have an extension mechanism, where non-standard information may be placed at the end of the defined syntax, with the length field increased to account for the extra information. Of course, with this form of extension, there is no expectation that a generic receiver will be able to understand the extra information.

For this reason, the length field should always be parsed and the information used to determine the offset to the next structure. A common implementation mistake is to assume that the bit stream being parsed contains only standardized usage and only account for the fields defined in the appropriate standard—especially for descriptors. Not correctly accounting for the length information could result in trying to interpret the remainder of the bit stream incorrectly.

5.5.3   References

1.   ATSC, "Guide to the Use of the Digital Television Standard," Advanced Television Systems Committee, Washington, D.C., Doc. A/54A, December 4, 2003

---

4. Note: This type of usage is more common in the MPEG-2 Standards than in the ATSC Standards.

5.6

DTV Program and System Information Protocol

5.6.1 Introduction[1]

The Program and System Information Protocol (PSIP) is a small collection of tables designed to operate within every Transport Stream for terrestrial broadcast of digital television. Its purpose is to describe the information at the system and event levels for all *virtual channels* carried in a particular *Transport Stream* (TS). Additionally, information for analog channels as well as digital channels from other Transport Streams may be incorporated.

5.6.2 System Overview

Under the ATSC Digital Television Standard, the 6 MHz channel used for analog TV broadcast supports about 19 Mbits/s of throughput for terrestrial broadcast [1]. Since audiovisual signals with standard resolution can be compressed using MPEG-2 to sustainable rates of around 6 Mbits/s, then approximately 3 or 4 digital TV channels can be safely supported in a single physical channel. Moreover, enough bandwidth remains within the same Transport Stream to provide several additional low-bandwidth non-conventional services such as:

- Weather reports
- Stock indices
- Headline news

---

1. This chapter is based on: ATSC, *Program and System Information Protocol*, Doc. A/65C, Advanced Television Systems Committee, Washington, D.C., January 2, 2006. For full details on this system, readers are encouraged to download the source document from the ATSC Web site (http://www.atsc.org). All ATSC Standards, Recommended Practices, and Information Guides are available at no charge.

- Software download (for games or enhanced applications)

- Image-driven classified ads

- Home shopping

- Pay-per-view information

It is therefore practical to anticipate that in the future, the list of services (virtual channels) carried in a physical transmission channel may reach ten or more. More importantly, the number and types of services may also change continuously, thus becoming a dynamic medium.

An important feature of terrestrial broadcasting is that sources follow a distributed information model rather than a centralized one. Unlike cable or satellite, service-providers are geographically distributed and have no interaction with respect to data unification or even synchronization. It was therefore necessary to develop a protocol for describing system information and event descriptions to be followed by every organization in charge of a physical transmission channel. System information allows navigation and access to each of the channels within the Transport Stream, whereas event descriptions give the user content information for browsing and selection.

5.6.2a   Elements of PSIP

PSIP is a collection of hierarchically-associated tables each of which describes particular elements of typical digital TV services [1]. Figures 5.6.1 and 5.6.2 show the different components and the notation used to describe them. The packets of the base tables are all labeled with the base PID (base_PID[2]) which has been chosen as 0x1FFB. The base tables are:

- The System Time Table (STT)

- Rating Region Table (RRT)

- Master Guide Table (MGT)

- Virtual Channel Table (VCT)

A second set of tables are the Event Information Tables (EIT) whose *Packet Identifiers* (PIDs) are defined in the MGT. A third set of tables are the Extended Text Tables (ETT), and similarly, their Packet Identifiers are defined in the MGT.

The System Time Table (STT) is a small data structure that fits in one Transport Stream packet and serves as a reference for time of day. Receivers can use this table as a reference for timing start times of advertised events.

It should be noted that, except for the MGT, PSIP Table Sections may start in any byte position within an MPEG-2 Transport Stream packet. The Master Guide Table is special in that the first byte always is aligned with the first byte of the packet payload. The ATSC A/65 Standard

---

2. ATSC Digital Television Standard documents contains symbolic references to syntactic elements used in the audio, video, and transport coding subsystems. These references usually contain the underscore character (e.g., sequence_end_code) and may consist of character strings that are not English words (e.g., dynrng). In this book they are identified by a distinctive font. Note also that certain syntax names are capitalized; e.g., Transport Stream.

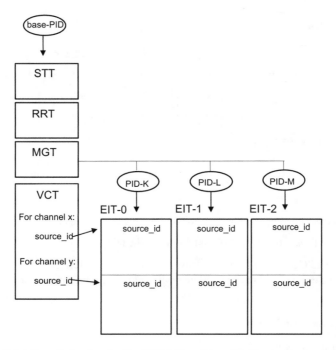

Figure 5.6.1 Overall structure for the PSIP tables. (*From* [1]. *Used with permission.*)

states this restriction as the pointer_field of the Transport Stream packet carrying the table_id field of the MGT section shall have the value 0x00 (section starts immediately after the pointer_field).

In general, Table Sections may span packet boundaries. Also, if the Table Sections are small enough, more than one PSIP Table Section may be present within a single Transport Stream packet. The MPEG-2 pointer_field mechanism is used to indicate the first byte of a Table Section within a packet payload. The starting byte of subsequent Table Sections that might be in the same payload is determined by processing successive section_length fields. The location of the section_length field is guaranteed to be consistent for any type of PSIP Table Section, as the format conforms to MPEG-2 defined Program Specific Information (PSI) tables.

If a specific packet payload does not include the start of a Table Section, then the payload_unit_start_indicator bit in the packet header is set to '0' and the pointer_field is not present.

Transmission syntax for the United States' voluntary program rating system is included in the A/65 standard. The Rating Region Table (RRT) has been designed to transmit the rating standard in use for each country using the standard. Provisions were made for different rating systems for different countries and multi-country regions as well.

The Master Guide Table (MGT) provides general information about all of the other tables that comprise the PSIP standard, specifically it:

• Defines table sizes necessary for memory allocation during decoding

• Defines version numbers to identify those tables that need to be updated

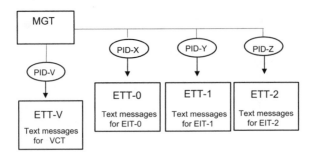

Figure 5.6.2 Extended text tables in the PSIP hierarchy. (*From* [1]. *Used with permission.*)

- Has a constrained header location to facilitate receiver acquisition

- Gives the Packet Identifiers (PIDs) that label the tables

The Virtual Channel Table (VCT), also referred to as the Terrestrial VCT (TVCT), contains a list of all the channels that are or will be on-line plus their attributes. Among the attributes are:

- Channel name

- Navigation identifiers

- Stream components and types

As part of PSIP there are several Event Information Tables, each of which describes the events or TV programs associated with each of the virtual channels listed in the VCT. Each EIT is valid for a time interval of 3 hours. Since the total number of EITs is 128, up to 16 days of programming may be advertised in advance. EIT-0 always denotes the current 3 hrs of programming, EIT-1 the next 3 hrs, and so on. As a minimum, the first four EITs must always be present in every Transport Stream.

Start times for EITs are constrained to be one of the following UTC times: 0:00 (midnight), 3:00, 6:00, 9:00, 12:00 (noon), 15:00, 18:00, and 21:00. Imposing constraints on the start times as well as the interval duration is necessary for the purpose of re-multiplexing. During re-multiplexing, EIT tables coming from several distinct Transport Streams may end up grouped together or vice-versa. If no constraints were imposed, re-multiplexing equipment would have to parse EITs by content in real time, which is a difficult task.

For example, consider a broadcast corporation operating in the Eastern time zone of the U.S. This corporation decides to carry 6 EITs (18 hrs of TV program information). If at present, the Eastern time is 15:30 EDT (19:30 UTC), then the coverage times for the EIT tables are as given in Table 5.6.1.

The abbreviation "nd" denotes *next day*. Before 17:00 EDT, the MGT will list the currently valid PIDs as: 123, 190, 237, 177, 295, and 221. At 17:00 EDT, table EIT-0 will become obsolete while the other ones will remain valid. At that time, the PID list can be changed to 190, 237, 177, 295, 221, maintaining the version number list as 4, 2, 7, 8, 15. Therefore, by simply shifting the listed PID values in the MGT, table EIT-1 can become EIT-0, table EIT-2 can become EIT-1, and so on.

Table 5.6.1 An Example of EIT Coverage Times (*After* [1].)

EIT number	Version Num.	Assigned PID	Coverage (UTC)	Coverage (EDT)
0	6	123	18:00 – 21:00	14:00 – 17:00
1	4	190	21:00 – 24:00	17:00 – 20:00
2	2	237	0:00 – 3:00	20:00 – 23:00
3	7	177	3:00 – 6:00	23:00 – 2:00 (nd)
4	8	295	6:00 – 9:00	2:00 (nd) – 5:00 (nd)
5	15	221	9:00 – 12:00	5:00 (nd) – 8:00 (nd)

However, it is also possible to regenerate one or several EITs at any time for correcting and/or updating the content (e.g., in cases where to be assigned events become known). Regeneration of EITs is flagged by updating version fields in the MGT. For example, if table EIT-2 needs to be updated at 16:17 EDT, then the new table must be transmitted with a version number equal to 3. Whenever the decoder monitoring the MGT detects a change in the version number of a table, it assumes that the table has changed and needs to be reloaded.

As illustrated in Figure 5.6.2, there can be several Extended Text Tables (ETTs), each of them having its PID defined in the MGT. Each Event Information Table (EIT) can have one ETT. Similarly, the Virtual Channel Table can have one ETT. As its name indicates, the purpose of an Extended Text Table (ETT) is to carry text messages. For example, for channels in the VCT, the messages can describe channel information, cost, coming attractions, and so on. Similarly, for an event such as a movie listed in the EIT, the typical message is a short paragraph that describes the movie itself. Extended Text Tables are optional.

5.6.2b   Application Example

For the purpose of this example, we assume that a broadcast group, here denominated NBZ, manages the frequency bands for RF channels 12 and 39 [1]. The first one is its analog channel whereas the second one will be used for digital broadcast. According to the premises established in PSIP Standard A/65, NBZ must carry the PSIP tables in the digital Transport Stream of RF channel 39. The tables must describe TV programs and other services provided on RF channel 39 but can also describe information for the analog RF channel 12.

Table 5.6.2 The First 3-Hour Segment to be Described in VCT and EIT-0 (*After* [1].)

		14:00 – 14:30	14:30 – 15:00	15:00 – 15:30	15:30 – 16:00	16:00 – 16:30	16:30 – 17:00
PTC 12 (12-0)	NBZ	City Life	City Life	Travel Show	Travel Show	News	News
PTC 39 (12-1)	NBZ	City Life	City Life	Travel Show	Travel Show	News	News
PTC 39 (12-2)	NBZ	Soccer	Golf Report	Golf Report	Car Racing	Car Racing	Car Racing
PTC 39 (12-3)	NBZ	Secret Agent	Secret Agent	Lost Worlds	Lost Worlds	Lost Worlds	Lost Worlds
PTC 39 (12-4)	NBZ	Headlines	Headlines	Headlines	Headlines	Headlines	Headlines

Table 5.6.3 Second 3-Hour Segment to be Described in VCT and EIT-1 (*After* [1].)

		17:00 – 17:30	17:30 – 18:00	18:00 – 18:30	18:30 – 19:00	19:00 – 19:30	19:30 –20:00
PTC 12 (12-0)	NBZ	Music Today	NY Comedy	World View	World View	News	News
PTC 39 (12-1)	NBZ	Music Today	NY Comedy	World View	World View	News	News
PTC 39 (12-2)	NBZ	Car Racing	Car Racing	Sports News	Tennis Playoffs	Tennis Playoffs	Tennis Playoffs
PTC 39 (12-3)	NBZ	Preview	The Bandit	The Bandit	The Bandit	The Bandit	Preview
PTC 39 (12-4)	NBZ	Headlines	Headlines	Headlines	Headlines	Headlines	Headlines

Assume that NBZ operates in the Eastern time zone of the U.S., and that the current time is 15:30 EDT (19:30 UTC). NBZ decides to operate in minimal configuration, therefore only the first four EITs need to be transmitted. As explained previously, EIT-0 must carry event information for the time window between 14:00 and 17:00 EDT, whereas EIT-1 to EIT-3 will cover the subsequent 9 hours. Tables 5.6.2 and 5.6.3 describe the scenario for the first 6 hours.

Similar tables can be built for the next 6 hours (for EIT-2 and EIT-3). According to this scenario, NBZ broadcasts four regular digital channels (also called virtual channels and denoted by their major and minor channel numbers), one with the same program as the analog transmission, another for sports, and a third one for movies. The fourth one supports a service displaying headlines with text and images.

The Master Guide Table (MGT)

The purpose of the MGT is to describe everything about the other tables, listing features such as version numbers, table sizes, and Packet Identifiers (PIDs) [1]. Figure 5.6.3 shows a typical Master Guide Table indicating, in this case, the existence in the Transport Stream of a Virtual Channel Table, the Rating Region Table, four EITs, one Extended Text Table for channels, and two Extended Text Tables for events.

The first entry of the MGT describes the version number and size of the Virtual Channel Table. The second entry corresponds to an instance of the Rating Region Table. If some region's policy makers decided to use more than one instance of an RRT, the MGT would list each PID, version number, and size. Notice that the base PID (0x1FFB) must be used for the VCT and the RRT instances as specified in PSIP.

The next entries in the MGT correspond to the first four EITs that must be supplied in the Transport Stream. The user is free to choose their PIDs as long as they are unique in the MGT list of PIDs. After the EITs, the MGT indicates the existence of an Extended Text Table for channels carried using PID 0x1AA0. Similarly, the last two entries in the MGT signal the existence of two Extended Text Tables, one for EIT-0 and the other for EIT-1.

Descriptors can be added for each entry as well as for the entire MGT. By using descriptors, future improvements can be incorporated without modifying the basic structure of the MGT. The MGT is like a "flag table" that continuously informs the decoder about the status of all the other tables (except the STT, which has an independent function). The MGT is continuously monitored at the receiver to prepare and anticipate changes in the channel/event structure. When tables are

MGT			
table_type	**PID**	**version_num.**	**table size**
VCT	0x1FFB (base_PID)	4	485 bytes
RRT – USA	0x1FFB (base_PID)	1	560 bytes
EIT-0	0x1FD0	6	2730 bytes
EIT-1	0x1FD1	4	1342 bytes
EIT-2	0x1DD1	2	1224 bytes
EIT-3	0x1DB3	7	1382 bytes
ETT for VCT	0x1AA0	21	4232 bytes
ETT-0	0x1BA0	10	32420 bytes
ETT-1	0x1BA1	2	42734 bytes

Figure 5.6.3 Contents of the Master Guide Table. (*From* [1]. *Used with permission.*)

changed at the broadcast side, their version numbers are incremented and the new numbers are listed in the MGT. Based on the version updates and on the memory requirements, the decoder can reload the newly defined tables for proper operation.

The Virtual Channel Table (VCT)

Figure 5.6.4 shows the structure of the VCT which essentially contains the list of channels available in the Transport Stream [1]. For convenience, it is possible to include analog channels and even other digital channels found in different Transport Streams.

The field number_of_channels_in_section indicates the number of channels described in one section of the VCT. In normal applications, all channel information will fit into one section. However, there may be rare times when most of the physical channel is used to convey dozens of low-bandwidth services such as audio-only and data channels in addition to one video program. In those cases, the channel information may be larger than the VCT section limit of 1 Kbyte and therefore VCT segmentation will be required.

For example, assuming that a physical channel conveys 20 low-bandwidth services in addition to a TV program, and assuming that their VCT information exceeds 1 Kbyte, then two or more sections may be defined. The first section may describe 12 virtual channels and the second 9 if such a partition leads to VCT sections with less than 1 Kbyte.

A new VCT containing updated information can be transmitted at any time with the version_number increased by one. However, since a VCT describes only those channels from a particular Transport Stream, virtual channels added to the VCT at arbitrary times will not be detected by the receiver until it is tuned to that particular TS. For this reason, it is highly recom-

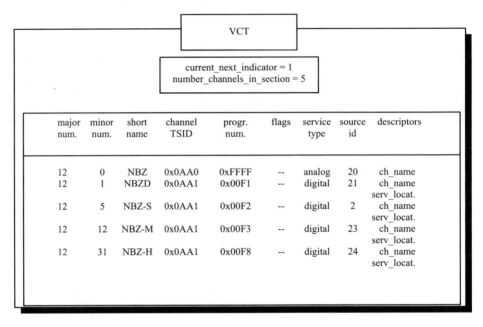

Figure 5.6.4 Content of the Virtual Channel Table. (*From* [1]. *Used with permission.*)

mended that channel addition be made in advance to give the receivers the opportunity to scan the frequencies and detect the channel presence.

The fields major_channel_number and minor_channel_number are used for identification. The first one, the major channel number, is used to group all channels that are to be identified as belonging to a particular broadcast corporation (or particular identifying number such as 12 in this case). The minor channel number specifies a particular channel within the group.

The field short_name is a seven-character name for the channel and may allow text-based access and navigation. The fields transport_stream_id and program_number are included to link the VCT with the PAT and sections of the PMT. A sequence of flags follows these fields. The flags indicate:

1) if the channel is hidden (e.g., for near-video-on-demand, NVOD, applications)

2) if the channel has a long text message in the VCT-ETT

3) if the channel is visible in general or has some conditional access constraints

After the flags, a description of the type of service offered is included, followed by the source_id. The source_id is simply an internal index for representing the particular logical channel. Event Information Tables and Extended Text Tables use this number to provide a list of associated events or text messages, respectively.

Two descriptors are associated with the logical channels in the example. The first one is extended_channel_name and, as its name indicates, it gives the full name of the channel. An example for channel NBZ-S could be: NBZ Sports and Fitness. The other one, the service_location descriptor, is used to list the available bit streams and their PIDs necessary to decode packets at the receiver. Assuming that NBZ-M offers bilingual transmission, then the following attributes are tabulated within its service_location descriptor:

PID_audio_1	AC-3 audio	English
PID_audio_2	AC-3 audio	Spanish
PID_video	MPEG-2 video	No language

Two VCTs may exist simultaneously in a TS: the current and the next VCT. The current VCT is recognized by having the flag current_next_indicator set to 1, while the next one has this flag set to 0. The next VCT should not be transmitted when it fits into a single Table Section, since delivery of the new current table will take effect as soon as the one (and only) section arrives. Multi-sectioned next VCTs may be sent, but should not be delivered until immediately before the point at which they are to become current. This recommendation arises because no mechanism is available to update the next tables without affecting the current table definition.

For multi-sectioned VCTs, delivery of the "next" table is helpful. Consider a TS containing the following Table Sections:

- Current VCT, version_number = 5, section 1 of 2
- Current VCT, version_number = 5, section 2 of 2
- Next VCT, version_number = 6, section 1 of 2
- Next VCT, version_number = 6, section 2 of 2

At the point in time when the "next" tables are to become current, the following Table Section may be placed into the Transport Stream:

- Current VCT, version_number = 6, section 1 of 2

At the moment this Table Section is processed, both sections of the version 6 VCT are understood to be the new "current" VCT, even before section 2 of 2 of VCT version 6 labeled "current" is received. As long as the "next" Table Sections have been cached, they can be taken as "current" as soon as the version number is seen to increment.

When the VCT refers to an analog service type, the channel_TSID cannot refer to the identifier of a "Transport Stream" in the MPEG-2 sense. Analog NTSC broadcast signals can, however, carry a 16-bit unique identifier called a "Transmission Signal Identifier."[3] For the example VCT in Figure 5.6.4, the Transmission Signal Identifier for channel 12-0 is 0x0AA0. Subsequently, receivers are expected to associate the NTSC broadcast identified by the Transmission Signal ID with the frequency tuned to acquire it. Given this association, a receiver can use the Transmission Signal ID to determine how to tune to the NTSC channel it identifies.

It is recommended that the broadcaster insert into the VCT any major-minor channel that would be used to carry any program announced in the EIT. This means if no current program was using 7-7, and if a program 16 days from now was going to use 7-7, that 7-7 would be in the VCT. This would enable receivers to include the channel number in a program guide presented to the consumer. If a program is announced in the EIT and the source ID for that program is not found in the VCT, the receiver cannot determine which "channel" to display for that program.

---

3. The 16-bit Transmission Signal ID for the NTSC VBI is specified in EIA/CEA-608-C [2].

Table 5.6.4 Receiver Behavior with Hidden and Hide Guide Attributes [*After* [1].)

hidden	hide_guide	Receiver Behavior		Description
		Surf	Guide	
0	x	✓	✓	Normal channel
1	1			Special access only
1	0		✓	Inactive channel

Any channels in the VCT which are not currently active must have the hidden attribute set to 1 and the hide_guide attribute set to 0.

Table 5.6.4 shows DTV behavior for the various combinations of the hidden and hide_guide attributes. In the table the "x" entry indicates "don't care." A check in the "surf" column indicates the channel is available by channel surfing and via direct channel number entry. A check in the "guide" column indicates that the channel may appear in the program guide listing.

The Event Information Tables (EITs)

The purpose of an EIT is to list all events for those channels that appear in the VCT for a given time window [1]. As mentioned before, EIT-0 describes the events for the first 3 hours, EIT-1 for the next 3 hours, and so on. EIT-i and EIT-j have different PIDs as defined in the MGT. In PSIP, tables can have a multitude of instances. The different instances of a table share the same table_id value and PID but use different table_id_extension values.

In PSIP, an instance of EIT-k contains the list of events for a single virtual channel with a unique source_id. For this reason, the table_id_extension has been renamed as source_id in the EIT syntax. Figure 5.6.5 shows, for example, the NBZ-S instance for EIT-0. Following similar procedures, the NBZD, NBZ-M, and NBZ-H instances of EIT-0 can be constructed. The process can be extended and repeated to obtain all of the instances for the other tables in the time sequence: EIT-1, EIT-2, and so on.

The three events programmed for the 3-hour period for NBZ-S are listed in Figure 5.6.5. The field event_id is a number used to identify each event. If an event time period extends over more than one EIT, the same event_id has to be used. The event_id is used to link events with their messages defined in the ETT, and therefore it has to be unique only within a virtual channel and a 3-hour interval defined by EITs. The event_id is followed by the start_time and then the length_in_seconds. Notice that events can have start times before the activation time (14:00 EDT in this example) of the table. The ETM_location specifies the existence and the location of an Extended Text Message (ETM) for this event. ETMs are simply long textual descriptions. The collection of ETMs constitutes an Extended Text Table (ETT).

An example of an ETM for the car racing event might be:

> "Live coverage from Indianapolis. This car race has become the largest single-day sporting event in the world. Two hundred laps of full action and speed."

Several descriptors can be associated with each event. One is the Content Advisory Descriptor which assigns a rating value according to one or more systems. Recall that the actual rating system definitions are tabulated within the RRT. Another is a closed caption descriptor, which signals the existence of closed captioning and lists the necessary parameters for decoding.

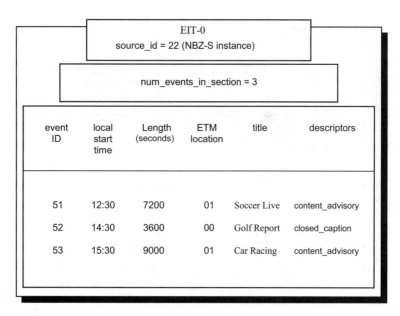

Figure 5.6.5 Content of EIT-0 for NBZ-S. (*From* [1]. *Used with permission.*)

The Rating Region Table (RRT)

The Rating Region Table is a fixed data structure in the sense that its content remains mostly unchanged [1]. It defines the rating standard that is applicable for each region and/or country. The concept of table instance introduced in the previous section is also used for the RRT. Several instances of the RRT can be constructed and carried in the TS simultaneously. Each instance is identified by a different table_id_extension value (which becomes the rating_region in the RRT syntax) and corresponds to one and only one particular region. Each instance has a different version number, which is also carried in the MGT. This feature allows updating each instance separately.

Figure 5.6.6 shows an example of one instance of an RRT, for a region called Tumbolia, assigned by the ATSC to rating_region 20. Each event listed in any of the EITs may carry a Content Advisory Descriptor. This descriptor is an index or pointer to one or more instances of the RRT.

Packetization and Transport

In the previous sections, we have described how to construct the MGT, VCT, RRT, and EITs based on the typical scenario described in Tables 5.6.1 and 5.6.2 [1]. The number of virtual channels described in the VCT is 5 and therefore, each EIT will have 5 instances.

For the example, the size of the MGT is less than a hundred bytes and the VCT ranges between 300 to around 1500 bytes depending on the length of the text strings. Similarly, each EIT instance can have from 1 to about 3 Kbytes depending again on the text length.

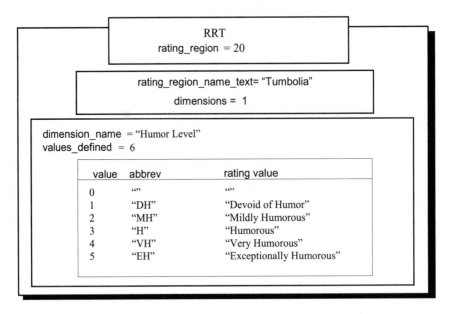

Figure 5.6.6 An instance of a Rating Region Table (RRT). (*From* [1]. *Used with permission.*)

Typically, the MGT, STT, VCT, and each instance of the RRT and EIT will have one or at most a few sections. For each table, the sections are appended one after the other, and then segmented into 184-byte packets. After adding the 4-byte MPEG-2 TS header, the packets are multiplexed with the others carrying audio, video, data, and any other components of the service. Figure 5.6.7 illustrates this process.

5.6.2c   Tuning Operations and Table Access

As described by the PSIP protocol, each TS will carry a set of tables describing system information and event description [1]. For channel tuning, the first step is to collect the VCT from the Transport Stream that contains the current list of services available. Figure 5.6.8 illustrates this process.

Once the VCT has been collected, a user can tune to any virtual channel present in the Transport Stream by referring to the major and minor channel numbers. Assuming that in this case, the user selects channel 5-11, then the process for decoding the audio and video components is shown in Figure 5.6.9.

For terrestrial broadcast, the existence of a Service Location Descriptor in the TVCT is mandatory. The PID values needed for acquisition of audio and video elementary streams may be found in either a service_location_descriptor() within a TVCT, or in a TS_program_map_section(). The service_location_descriptor() has been included in PSIP to minimize the time required for changing and tuning to channels. However, PAT and PMT information is required to be present in the Transport Stream to provide MPEG-2 compliance. Access to data or other supplemental

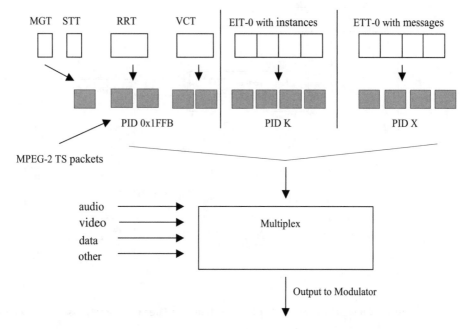

Figure 5.6.7 Packetization and transport of the PSIP tables. (*From* [1]. *Used with permission.*)

services may require access to the PAT or TS_program_map_section(). Cable systems may or may not carry the Service Location Descriptor, and the information contained therein will be found in the TS_program_map_section().

The PMT should always be processed and monitored for changes because, in some instances, the structure of a service may exceed the capability of the service_location_descriptor() to describe it. One example is a service that includes multiple audio tracks for the same language. In this case, the TS_program_map_section() carries the textual name for each of the tracks to help in user selection.

5.6.2d   GPS Time

The System Time Table provides time of day information to receivers [1]. In PSIP, time of day is represented as the number of seconds that have elapsed since the beginning of GPS time, 00:00:00 UTC January 6, 1980. GPS time is referenced to the master clock at the US Naval Observatory and steered to Coordinated Universal Time (UTC). UTC is the time source we use to set our clocks.

UTC is occasionally adjusted by one-second increments to ensure that the difference between a uniform time scale defined by atomic clocks does not differ from the Earth's rotational time by more than 0.9 seconds. The timing of occurrence of these leap seconds is determined by careful observations of the Earth's rotation; each is announced months in advance. On the days it is scheduled to occur, the leap second is inserted just following 12:59:59 p.m. UTC.

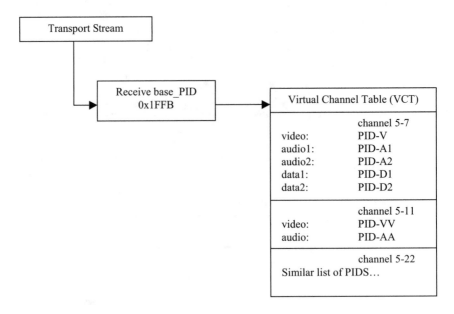

Figure 5.6.8 Extraction of the VCT from the Transport Stream. (*From* [1]. *Used with permission.*)

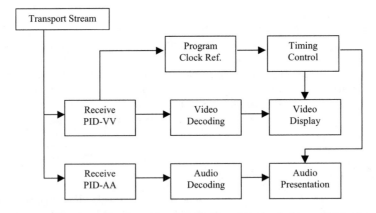

Figure 5.6.9 Acquisition of audiovisual components. (*From* [1]. *Used with permission.*)

UTC can be directly computed from the count of GPS seconds since January 6, 1980, by subtracting from it the count of leap seconds that have occurred since the beginning of GPS time.

In the A/65 protocol, times of future events (such as event start times in the EIT) are specified the same as time of day, as the count of seconds since January 6, 1980. Converting an event start time to UTC and local time involves the same calculation as the conversion of system time to local time. In both cases, the leap seconds count is subtracted from the count of GPS seconds to derive UTC.

GPS time is used to represent future times because it allows the receiver to compute the time interval to the future event without regard for the possible leap second that may occur in the meantime. Also, if UTC were to be used instead, it would not be possible to specify an event time that occurred right at the point in time where a leap second was added. UTC is discontinuous at those points.

Around the time a leap second event occurs, program start times represented in local time (UTC adjusted by local time zone and [as needed] daylight savings time) may appear to be off by plus or minus one second. PSIP generating equipment may use one of two methods to handle leap seconds:

- **Method A:** PSIP generating equipment does not anticipate the future occurrence of a leap second. In this case, prior to the leap second, program start times will appear correct. An event starting at exactly 10 a.m. will be computed as starting at 10:00:00. But just following the leap second, that same event time will be computed as 9:59:59. The PSIP generating equipment should re-compute the start times in all the EITs and introduce the leap second correction. Once that happens, and receivers have updated their EIT data, the computed time will again show as 10:00:00. In this way the disruption can be limited to a matter of seconds.

- **Method B:** PSIP generating equipment does anticipate the occurrence of a leap second, and adjusts program start times for events happening after the new leap second is added. If the leap second event is to occur at midnight tonight, an event starting at 10 a.m. tomorrow will be computed by receiving equipment as starting at 10:00:01.

For certain types of events, the precision of method B is necessary. By specifying events using a time system that involves no discontinuities, difficulties involving leap seconds are avoided. Events such as program start times do not require that level of precision. Therefore, method A works well.

Consider the following example. Times are given relative to UTC, and would be corrected to local time zone and daylight savings time as necessary.

- Time of day (UTC): 1:00 p.m., December 30, 1998

- Event start time (UTC): 2:00 p.m., January 2, 1999

- A leap second event will occur just after 12:59:59 p.m. on December 31, 1998

- Leap seconds count on December 30 is 12

The data in the System Time message is:

- GPS seconds = 599,058,012 = 0x23B4E65C

- GPS to UTC offset = 12

Using method A (upcoming leap second event is not accounted for):

- Event start time in EIT: 599,320,812 = 0x23B8E8EC

- Converted to UTC: 2:00:00 p.m., January 2, 1999

- Number of seconds to event: 262,800 = 73 hours, 0 minutes, 0 seconds

Using method B (upcoming leap second event is anticipated):

- Event start time in EIT: 599,320,813 = 0x23B8E8ED

- Converted to UTC: 2:00:01 p.m., January 2, 1999

- Number of seconds to event: 262,801 = 73 hours, 0 minutes, 1 second

Note that using method B, the number of seconds to event is correct, and does not need to be recomputed when the leap seconds count moves from 12 to 13 at year-end.

5.6.2e   Interpretation of MGT Table Version Numbers

On first glance, it may appear that the MGT simply provides the version numbers for Table Sections that make up the EIT/ETT tables for each timeslot. For example, the MGT may indicate a table_type_version_number of 5 for a table_type value of 0x0100 (EIT-0), which could lead one to say EIT-0 is at version 5. In fact, the MGT does give table version information for all transmitted tables, but a careful and correct interpretation of the data provided, including table_type_PID, must be made to avoid errors in processing.

The proper interpretation of table_type_version_number is to consider it to reflect the version_number field in the referenced table. In accordance with the MPEG-2 Systems Standard, the scope of table version_number is limited to Table Sections delivered in transport packets with a common PID value. For example, for Table Sections with a given value of table_ID, a Table Section delivered in transport packets with PID value 0x1E00 and version_number 6 must be interpreted as a separate and distinct table from a Table Section delivered in transport packets with PID value 0x1E01 and version_number 6.

The following example is designed to illustrate the distinction between the simple (incorrect) interpretation and the correct one. In the illustration, the incorrect interpretation leads to processing errors, which involve re-loading tables that have not in fact changed, or (more seriously) not updating tables that *have* changed.

For the following example, the time zone offset is 0. Each EIT table instance is associated with a separate PID (as per A/65 rules).

1)   Assume that it is noon. From noon to 3:00 p.m. the following is true:

a) The EIT describing noon to 3:00 p.m. is in PID[4] 0x1000; version number is 0

b) The EIT describing 3:00 p.m. to 6:00 p.m. is in PID 0x1001; version number is 1

c) The EIT describing 6:00 p.m. to 9:00 p.m, is in PID 0x1002; version number is 0

d) The EIT describing 9:00 p.m. to midnight is in PID 0x1003; version number is 0

e) The MGT is at version 7 and indicates:

- EIT-0, PID 0x1000, version number 0

- EIT-1, PID 0x1001, version number 1

- EIT-2, PID 0x1002, version number 0

- EIT-3, PID 0x1003, version number 0

---

4.   The expression in PID as used here is a shorthand way of saying that the indicated Table Section is carried in transport packets with a PID value equal to the indicated value.

2) The time moves to 3:00 p.m., crossing a timeslot boundary. Assume the EIT describing 6:00 p.m. to 9:00 p.m. is changed now too.

   a) The EIT for noon to 3:00 p.m. is no longer sent, since its time has passed

   b) The EIT for 3:00 p.m. to 6:00 p.m. is still in PID 0x1001; version number is still 1

   c) The EIT for 6:00 p.m. to 9:00 p.m. is still in PID 0x1002; but its content changed at the same time, so its version number is moved to 1

   d) The EIT for 9:00 p.m. to midnight is still in PID 0x1003; version number is still 0

   e) MGT moves to version 8 and indicates:

   - EIT-0, PID 0x1001, version number 1

   - EIT-1, PID 0x1002, version number 1

   - EIT-2, PID 0x1003, version number 0

What is now EIT-0 did not change. What is now EIT-1 *did* change.

For this case, if the MGT is interpreted to give the version numbers of EIT-*n* for each value of *n*, the receiver will see the version of EIT-0 change from 0 to one and refresh it. It will decide the version of EIT-1 has not changed, and not refresh it. But both inferences are incorrect: in this example, EIT-0 has not changed, and EIT-1 *has* changed.

The correct interpretation involves processing version numbers with respect to the associated PID values. Looking at the same example, the MGT indicates that the table associated with PID 0x1001 did not change versions. Likewise, the table associated with PID value 0x1002 changed from version 0 to 1 and should be refreshed.

5.6.2f   Use of Analog Transmission Signal ID

The Virtual Channel Table in PSIP associates a user-friendly definition of a service (a channel name and number) with the physical location of that service [1]. Both digital and analog services are accommodated. For digital services, the Transport Stream ID (TSID) parameter defined in ISO/IEC 13818-1 (MPEG-2 Systems) is used as a unique identifier at the TS level. For analog services, an identifier called the Transmission Signal ID (the acronym is also TSID) may be used.

The analog TSID, like its digital counterpart, is a 16-bit number that uniquely identifies the NTSC signal within which it is carried. EIA/CEA-608-C [2] defines the data format for carriage of the Transmission Signal ID within eXtended Data Service (XDS) packets in the NTSC vertical blanking Interval.

In the U.S., the DTV system is designed with the expectation that the analog TSID will be included in any NTSC broadcast signal referenced by PSIP data. Whenever PSIP data provides a reference to an analog service, the receiver is expected to use that service's analog TSID to make a positive identification. The receiver is expected to not associate any channel or program information data with an NTSC service that does not broadcast its analog TSID.

5.6.2g   Use of Component Name Descriptor

The component_name_descriptor() provides a mechanism to associate a multilingual textual label with an Elementary Stream component of any MPEG-2 Program. If the program consists of one video stream and one audio track, such a label would not give much value. A program may be offered multilingually, for example with separate French and English tracks. In that case, a receiving device may choose, without need for user intervention, the track corresponding to the language set up as the user's preferred language.

It may be, however, that the service happens to have two English-language audio tracks of the same audio type (for example both may be Complete Main audio tracks). In another case, one or more of the audio tracks may not be associated with a spoken language. An example of such a track, sometimes called *clean effects*, is ambient sound such as crowd noise from a sporting event. In both of these cases, use of the component_name_descriptor() is mandatory by the rules established in the A/65 Standard. The net result is that a display device will always have sufficient information to either choose an audio track by its language, by its type, or will have text describing each track that can be used to create an on-screen user dialog to facilitate the user's choice.

5.6.3   Directed Channel Change

A Directed Channel Change request is a trigger event sent within the PSIP stream of the DTV multiplex that will cause a DCC-capable DTV reference receiver (DCCRR) to select a different virtual channel from that to which it is already tuned [3]. Depending upon the kind of DCC request, the change to a different virtual channel can occur without intervention by the viewer, if the viewer has enabled the capability by providing required information during a setup process or during operation of the display. Alternatively, the change can take place under direct control of the viewer, for instance using a remote control device.

To enable the automatic operation of a DCC request within a DCCRR, the DTV viewer will be required to provide information to the display system. This may be accomplished through an interactive setup session or may be done during operation of the receiver, as different viewing options become available. The information provided by the viewer to the DCCRR will permit the unit to determine which, if any, alternate virtual channel the DCCRR should display upon receipt of a DCC request. This selection will take place through the DCCRR matching the viewer information with categorization information or other selection criteria sent by the broadcaster or system operator. There are also forms of DCC request that enable real time viewer selections among alternate program streams, such as, for example, alternate camera views of a sporting event.

5.6.3a   DCC Switch Criteria

A switch from a currently viewed virtual channel to another virtual channel may be accomplished upon occurrence of any of the following selection criteria, some of which may be used in combination [3]:

• Unconditional switch

• Postal code, zip code, or location code

- Program Identifier
- Demographic category
- Content subject category
- Authorization level
- Content advisory value
- One of eight user categories

In addition to the criteria listed above that serve to include groups of viewers into a DCC request, several of the criteria may also be used to exclude viewers from inclusion in a request. For example, instead of listing many zip codes for inclusion of viewers in a DCC request, the inverse logic may be used to specify the group of all viewers not included within a group of zip codes.

Two other DCC request actions have also been defined: an action to be taken upon a viewer switching away from a channel, and an action to be taken upon a viewer switching into a channel.

The broadcaster may specify more than one type of selection criteria within a single DCC request. For example, it is possible to specify several individual zip codes by employing the loop structure within the DCC Table.

Unconditional Switch

An unconditional switch would cause all viewers' channel, regardless of any other DCC selection criteria selected within the DCCRR, to switch to a specified virtual channel [3]. A potential use of this criteria would be to aggregate viewers on different virtual channels to a single channel.

Postal Code, Zip Code, or Location Code

A channel change based upon the viewer's location may be accomplished by using one or more of these criteria [3]. Broadcasters may use these capabilities to provide targeted programming based upon a viewer's location within the viewing area.

Program Identifier

A channel change based upon a program's episode and version number may be accomplished with this criteria [3]. Use of this function would permit a broadcaster to direct a viewer's attention to a broadcast of a particular program having a certain episode and version number. If the viewer had previously enabled a function within the DCCRR that would "remember" a particular program's episode/version number, the viewer could be directed to that program again upon detection of this criteria within the multiplex.

Demographic Category

A Directed Channel Change may be accomplished using demographic categories as a switching criteria [3]. Demographics such as age group and gender can be selected.

Content Subject Category

A channel change based upon the subject matter of the content of the program can be accomplished [3]. Nearly 140 categories of subject matter have been tabulated that can be assigned to describe the content of a program. A broadcaster may use this category of DCC request switching to direct a viewer to a program based upon the viewer's desire to receive content of that subject matter. Although nearly 140 subject categories have been identified for inclusion in the current version of the specification, additional categories may be determined in the future and may be transmitted to the DCCRR through a table revision mechanism (the Directed Channel Change Selection Code Table).

Authorization Level

A channel change may be accomplished using this mechanism in the event that the viewer attempts to switch to a virtual channel that he or she is not authorized to view [3]. This category of DCC would permit the DCCRR to be directed to an alternative channel (such as a barker channel informing the viewer of ineligibility to view the channel) instead of the channel the viewer attempted to tune.

Content Advisory Level

This category of DCC is similar to that described for the Authorization Level but would redirect the viewer to another channel based on the content advisory level in the DCCRR [3].

User Specified Category

This category of DCC would allow a broadcaster to specify one of eight classifications of a program so that if a viewer pressed one of eight viewer-direct-select buttons on a remote control, he or she would be directed to a virtual channel airing a program having that classification [3]. This function would permit a broadcaster to define classifications not anticipated by this standard and then permit the viewer to be directed to programs or segments having those classifications.

Arriving/Departing Request

Two additional Directed Channel Change request actions are specified [3]. A Departing Request descriptor may be used within a DCC to signal the occurrence of a departing request or to cause a text box to appear for a definable amount of time prior to performing a channel change requested by the viewer. The text box may be used by the broadcaster to present information to the viewer, such as plot elements remaining in the program or upcoming segment schedule information.

An Arriving Request descriptor may be used to signal the occurrences of an arriving request or cause a text box to appear for a definable amount of time upon arrival at a newly tuned virtual channel. For example, the text box may be used by the broadcaster to convey story line information up to this point in time, or even program schedule change or preemption information such as a program delayed announcement due to the previous or current program running long.

5.6.4   References

1.   ATSC: "Program and System Information Protocol for Terrestrial Broadcast and Cable—Revision B," Advanced Television Systems Committee, Washington, D.C., Doc. A/65B, March 18, 2003.

2.   EIA/CEA-608-C: "Line 21 Data Service," Electronic Industries Alliance, Arlington, VA.

3.   ATSC: "Recommended Practice: Program and System Information Protocol Implementation Guidelines for Broadcasters," Advanced Television Systems Committee, Washington, D.C., Doc. A/69, June 25, 2002.

5.7

Multiple Transmitter Networks

5.7.1   Introduction[1]

Many of the challenges of RF transmission, especially as they apply to digital transmission, can be addressed by using multiple transmitters to cover a service area [1]. Because of the limitations in the spectrum available, many systems based on the use of multiple transmitters must operate those transmitters all on the same frequency, hence the name *single frequency network* (SFN). At the same time, use of SFNs leads to a range of additional complications that must be addressed in the design of the network.

SFNs for single-carrier signals such as 8-VSB become possible because of the presence of adaptive equalizers in receivers. When signals from multiple transmitters arrive at a receiver, under the right conditions, the adaptive equalizer in that receiver can treat the several signals as echoes of one another and extract the data they carry. The conditions are controlled by the capabilities of the adaptive equalizer and will become less stringent as the technology of adaptive equalizers improves over time.

5.7.1a   Benefits of Multiple Transmitters

A number of benefits may accrue to the use of multiple transmitters to cover a service area [1]. Among these can be the ability to obtain more uniform signal levels throughout the area being served and the maintenance of higher average signal levels over that area. These results come from the fact that the average distance from any point within the service area to a transmitter is reduced. Reducing the distance also reduces the variability of the signal level with location and time and thereby reduces the required fade margin needed to maintain any particular level of reliability of service. These reductions, in turn, permit operation with less overall *effective radiated power* (ERP) and/or antenna height.

---

1. This chapter is based on: ATSC, "ATSC Recommended Practice: Design Of Synchronized Multiple Transmitter Networks," Advanced Television Systems Committee, Washington, D.C., Doc. A/111, September 3, 2004. Readers are encouraged to download the entire Recommended Practice from the ATSC Web site (http://www.atsc.org). All ATSC Standards, Recommended Practices, and Information Guides are available at no charge.

When transmitters can be operated at lower power levels and/or elevations, the interference they cause to their neighbors is reduced. Using multiple transmitters allows a station to provide significantly higher signal levels near the edge of its service area without causing the level of interference to its neighbor that would arise if the same signal levels were delivered from a single, central transmitter. The interference reductions come from the significantly smaller interference zones that surround transmitters that use relatively lower power and/or antenna heights.

With the use of multiple transmitters comes the ability to overcome terrain limitations by filling in areas that would otherwise receive insufficient signal level. When the terrain limitations are caused by obstructions that isolate an area from another (perhaps the main) transmitter, advantage may be taken of the obstructions in the design of the network. The obstructions can serve to help isolate signals from different transmitters within the network, making it easier to control interference between the network's transmitters. When terrain obstructions are used in this way, it may be possible to place transmitters farther apart than if such obstructions were not utilized for isolation.

Where homes are illuminated by sufficiently strong signals from two or more transmitters, it may be possible to take advantage of the multiple signals to provide more reliable indoor reception. When a single transmitter is used, standing waves within a home sheathed in metal likely will result in areas within that home having signal levels too low to use. Signals arriving from different directions will enter the resonant cavity of the home through different ports (windows) and set up standing waves in different places. The result often may be that areas within the home receiving low signal levels from one transmitter will receive adequate signal levels from another transmitter, thereby making reliable reception possible in many more places within the home.

5.7.1b   Limitations of SFNs

While they have the potential to solve or at least ameliorate many of the challenges of RF transmission, SFNs also have limitations of their own [1]. Foremost among these is there will be interference between the signals from the several transmitters in a network. This "system-internal" interference must be managed so as to bring it within the range of capabilities of the adaptive equalizers of the largest number of receivers possible. Where the interference falls outside the range that can be handled by a given adaptive equalizer, other measures, such as the use of an outdoor directional antenna, must be applied.

The characteristics of adaptive equalizers that are important for single frequency network design are:

• The ability to deal with multiple signals (echoes) with equal signal levels.

• The length of echo delay time before and after the main (strongest) signal that can be handled by the equalizer (or the total *delay spread* of echoes that can be handled when an equalizer design does not treat one of the signals as the main signal).

• The interfering signal level below which the adaptive equalizer is not needed because the interference is too low in level to prevent reception.

The last of these characteristics, which is somewhat lower in level than the co-channel interference threshold, determines the areas where the performance of the adaptive equalizer matters. In places with echo interference below the point at which the adaptive equalizer needs to correct the signal, it is not necessary in design of the SFN to consider the differential arrival times of the signals from the various transmitters. In places with echo interference above that threshold, the arrival times of the signals matter if the adaptive equalizer is to be able to correct for the echo interference. Differential delays between signals above the echo interference threshold must fall within the time window that the adaptive equalizer can correct if the signals are to be received.

Current receiver designs have a fixed time window inside which echoes can be equalized. The amplitudes of correctable echoes also are a function of their time displacement from the main signal. The closer together the signals are in time, the closer they can be in amplitude. The further apart they are in time, the lower in level the echoes must be for the equalizer to work. These relationships are improving dramatically in newer receiver front-end designs, and they can be expected to continue improving at least over the next several generations of designs. As they improve, limitations on SFN designs will be reduced.

The handling of signals with equal signal level echoes is a missing capability in early receiver front-end designs, but it is now recognized as necessary for receivers to work in many situations that occur naturally, without even considering their generation in SFNs. The reason for this is that any time there is no direct path from the transmitter to the receiver, the receiver will receive all of its input energy from reflections of one sort or another. When this happens, there may be a number of signals (echoes) arriving at the receiver that are about equal in amplitude, and they may vary over time, with the strongest one changing from time-to-time. This is called a Rayleigh channel when it occurs, and it is now recognized that Rayleigh channels are more prevalent than once thought. For example, they often exist in city canyons and mid-rise areas, behind hills, and so on. They also exist in many indoor situations. If receivers are to deal with these cases, adaptive equalizers will have to be designed to handle them. Thus, SFNs will be able to take advantage of receiver capabilities that are needed in natural circumstances.

Radio frequency signals travel at a speed of about 3/16 mile per microsecond. Another way to express the same relationship is that radio frequency signals travel a mile in about 5-1/3 μs. If a pair of transmitters emits the same signal simultaneously and a receiver is located equidistant from the two transmitters, the signals will arrive at the receiver simultaneously. If the receiver is not equidistant from the transmitters, the arrival times at the receiver of the signals from the two transmitters will differ by 5-1/3 μs for each mile of difference in path length. In designing the network, the determination of the sizes of the cells and the related spacing of the transmitters will be dependent on this relationship between time and distance and on the delay spread capability of the receiver adaptive equalizer.

Because receivers have limited delay-spread capability, there is a corresponding limit on the sizes of cells and spacing of transmitters in SFNs. As receiver front-end technology improves over time, this limitation can be expected to be relaxed. As the limitation on cell sizes is relaxed and cells become larger, it can be helpful to network design to adjust the relative emission times of the transmitters in the network. This allows putting the locus of equidistant points from various transmitters where needed to maximize the audience reached and to minimize internal interference within the network. When such time offsets are used, it becomes desirable to be able to measure the arrival times at receiving locations of the signals from the transmitters in the network. Such measurements can be difficult since the transmitters are intentionally transmitting exactly the same signals in order to allow receivers to treat them as echoes of one another. Somehow the transmitters have to be differentiated from one another if their respective contributions to the aggregate signal received at any location are to be determined. To aid in the identification of individual transmitters in a network, a buried spread spectrum pseudorandom "RF Watermark" signal is included in ATSC Standard A/110, "Synchronization Standard for Distributed Transmission" [2].

Table 5.7.1 Comparison of Different Distributed Transmitter Networks (*After* [1])

SFN Configuration	Distributed Transmitter	DOCR	Distributed Translator
Complexity/cost	High	Low	Low
Requirement for feeder link	Yes	No	No
Delay adjustment capability	Yes	No	Yes
Output power level	No limit	Low/moderate	No limit
DTV RF channels required	One	One	At least two
Recommended implementation	Large area SFN	Cover small area: gap filler, and coverage extender	Large area SFN, gap filler, and coverage extender

5.7.2   Single Frequency Network Concepts

Table 5.7.1 lists the most common forms of distributed transmitter networks and some of their primary characteristics.

5.7.2a   Digital On-Channel Repeaters

In block diagram form, a *digital on-channel repeater* (DOCR) looks just like a translator or booster (Figure 5.7.1) [1]. It consists of a receiving antenna, a receiver, some signal processing, a transmitter, and a transmitting antenna. The DOCR is closest in nature to a boosters in that it receives the off-the-air DTV signal, processes it, and retransmits it on the same frequency. This greatly simplifies the single frequency network, but also leads to limitations in how the system can be applied. For example, due to the internal signal processing and filtering delays of a DOCR, the signal from the main transmitter will arrive at the DOCR coverage area first, which acts as a *pre-echo*, relative to the output signal from the DOCR. This type of multipath distortion is very harmful to reception by ATSC legacy receivers. With the anticipated performance improvements of newer receivers, this problem is expected to be reduced or eliminated. The main advantage for the DOCR is its simplicity and low cost. With a DOCR, there is no need for a separate *studio-to-transmitter link* (STL).

There are several configurations that can be used in a DOCR. Shown in Figures 5.7.2 through 5.7.5, they can be designated as:

- RF processing DOCR (Figure 5.7.2)

- IF processing DOCR (Figure 5.7.3)

- Baseband decoding and re-generation DOCR (Figure 5.7.4)

- Baseband demodulation and equalization DOCR (Figure 5.7.5)

Figure 5.7.2 shows the simplest form of DOCR, the RF processing DOCR, in which the receiver comprises a preselector and low-level amplifier, the signal processing comprises an RF bandpass filter at the channel of operation, and the transmitter comprises a power amplifier. There is no frequency translation of any sort in this arrangement. It has the shortest processing delay of any of the DOCR configurations—usually a fraction of a microsecond. This configuration, however, has very limited first adjacent channel interference rejection capability. Such a configuration can result in the generation of intermodulation products in the amplifier and in degradation of the re-transmitted signal quality. Meanwhile, due to limited isolation (i.e., too much coupling) between transmitting and receiving antennas, the re-transmitted signal could

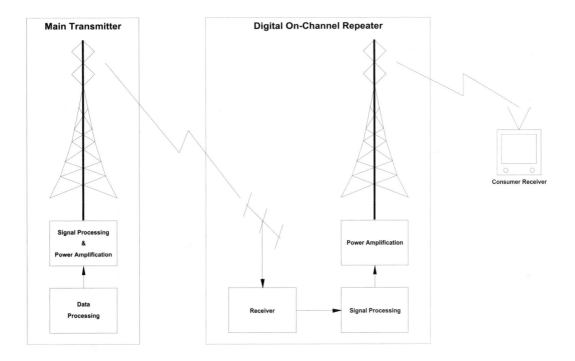

Figure 5.7.1 Digital on-channel repeater (DOCR) generic block diagram. (*From* [1]. *Used with permission.*)

loop back and re-enter the receiving antenna. This can also degrade the signal quality, causing *spectrum ripple* and other distortions. The only way to limit or avoid the signal loopback in this type of DOCR is to increase antenna isolation, which is determined by the site environment, or to limit the DOCR output power. Usually, the RF processing DOCR transmitter output power is less than 10 W, resulting in an effective radiated power (ERP) on the order of several dozen watts.

Conversion to an intermediate frequency (IF) for signal processing is the principal feature that differentiates Figure 5.7.3 from the RF processing DOCR. In this arrangement, a local oscillator and mixer are used to convert the incoming signal to the IF frequency, where it can be more easily amplified and filtered. The same local oscillator used for the downconversion to IF in the receiver can be used for upconversion in the transmitter, resulting in the signal being returned to precisely the same frequency at which it originated (with some amount of the local oscillator phase noise added to the signal). The delay time through the IF processing DOCR will be decided mostly by the IF filter implemented. A SAW filter can have much sharper passband edges, better control of envelope delay, greater attenuation in the stopband, and generally more repeatable characteristics than most other kinds of filters. Transit delay time for the signal can be on the order of 1–2 μs, so the delay through an IF processing DOCR, Figure 5.7.3, will be from a fraction of a microsecond to about 2 μs—somewhat longer than the RF processing DOCR. The IF processing DOCR has better first adjacent channel interference rejection capability than does

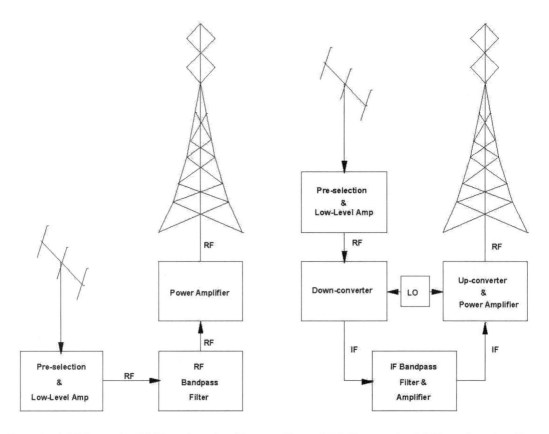

Figure 5.7.2 RF Processing DOCR configuration. (*From* [1]. *Used with permission.*)

Figure 5.7.3 IF processing DOCR configuration. (*From* [1]. *Used with permission.*)

the RF processing DOCR, but it retains the signal loopback problem, which limits its output power.

Figure 5.7.4 shows a receiver that demodulates the incoming signal to a digital baseband signal in which *forward error correction* (FEC) can be applied. This restores the bit stream to perfect condition, correcting all errors and eliminating all effects of the analog channel through which the signal passed in reaching the DOCR. The bit stream then is transmitted, starting with formation of the bit stream into the symbols in an exciter, just as in a normal transmitter. If no special steps are taken to set the correct trellis encoder states, the output of the DOCR of Figure 5.7.4 would be incoherent with respect to its input. This would result in the signal from such a repeater acting like noise when interfering with the signal from another transmitter in the network rather than acting like an echo of the signal from that other transmitter. Thus, additional data processing is required to establish the correct trellis states for retransmission. It should also be noted that this form of DOCR has a very long delay through it, measured in milliseconds, mostly caused by the de-interleaving process. This delay is well outside the ATSC receiver equalization range. Although by regenerating the DTV signal it totally eliminates the adjacent channel

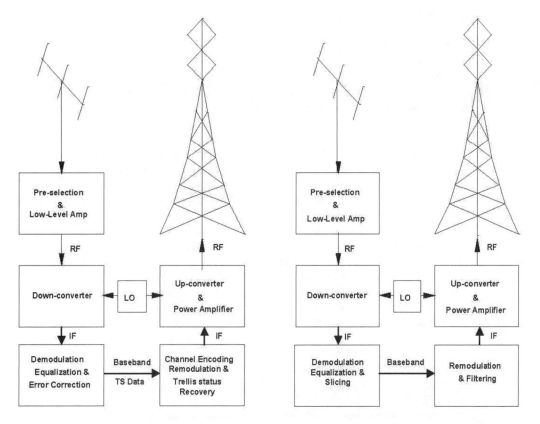

Figure 5.7.4 Baseband decoding DOCR configuration. (*From* [1]. *Used with permission.*)

Figure 5.7.5 Baseband equalization DOCR configuration. (*From* [1]. *Used with permission.*)

interference and signal loopback problems, this type of DOCR has very little practical use, unless the intended DOCR coverage area is totally isolated from the main DTV transmitter.

A more practical intermediate method is the baseband equalization DOCR (or EDOCR) as shown in Figure 5.7.5. It fits between the techniques of Figures 5.7.2 or 5.7.3 and 5.7.4. This type of DOCR demodulates the received signal and applies adaptive equalization in order to reduce or eliminate adjacent channel interference, signal loopback, and multipath distortion occurring in the path from the main transmitter to the EDOCR. In determining the correct 3-bit symbol levels, it also carries out symbol level slicing or trellis decoding, which can achieve several dB of noise reduction and minimizes the impact of channel distortions. The baseband output of the equalizer and slicer is re-modulated, filtered, frequency shifted, and amplified for re-transmission. The delay of the baseband processing is in the order of a few dozen VSB symbol times. The total EDOCR internal delay is in the order of a few microseconds once the time delays of the adaptive equalizer and the pulse shaping (*root raised cosine*) filters (one each for receive and transmit) are taken into account. This amount of delay could have an impact on ATSC legacy receivers. The baseband equalization DOCR allows retransmission of a clean signal without the lengthy delays inherent in the baseband decode/regeneration method of Figure 5.7.4. It can also transmit at higher power than the RF and IF processing DOCRs. The shortcoming of this method is that since there is not complete error correction, any errors that occur in interpretation of the received data will be built into the retransmitted signal. This makes it important in designing the EDOCR installation to include sufficient receiving antenna gain and received signal level that errors are minimized in the absence of error correction. If a fairly clean received signal cannot be obtained, it may be better to use the RF or IF processing DOCR and allow viewers' receivers to apply full error correction to the relayed signal rather than retransmitting processed signals that contain data errors.

All of the forms of DOCR described have a number of limitations in common. Most of the limitations arise from the facts that a DOCR receives and re-transmits on the same frequency, and that obtaining good isolation between transmitting and receiving antennas is a difficult proposition. The result is coupling from the DOCR output back to its input. This coupling leads to feedback around the amplifiers in the DOCR. Such feedback can result in oscillation in the cases of the RF and IF processing DOCR. Short of oscillation, it can result in signal distortions in amplitude and group delay ripple similar to those suffered in a propagation channel. These two designs will also suffer from the accumulation of noise along the cascade of transmitters and propagation channels from the signal source to the ultimate receiver. The application of adaptive equalizers to the feedback path around the DOCR holds some promise to mitigate the distortions, but it cannot help with the noise accumulation.

The feedback around a DOCR puts a limitation on the power that can be transmitted by such a device. A margin must be provided to keep the system well below the level of oscillation, and the point of oscillation will be determined by the isolation between the transmitting and receiving antennas. All of this tends to make high power level operation of a DOCR problematic.

Similarly, the time delay through a DOCR is significant for network design. As one goes from design-to-design from Figure 5.7.2 to 5.7.3, 5.7.5, and then 5.7.4, the time delay gets longer. The time delay determines over what area the combination of signals from the transmitters in the network stay within the capability of the receiver adaptive equalizer to correct the apparent echoes caused by receiving signals from multiple transmitters. The geometry between the source transmitter, the DOCR, and the receiver determine the delay spread actually seen by the receiver. To this must be added the delay of the DOCR. Additional delay in the DOCR can only push the delay spread in the wrong direction (extending pre-echo), further limiting the area in which a receiver having a given adaptive equalizer capability will find the delay spread within its ability to correct.

The relative merits of different DOCR configurations are summarized in Table 5.7.2.

5.7.2b   Distributed Transmission Systems

Distributed transmission networks differ from DOCRs in that each transmitter in the network is fed over a link separate from the on-air signal delivered to viewers (although this may be over a different broadcast channel in the case of distributed translators) [1]. This separation of the delivery and emission channels provides complete flexibility in the design of individual transmitters and the network, subject only to the limitations of consumer receiver adaptive equalizer capabilities. From the standpoint of the network and its transmitters alone, any power level and any relative emission timing are possible. The important requirement is to assure that all transmitters in

Table 5.7.2 Performance Comparison of Different DOCR (*After* [1].)

DOCR Configurations	RF Processing DOCR	IF Processing DOCR	Baseband Equalization DOCR	Baseband Decode/ Regeneration DOCR
Complexity/cost	Very low	Low	High	High
DOCR internal delay (extending pre-echo)	Few tenths μs very small	Around 1 – 2 μs small	Several μs medium	mS level, beyond equalizer range
Input adjacent channel suppression capability	Very weak	Medium	High	High
Output power level	Low	Low	Moderate	Moderate
Multihop noise/error accumulation	Yes	Yes	Maybe	No

the network emit the same symbols for the same data delivered to their inputs and at the same time (plus whatever time offsets may be designed into the network).

There are a number of ways in which signals in a distributed transmission network could be delivered to the several transmitters. These include methods in which fully modulated signals are delivered to the various transmitter locations and either up-converted or used to control the re-modulation of the data onto a new signal, a method in which data representative of the symbols to be broadcast are distributed to the transmitters, and a method in which standard MPEG-2 transport stream packets are delivered to the transmitters together with the necessary information to allow the various transmitters to emit their signals in synchronization with one another. It is the latter method that has been selected for use in ATSC Standard A/110 [2].

In the method documented in A/110, the data streams delivered to the transmitters are the same as now delivered over standard STLs for the single transmitters currently in use, with information added to those streams to allow synchronizing the transmitters. This method utilizes a very small portion of the capacity of the channel but allows continued use of the entire existing infrastructure designed and built around the 19.39 Mbps data rate. This technique permits complete flexibility in setting the power levels and relative emission timing of the transmitters in a network while assuring that they emit the same symbols for the same data inputs. While originally intended for use in SFNs, the selected method also permits extension to *multiple frequency networks* (MFNs), using a second broadcast channel as an STL to deliver the data stream to multiple distributed translators that themselves operate in an SFN. Various combinations of distributed transmitters and distributed translators are possible, and, in some cases, whether a given configuration constitutes an SFN or an MFN will depend only upon whether there are viewers in a position to be able to receive the signals that are also relaying the data stream to successive transmitters in the network.

5.7.3   Distributed Transmission Architecture

At the most basic level, the application of a distributed transmission (DTx) network can be divided into two categories [2]:

• Simple, where a second transmitter is added to a main system.

- Complex, where multiple (three or more) transmitters are used in the system. Within the "complex" realm, systems can be further distinguished by "cell size."

5.7.3a   Simple DTx Network

A DTx network has many advantages when compared to on-channel repeaters and translators [2]. In a DTx network, the digital bit stream is distributed over a separate path to each transmitter in the network. This eliminates both any errors that might have arisen using over-the-air distribution and the need for isolation between the transmitter and the over-the-air receiver. The result is that higher output power can be achieved. The transmitters also are synchronized, which allows control of the time delay between the two transmitters. Moreover, a second broadcast channel is not required, as it is when translators are used.

The major limitations of DTx networks are the network designs necessary to minimize mutual interference to consumer receivers (network internal interference) generated by the multiple transmitters and the limitations created by the operating ranges of consumer receiver adaptive equalizers. The impact of these limitations is in the distance by which the transmitters in a network can be separated from one another while still permitting the best possible improvement in reception.

The application of DTx networks can be illustrated by two examples related to terrain and coverage. In the first example, coverage is improved in terrain-shielded areas within a station's *noise limited contour* (NLC). This use of a DTx network recognizes known coverage limitations by locating a distributed transmitter where it best can fill in an area receiving low signal levels because of terrain shielding. Because the distributed transmitters operate on the same channel, the spectrum allocations limitations on the use of translators do not preclude such service improvements.

The second example is the use of a second transmitter to maximize service in areas beyond the NLC of a single transmitter. In this case, the population to be served cannot receive the signal from the first transmitter, and the second transmitter is located so as to extend service to areas that otherwise could not be reached.

5.7.3b   Complex DTx Network

Complex DTx Networks consist of multiple distributed transmitters and can use all of the concepts of the simple types discussed previously [2]. Such systems have the ability to distribute power more uniformly throughout a total service area and to provide reception at any location from multiple directions. Three concepts are closely related to the use of complex, multiple-transmitter applications: *large cell* systems, *small cell* systems, and *micro-cell* systems.

Large Cell System

Distributed transmission networks that use powerful transmitters on tall towers to cover large areas are called large cell systems [2]. Such systems might have, for example, five cells each covering its own service area with a radius of 20–30 miles (30–50 km). The service areas of the individual cell transmitters would have large areas of overlap, and the network would require careful design to minimize the effects of delay spread on receivers in the overlap regions. Consideration of transmitting antenna patterns, antenna elevations, transmitter power levels, and transmitter timing would be required. As DTV receiver adaptive equalizer designs improve over time, it can be expected that the constraints required on large cell designs and the difficulty of their implementation will be eased by the improved ability of receivers to handle the longer delay spreads that result from large cell networks.

Small Cell System

Distributed transmission networks that use lower power transmitters on shorter towers to cover smaller regions are called small cell systems [2]. Such systems might have, for example, a dozen cells, each covering its own service area with a radius of 5–10 miles (8–16 km). The service areas of the individual cell transmitters would have relatively small areas of overlap, and the network would allow a simpler design while still minimizing the effects of delay spread on receivers in the overlap regions. Transmitting antenna patterns, antenna elevations, transmitter power levels, and transmitter timing would be designed to adjust the service areas to closely match the populations intended to be served. As DTV receiver adaptive equalizer designs improve over time, it can be expected that small cell designs can be further simplified due to the improved ability of receivers to handle the reception of multiple signals that result from small cell networks.

Micro-Cell System

Micro-cell systems use very low power transmitters to cover very small areas [2]. They may be intermixed with either large cell or small cell networks to fill in coverage in places like city canyons, tunnels, or small valleys. Cities with tall buildings that emulate the effect of mountains and valleys with river gorges (e.g., New York and Chicago) can benefit from the use of microcells. Design of micro-cell systems will require the use of different design tools than typically used for broadcast applications—for example, those used for design of cellular telephone networks as opposed to the Longley-Rice methods used for longer range propagation modeling.

Ultimately, the application of distributed transmission networks can use any or all of these concepts, integrated together in a master plan. This will allow evolution from a single, high power, tall tower system to a hybrid system based upon combinations of these concepts and other unique solutions created for each individual broadcast market.

5.7.3c   DTx Synchronization Mechanisms

The synchronization requirements outlined in previously lead to the reference top-level system configuration shown in Figure 5.7.6 [2]. The system comprises three elements:

- An external time and frequency reference (shown as GPS)

- A *Distributed Transmission Adapter* (DTxA) situated at the source end of the distribution (studio-to-transmitter link, STL) subsystem

- A slave synchronization subsystem included in each of the transmitters.

The heavy lines in Figure 5.7.6 show the paths taken by synchronization signals generated in the DTxA, and a bar across the top of the figure shows the area of applicability of ATSC standard A/110.

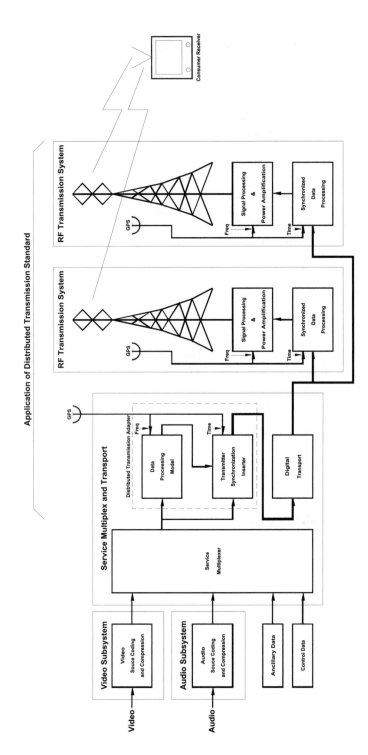

Figure 5.7.6 Synchronized DTV transmitter block diagram. (*From* [2]. *Used with permission.*)

Distributed Transmission Adapter at Source

The Distributed Transmission Adapter is used to create a pair of synchronization signals that are multiplexed into the transport stream prior to distribution over the STL system [2]. The signals produced by the DTxA are a *Cadence Signal* (CS), which establishes the phase of the data frames relative to the TS packets, and a Distributed Transmission Packet (DTxP), which carries information for slaving the pre-coders and trellis coders in the transmitters and carries command information specifying the necessary time offset for each transmitter. In addition, the DTxA indicates operating mode to the transmitters and provides information to be transmitted in the data field sync data segment through a *Field Rate Side Channel*, which carries information updated regularly at a data field rate. To accomplish these functions, the DTxA includes a data processing model equivalent to the data processing subsection of an A/53 modulator to serve as a master reference to which the Slave Synchronizers at the transmitters are slaved.

Slave Synchronization of Transmitters

At each transmitter, a *Slave Synchronizer* is employed to capture the Cadence Signal and the distributed transmission packet, to slave the data frame phasing to the Cadence Signal, and to slave the pre-coder and trellis encoder to the data in the DTxP [2]. The Slave Synchronizer extracts mode information from the Field Rate Side Channel to set the transmitter to the desired mode. It also extracts time offset command information addressed to its associated transmitter and uses it to adjust the emission time of the output symbols.

External Time and Frequency Reference

A common time and frequency reference (such as GPS) is required at several locations in the system [2]. The time component of the external reference is used by the Distributed Transmission Adapter to produce the time-offset information to be sent to the Slave Synchronizers to adjust the emission times of their associated transmitters. The DTxA uses the frequency component to precisely maintain its output transport stream data rate to tight tolerances. The time component also is used by the Slave Synchronizers at the transmitters to adjust the emission times of the associated signals to the time offsets sent from the Distributed Transmission Adapter. The frequency component also is used by the Slave Synchronizers at the transmitters to precisely set the frequencies of the transmitters in order to minimize the apparent creation of Doppler shift and the consequent burdening of receiver adaptive equalizers by frequency differences between transmitters. The Slave Synchronizers can also use the frequency component to reestablish the precise bit rate, and thereby stabilize the timing, of the Transport Stream after its transmission through STLs having some amount of time variation in their delivery of the signal, as in satellite relay, some over-the-air receivers, and some microwave systems.

5.7.4 Multiple Frequency Network Concepts

A *multiple frequency network* (MFN) uses more than one channel for transmission [1]. In the purest case, for N transmitters, N channels are used. But where distributed transmission technology is being applied, channels may be shared among a number of transmitters. For N transmitters, the number of channels used will be less than N. Some of the transmitters will be synchronized, operating on the same channel. In this situation, the network is actually a hybrid of multiple frequency and single frequency techniques.

5.7.4a   Translators

A translator is part of a multiple frequency network [1]. It receives an off-air signal on one channel and retransmits it on another channel.

Even in some relatively unpopulated areas, especially where NTSC translator systems are already deployed, there are not enough additional channels to accommodate traditional translator networks for ATSC signals during the transition phase. In these situations, use of distributed translators allows ATSC translator systems to be built using fewer channels.

A distributed translator system applies distributed transmission technology to create a network of synchronized transmitters on one channel, which retransmits a signal received off-the-air from a main transmitter, distributed transmitter, another translator, or a translator network. The advantages of a distributed translator system over conventional translators include signal regeneration and conservation of spectrum.

The distributed transmission system initially was designed to use STLs to convey MPEG-2 transport stream streams to slave transmitters in a network. This certainly could be done for distributed translator systems, but where an off-air signal is available, use of STLs would be costly and redundant. The advantage of a distributed translator system over a distributed transmission system is that STLs are not required—the signal may be taken off the air.

5.7.5   Implementation Considerations

A generalized distributed transmission system may include different combinations of DTx and DOCR elements [1]. Nothing inherently prohibits using one in the presence of another. System architects may tailor the deployment of the various technologies to the particular system being designed, considering such issues as power, cost, and terrain shielding.

For example, where low power is adequate, particularly where there is terrain shielding, a DOCR may be the most cost-effective solution. If higher power is necessary then a distributed transmission translator might be better.

5.7.6   References

1.   ATSC, "ATSC Recommended Practice: Design Of Synchronized Multiple Transmitter Networks," Advanced Television Systems Committee, Washington, D.C., Doc. A/111, September 3, 2004.

2.   ATSC: "Synchronization Standard for Distributed Transmission," Advanced Television Systems Committee, Washington, D.C., Doc. A/110A, July 19, 2005.

5.8

DTV Satellite Transmission

5.8.1 Introduction[1]

In recognition of the importance of satellite transmission in the distribution of digital video and audio programming, the ATSC developed two standards optimized for use with the ATSC digital television suite of standards. The first, document A/80, defines the parameters necessary to transmit DTV signals (transport, video, audio, and data) over satellite to one or more production and/or transmission centers. Although the ATSC had identified the particulars for terrestrial transmission of DTV (using FEC-encoded 8-VSB modulation), this method is inappropriate for satellite transmission because of the many differences between the terrestrial and satellite transmission environments. The second standard, document A/81, describes a direct-to-home satellite transmission system intended for distribution of programming directly to consumers.

5.8.2 ATSC A/80 Standard

The ATSC, in document A/80, defined a standard for modulation and coding of data delivered over satellite for digital television applications [1]. These data can be a collection of program material including video, audio, data, multimedia, or other material generated in a digital format. They include digital multiplex bit streams constructed in accordance with ISO/IEC 13818-1 (MPEG-2 Systems), but are not limited to these. The standard includes provisions for arbitrary types of data as well.

Document A/80 covers the transformation of data using error correction, signal mapping, and modulation to produce a digital carrier suitable for satellite transmission. In particular, quadrature phase shift modulation (QPSK), eight phase shift modulation (8PSK), and 16 quadrature amplitude modulation (16QAM) schemes are specified. The main distinction between QPSK, 8PSK, and 16QAM is the amount of bandwidth and power required for transmission. Generally,

---

1. *Editor's note*: This chapter provides an overview of ATSC DTV satellite services based on ATSC A/80 and A/81. Readers are encouraged to download the entire standards from the ATSC Web site (http://www.atsc.org). All ATSC Standards, Recommended Practices, and Information Guides are available at no charge.

for the same data rate, progressively less bandwidth is consumed by QPSK, 8PSK, and 16QAM, respectively, but the improved bandwidth efficiency is accompanied by an increase in power to deliver the same level of signal quality.

A second parameter, coding, also influences the amount of bandwidth and power required for transmission. Coding, or in this instance forward error correction (FEC), adds information to the data stream that reduces the amount of power required for transmission and improves reconstruction of the data stream received at the demodulator. While the addition of more correction bits improves the quality of the received signal, it also consumes more bandwidth in the process. So, the selection of FEC serves as another tool to balance bandwidth and power in the satellite transmission link. Other parameters exist as well, such as transmit filter shape factor (α), which have an effect on the overall bandwidth and power efficiency of the system.

System operators optimize the transmission parameters of a satellite link by carefully considering a number of trade-offs. In a typical scenario for a broadcast network, material is generated at multiple locations and requires delivery to multiple destinations by transmitting one or more carriers over satellite, as dictated by the application. Faced with various size antennas, available satellite bandwidth, satellite power, and a number of other variables, the operator will tailor the system to efficiently deliver the data payload. The important tools available to the operator for dealing with this array of system variables include the selection of the modulation, FEC, and α value for transmission.

5.8.2a   Services and Applications

The ATSC satellite transmission standard includes provisions for two distinct types of services [1]:

• *Contribution service*—the transmission of programming/data from a programming source to a broadcast center. Examples include such material as digital satellite news gathering (DSNG), sports, and special events.

• *Distribution service*—the transmission of material (programming and/or data) from a broadcast center to its affiliate or member stations.

The A/80 standard relies heavily upon previous work done by the Digital Video Broadcasting (DVB) Project of the European Broadcast Union (EBU) for satellite transmission. Where applicable, the ATSC standard sets forth requirements by reference to those standards, particularly EN 300 421 (QPSK) and EN 301 210 (QPSK, 8PSK, and 16QAM).

The modulation and coding defined in the standard have mandatory and optional provisions. QPSK is considered mandatory as a mode of transmission, while 8PSK and 16QAM are optional. Whether equipment implements optional features is a decision for the manufacturer. However, when optional features are implemented they must be in accordance with the standard in order to be compliant with it.

5.8.2b   System Overview

A digital satellite transmission system is designed to deliver data from one location to one or more destinations. A block diagram of a simple system is shown in Figure 5.8.1 [1]. The drawing depicts a *data source* and *data sink*, which might represent a video encoder/multiplexer or decoder/demultiplexer for ATSC applications, but can also represent a variety of other sources

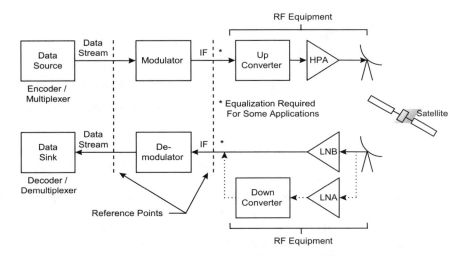

Figure 5.8.1 Overall system block diagram of a digital satellite system. The ATSC standard described in document A/80 covers the elements noted by the given reference points. (*From* [1]. *Used with permission.*)

that produce a digital data stream. This particular point, the accommodation of arbitrary data streams, is a distinguishing feature of the A/80 Standard.

The subject of the A/80 standard is the segment between the dashed lines designated by the reference points on Figure 5.8.1; it includes the modulator and demodulator. Only the modulation parameters are specified; the receive equipment is designed to recover the transmitted signal. The ATSC standard does not preclude combining equipment outside the dashed lines with the modulator or demodulator, but it sets a logical demarcation between functions.

In the figure, the modulator accepts a data stream and operates upon it to generate an intermediate frequency (IF) carrier suitable for satellite transmission. The data are acted upon by forward error correction (FEC), interleaving and mapping to QPSK, 8PSK or 16QAM, frequency conversion, and other operations to generate the IF carrier. The selection of the modulation type and FEC affects the bandwidth of the IF signal produced by the modulator. Selecting QPSK, 8PSK, or 16QAM consumes successively less bandwidth as the modulation type changes. It is possible, then, to use less bandwidth for the same data rate or to increase the data rate through the available bandwidth by altering the modulation type.

Coding or FEC has a similar impact on bandwidth. More powerful coding adds more information to the data stream and increases the occupied bandwidth of the IF signal emitted by the modulator. There are two types of coding applied in the modulator. An outer Reed-Solomon code is concatenated with an inner convolutional/trellis code to produce error correction capability exceeding the ability of either coding method used alone. The amount of coding is referred to as the *code rate*, quantified by a dimensionless fraction (k/n) where n indicates the number of bits out of the encoder given k input bits (e.g., rate 1/2 or rate 7/8). The Reed-Solomon code rate is fixed at 204,188 but the inner convolutional/trellis code rate is selectable, offering the opportunity to modify the transmitted IF bandwidth.

One consequence of selecting a more bandwidth-efficient modulation or a higher inner code rate is an increase in the amount of power required to deliver the same level of performance. The

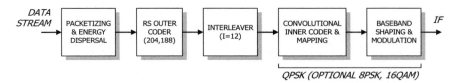

Figure 5.8.2 Block diagram of the baseband and modulator subsystem. (*From* [1]. *Used with permission.*)

key measure of power is the E_b/N_o (energy per useful bit relative to the noise power per Hz), and the key performance parameter is the bit error rate (BER) delivered at a particular E_b/N_o. For digital video, a BER of about 10^{-10} is necessary to produce high-quality video. Thus, noting the E_b/N_o required to produce a given BER provides a way of comparing modulation and coding schemes. It also provides a relative measure of the power required from a satellite transponder, at least for linear transponder operation.

The basic processes applied to the data stream are illustrated in Figure 5.8.2. Specifically,

• Packetizing and energy dispersal

• Reed-Solomon outer coding

• Interleaving

• Convolutional inner coding

• Baseband shaping for modulation

• Modulation

The input to the modulator is a data stream of specified characteristics. The physical and electrical properties of the data interface, however, are outside the scope of this standard. The output of the modulator is an IF signal that is modulated by the processed input data stream. This is the signal delivered to RF equipment for transmission to the satellite. Table 5.8.1 lists the primary system inputs and outputs.

The data stream is the digital input applied to the modulator. There are two types of packet structures supported by the standard, as given in Table 5.8.2.

5.8.3   ATSC DTH Satellite Broadcast Standard

ATSC Standard A/81 describes the emission system for a direct-to-home (DTH) satellite broadcast system [2]. This specification defines extensions to the audio, video, transport, and PSIP subsystems as defined in ATSC standards A/52, A/53, and A/65. The emission system defined in this document includes carriage of data broadcasting as defined in ATSC standard A/90, without requiring extensions. Transmission and conditional access subsystems are not defined, allowing service providers to use existing subsystems.

The ATSC DTH satellite broadcast system consists of two major elements:

• The transmission system

• Integrated receiver decoder, commonly referred as a set top box (STB).

Table 5.8.1 System Interfaces Specified in A/80 (*After* [1].)

Location	System Inputs/Outputs	Type	Connection
Transmit station	Input	MPEG-2 transport (Note 1) or arbitrary	From MPEG-2 multiplexer or other device
	Output	70/140 MHz IF, L-band IF, RF (Note 2)	To RF devices
Receive installation	Input	70/140 MHz IF, L-band IF (Note 2)	From RF devices
	Output	MPEG-2 transport (Note 1) or arbitrary	To MPEG-2 de-multiplexer or other device

1 In accordance with ISO/IEC 13838-1
2 The IF bandwidth may impose a limitation on the maximum symbol rate.

Table 5.8.2 Input Data Stream Structures (*After* [1].)

Type	Description
1	The packet structure shall be a constant rate MPEG-2 transport per ISO/IEC 13818-1 (188 or 204 bytes per packet including 0x47 sync, MSB first).
2	The input shall be a constant rate data stream that is arbitrary. In this case, the modulator takes successive 187 byte portions from this stream and prepends a 0x47 sync byte to each portion, to create a 188 byte MPEG-2 like packet. (The demodulator will remove this packetization so as to deliver the original, arbitrary stream at the demodulator output.)

5.8.3a   Transmission System

The transmission system comprises an emission multiplex, a modulator/encoder, and a transmitter [2]. The emission multiplexer requirements are discussed in this standard. Specifications for the modulator/encoder and the transmitter are left for service provider to develop.

Figure 5.8.3 shows a functional block diagram of a DTH satellite transmission system. The emission multiplexer accepts and combines:

- ATSC multi-program Transport Streams (A/53, A/65, A/70, and A/90 protocols) from different sources

- Satellite extensions to PSIP

Additionally, the emission multiplexer may accept:

- MPEG-compliant (non-ATSC) Transport Streams

- Data streams such as A/90 and DVB data broadcast

Each multi-program Transport Stream output from the emission multiplexer to a modulator must conform with:

- The transport, audio, and video format extensions defined for satellite delivery in standard A/81.

- System information with all the normative elements from the ATSC PSIP standard (A/65) and satellite extensions, such as the satellite Virtual Channel Table (defined in A/81).

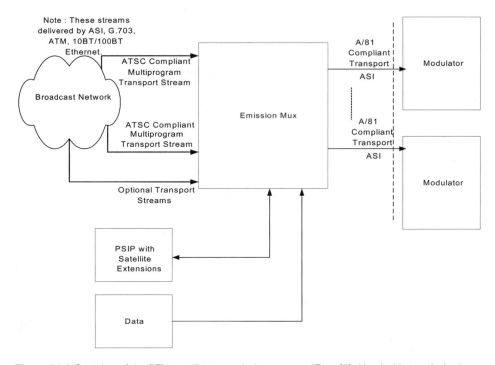

Figure 5.8.3 Overview of the DTH satellite transmission system. (*From* [2]. *Used with permission.*)

Transport Streams at the output of the emission multiplex may also carry additional information to support delivery of system-specific data (such as DVB-SI, ATSC A/56, control data, EIA-608B captions using ANSI/SCTE 20, and MPEG-1 Layer 2 audio).

5.8.3b   Integrated Receiver Decoder System

A functional block diagram of an integrated receiver decoder (IRD) system is depicted in Figure 5.8.4 [2]. This system demodulates and decodes audio, video, and data streams compatible with the transmission system described in A/81.

5.8.3c   Compression Format Constraints

The allowed compression formats are listed in Table 5.8.3, and for 25/50 Hz video, Table 3 in ATSC A/63 (see [3]).

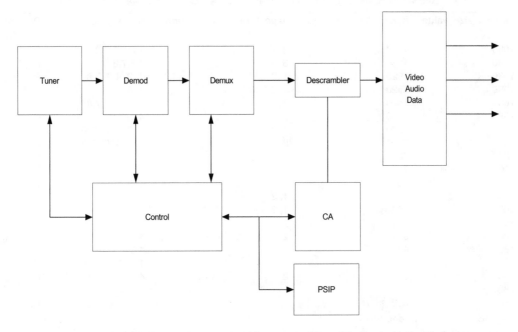

Figure 5.8.4 Functional block diagram of an IRD system. (*From* [2]. *Used with permission.*)

5.8.3d PSIP

Some Program and System Information Protocol (PSIP) tables used in the DTH satellite broadcast standard are common with terrestrial broadcast and/or cable systems defined in ATSC document A/65 [2].

The following tables must be included in all ATSC-compliant Transport Streams to be transmitted via satellite broadcast:

- The satellite *Virtual Channel Table* (SVCT) defining, at a minimum, the virtual channel structure for the collection of MPEG-2 programs embedded in the Transport Stream in which the SVCT is carried.

- The *Master Guide Table* (MGT) defining the type, packet identifiers, and versions for all of the other satellite PSIP tables included in the Transport Stream, except for the system time table (STT).

- The *Rating Region Table* (RRT) defining the TV parental guideline system referenced by any content advisory descriptor carried within the Transport Stream.

- The *System Time Table* (STT), defining the current date and time of day and daylight savings time transition timing.

- The first four *Aggregate Event Information Tables* (AEIT-0, AEIT-1, AEIT-2, and AEIT-3), which deliver event title and schedule information that may be used to support an electronic program guide application.

Table 5.8.3 Allowed Compression Formats Under ATSC DTH Satellite Standard (*After* [2].)

vertical_size_value	horizontal_size_value	aspect_ratio_information	frame_rate_code	Progressive/ Interlaced
1080	1280	3	1,2,4,5,7,8	P
1080	1280	3	4,5,7,8	I
1080	1920	1, 3	1, 2, 4, 5,7,8	P
1080	1920	1, 3	4, 5,7,8	I
1080	1440	3	1, 2, 4, 5,7,8	P
1080	1440	3	4, 5,7,8	I
720	1280	1, 3	1, 2, 4, 5, 7, 8	P
480	720	2, 3	1, 2, 4, 5, 7, 8	P
480	720	2, 3	4, 5	I
480	704	2, 3	1, 2, 4, 5, 7, 8	P
480	704	2, 3	4, 5	I
480	640	1, 2	1, 2, 4, 5, 7, 8	P
480	640	1, 2	4, 5	I
480	544	2	1	P
480	544	2	4	I
480	480	2	4,5	I
480	528	2	1	P
480	528	2	4	I
480	352	2	1	P
480	352	2	4	I

Legend for MPEG-2 Coded Values

aspect_ratio_information: 1 = square samples, 2 = 4:3 display aspect ratio, 3 = 16:9 display aspect ratio
frame_rate_code: 1 = 23.976 Hz, 2 = 24 Hz, 4 = 29.97 Hz, 5 = 30 Hz, 7 = 59.94 Hz, 8 = 60 Hz
Progressive/Interlace: I= interlaced scan, P = progressive scan

Satellite Virtual Channel Table (SVCT)

The Satellite Virtual Channel Table (SVCT), like its cable and terrestrial broadcast counterparts, contains a list of attributes for virtual channels carried in the Transport Stream [2]. Any changes in the virtual channel structure are conveyed with a new "version number." The basic information contained in the body of the SVCT includes:

- Transport Stream ID

- Major and minor channel number

- Short channel name

- Carrier frequency

- Program number

- Location field for extended text messages

- Service type

- Modulation parameters

Additional information may be carried by *descriptors*, which may be placed in the descriptor loop after the basic information.

Unlike cable and terrestrial applications where just one Virtual Channel Table is present in any given Transport Stream, in the satellite application more than one may be present. A receiver is expected to capture and record one or more SVCTs for use in navigation and service acquisition.

5.8.4   References

1.   ATSC: "Modulation And Coding Requirements For Digital TV (DTV) Applications Over Satellite," Doc. A/80, Advanced Television Systems Committee, Washington, D.C., July, 17, 1999.

2.   ATSC: "Direct-to-Home Satellite Broadcast Standard," Doc. A/81, Advanced Television Systems Committee, Washington, D.C., July 20, 2003.

3.   ATSC: "Standard for Coding 25/50 Hz Video," Doc. A/63, Advanced Television Systems Committee, Washington, D.C., May 2, 1997.

5.9
DTV Data Broadcasting

5.9.1   Introduction[1]

Unlike NTSC VBI-based data services, DTV Data Services are an integral part of the broadcast signal. Data shares the same multiplex with the video and audio, and the same fundamental MPEG-2 acquisition mechanisms are used to acquire data, video, and audio signals. Data services may also be announced in a program guide like video/audio programming.

The ATSC Data Broadcast Standard, described in ATSC document A/90, specifies the following key elements:

- Data Services announcement

- Data delivery models

- Application signaling

- MPEG-2 systems tools

- Protocols

The Data Broadcast Standard can be used for wide range of applications, including:

- Delivery of declarative data such as HTML code

- Delivery of procedural data such as Java® code

- Delivery of software, images, graphics, and other files

Building upon the foundation of A/90, the ATSC developed a suite of standards that extend the basic capabilities of the Data Broadcast Standard. Applications taking advantage of these features are designed to be flexible with regard to efficient utilization of the available DTV bandwidth. While the instantaneous bandwidth required to carry the video program will vary, the bandwidth of the transport pipe is constant. Rather than transmit null packets, *opportunistic data*

---

1. This chapter is based on the A/90 suite of ATSC Data Broadcast Standards. Readers are encouraged to download the source documents from the ATSC Web site (http://www.atsc.org). All ATSC Standards, Recommended Practices, and Information Guides are available at no charge.

packets may be sent instead. The concept of opportunistic bandwidth for the transmission of data is illustrated in Figure 5.9.1

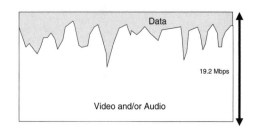

Figure 5.9.1 The concept of opportunistic broadcasting of data over the DTV system.

5.9.2   ATSC A/90 Standard

The ATSC A/90 Data Broadcast Standard describes the available encapsulation protocols used to transport data within the ATSC MPEG-2 transport multiplex [3]. The specification explains the Service Description Framework used for the discovery of Program Elements and the binding of applications to their Program Elements. Additionally, the specification describes the additions to the ATSC Program and System Information Protocol standard such that a Data Service may be announced within the existing Electronic Program Guide (EPG) mechanisms. The specification also explains the use of elements in MPEG-2 PSI to aide the SDF for binding of applications to their Program Elements. It must also be noted that the PSIP Standard and MPEG-2 PSI (PAT and PMT) aid the SDF with the discovery of Program Elements. Figure 5.9.2 provides an overview of the Data Broadcast Standard.

The basis of the Data Broadcast Standard is formed by the MPEG-2 Transport Stream (TS) as defined in ISO/IEC 13818-1 (MPEG-2 Systems) and Amendments 1 and 2 specifying registration procedures for the copyright identifier and the format identifier, respectively. Data information can be transported within this MPEG-2 TS by means of the following encapsulation protocols:

• Data Piping

• Data Streaming

• Addressable Sections

• Data Download

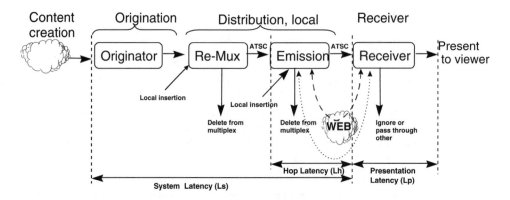

Figure 5.9.2 Overall system diagram of the ATSC data broadcast standard. (*After* [3].)

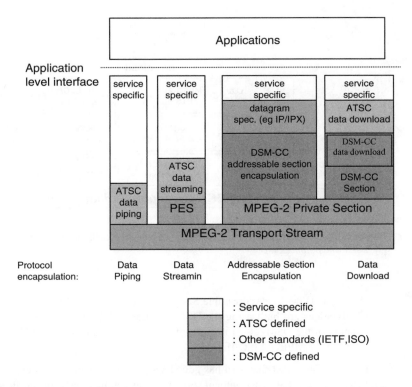

Figure 5.9.3 ATSC data broadcast standard graphical encapsulation overview and relation to other standards. (*From* [4]. *Used with permission.*)

Figure 5.9.3 identifies what is standardized and by which body. ISO has standardized the MPEG-2 TS in ISO/IEC 13818-1 and the DSM-CC framework in ISO/IEC 13818-6. The IETF has standardized the Internet Protocol (IP) in RFC 791. The ATSC has specified within the Data Broadcast Standard the ATSC Data Piping protocol, the ATSC Data Streaming Encapsulation, the ATSC DSM-CC Addressable Section Encapsulation, and the ATSC Download Protocol. Within Figure 5.9.3, the encapsulation of the Internet Protocol (IP) is just an example. Other network protocols can also be encapsulated.

As shown in Figure 5.9.3, the Data Broadcast Standard specifies different encapsulation protocols for different application areas:

- The Data Piping specification provides minimal information on how to acquire and assemble the data bytes from the MPEG-2 TS. It only specifies how to put data into the MPEG-2 Transport Stream packets.

- The Data Streaming specification provides additional functionality, especially for timing. It is possible to achieve synchronous data broadcast or synchronized data broadcast. The data streaming specification is based on PES packets as defined in MPEG-2 ISO/IEC 13818-1.

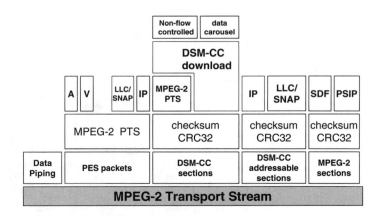

Figure 5.9.4 ATSC data broadcast protocol packetization, synchronization, and protection layers. (*From* [4]. *Used with permission.*)

- The ATSC DSM-CC Addressable Section Encapsulation and the ATSC Download Protocol specifications are built using the DSM-CC framework of MPEG-2 ISO/IEC 13818-6. The mapping of the DSM-CC protocols onto MPEG-2 TS utilizes the MPEG-2 Private Sections as defined in MPEG-2 ISO/IEC 13818-1.

The Data Broadcast Standard has added specific information in order for the framework to work within the ATSC environment, especially in conjunction with the Program and System Information Protocol.

In the Data Broadcast standard, each Data Service may be composed of one or more applications integrated to the remaining ATSC infrastructure by means of Announcement, Discovery, and Binding functions.

The announcement and discovery specification is part of the PSIP specification, along with the new elements documented in the Data Broadcast Standard. Additionally, further discovery definition and application binding are part of the Service Description Framework (SDF) that is described in the Data Broadcast Standard. Finally, data delivery is part of the Data Broadcast Standard. Figure 5.9.4 illustrates the packetization, synchronization, and protection layers of the Data Broadcast protocols.

5.9.2a   Data Service Definition

A Data Service is the collection of applications and associated resources signaled in the Data Service Table (DST) of the Service Description Framework [4]. A Data Service is required to have one Program Element (data Elementary Stream) conveying the DST and optionally the Network Resources Table (NRT). A virtual channel, as described by the PSIP Virtual Channel Table (VCT), may include a maximum of a single Data Service. The minor_channel_number of a data-only channel (service_type[2] 0x04 in the VCT) must be in the range 100–999. A Data Service may be composed of any number of Program Elements and other resources, and the Program Elements may include any combination of the encapsulation protocols specified in the Data Broadcast Standard.

The discovery of a Data Service in an ATSC transport multiplex is independent of the announcement of a Data Service. The announcement procedure formalizes the technique used to signal the Data Service event (i.e., the start time and the duration) during which the Data Service is active. The discovery procedure formalizes the mechanism used for identifying the presence of a Data Service and for identifying the components comprising the Data Service. The binding mechanism describes a way to provide application specific parameters and resource signaling.

5.9.2b   Data Encapsulations

As illustrated in Figure 5.9.2 and Figure 5.9.4, there are multiple ways to encapsulate data within the transport multiplex [4]. The mechanisms have different characteristics concerning timing, filtering, overhead, size, and so on. The selection of the appropriate mechanism is based on the specific requirements of the target application.

Packetization of Data Entities

Generally, data of any protocol is transmitted in a packetized form (*Data Entities*) [4]. These Data Entities may have different lengths. If the data is not packetized or the packetization method is irrelevant or hidden to the ATSC transmission chain the most appropriate way of transmission is Data Piping.

At the MPEG-2 Transport Stream layer, data is transmitted within MPEG-2 TS packets with a fixed length of 188 bytes (184 bytes payload); therefore, Data Entities of higher layers must often be split at the transmission side and must be re-assembled at the reception device. For the splitting of the Data Entities, there are several possible packetization mechanisms:

- Private mechanisms based on the Data Piping

- MPEG-2 Packetized Elementary Streams (PES)

- MPEG-2 Private Sections

- DSM-CC Data Download Blocks

MPEG-2 PES provides a mechanism to transmit Data Entities of variable size with a maximum length of 64 Kbytes. Additionally, it provides the facility to synchronize different data streams accurately (as used in MPEG-2 for synchronization of Video and Audio). MPEG-2 PES was chosen by ATSC for the Data Broadcast Standard for the transmission of all synchronous data streams and for synchronized data streaming.

MPEG-2 Private Sections can be used to transmit Data Entities of variable size with a maximum length for each data entity of just less than 4 Kbytes (encapsulation specific). The transmission is asynchronous. Data Entities can be recombined to form larger data objects (for example,

---

2. ATSC Data Broadcast Standards documents contain symbolic references to syntactic elements used in the audio, video, and transport coding subsystems. These references usually contain the underscore character (e.g., sequence_end_code) and may consist of character strings that are not English words (e.g., dynrng). In this book they are identified by a distinctive font. Note also that certain syntax names are capitalized; e.g., Transport Stream.

Table 5.9.1 Encapsulation Protocol Selection Matrix (*After* [4].)

	Bounded	Unbounded/Streaming	Network Datagram
Asynchronous	Asynchronous Module Download Protocol	Asynchronous Data Streaming Download Protocol	DSM-CC Addressable Sections
Synchronous		Synchronous Data Streaming (PES)	Synchronous Data Streaming (PES)
Synchronized	Synchronized Data Download Protocol	Synchronized Data Streaming (PES)	Synchronized Data Streaming (PES)

modules reassembled from Download Data Blocks). MPEG-2 Private Sections are built in a way that MPEG-2 demultiplexers can filter out single sections in hardware, thereby reducing the required software processing power of the receiver. This concept is the main reason why the MPEG-2 Private Sections have been chosen as the mechanism for the transmission of asynchronous network protocols and data downloads (*Download Protocol*). The Download Protocol provides a means of re-assembling sections (DownloadDataBlocks) into modules, where the section data may be up to 4066 bytes in size. DSM-CC Addressable Sections are used for the carriage of Data Entities like IP Datagrams. Again, these sections are designed to engage the section filtering capability of receivers.

Encapsulation Protocol Selection Guidance

Table 5.9.1 provides a generalized guide for determining the recommended encapsulation method [4]. The column headings represent the data size characteristics. The row headings represent the data timing or synchronization requirements.

The delivery of asynchronous data Elementary Streams is not subject to any timing constraints derived from the presence of MPEG-2 Systems time stamps in the ATSC Transport Stream. In essence, asynchronous data Elementary Streams are not time sensitive streams and the bytes of such streams are delivered at a fixed leak rate from a well-defined Transport System Target Decoder (T-STD) buffer model to the data receiver. Asynchronous data Elementary Streams have stream_type value 0x0B, 0x0D or 0x95.

Synchronous data Elementary Streams are used for applications requiring continuous streaming of data to the receiver at a regular and piece-wise constant data rate. Synchronous data is delivered as a stream of 2-byte Data Access Units, each Data Access Unit being associated with a precise delivery time derived directly from the PTS field in the PES packet header and a leak rate specified either in the PES packet header or the synchronous data header structure at the beginning of the PES packet payload. The Presentation Time Stamps in the PES packets specifies the time of delivery of the first synchronous Data Access Unit in the PES packet relative to the System Time Clock (STC) reconstructed from the PCR fields in the ATSC Transport Stream. The leak rate is then used to infer the delivery time for each of the following synchronous access units within the same PES packet. Delivery of the synchronous Data Access Units in the next PES packet starts at the time specified by the PTS of that next PES packet. The PCR time stamps may be delivered in the Transport Packets conveying the synchronous data Elementary Stream. The delivery of synchronous Data Access Units is governed by a well-defined T-STD buffer model. The difference between synchronous data and asynchronous data is the fact that synchronous Data Access Units have been defined for the sole purpose of delivering bytes out of the T-

STD buffer model according to a strict timing tied to the 27 MHz System Time Clock of the receiver. Synchronous data Elementary Streams have stream_type value 0xC2.

Synchronized data Elementary Streams are used for applications requiring presentation of data at precise but not necessarily regular instants. The presentation times are typically, but not necessarily, associated with some other reference video, audio, or data Elementary Stream. Synchronized data is delivered as a series of Data Access Units spanning one or several PES packets. The time separating two consecutive synchronized Data Access Units is arbitrary. A synchronized data Elementary Stream includes Presentation Time Stamps. The Program Clock Reference time stamps are typically delivered in the MPEG-2 Transport Stream packets of a reference Elementary Stream, but they may be in the same Elementary Stream as the synchronized data. The purpose of the PTS time stamps is to specify the instant in time, relative to the STC, at which a Data Access Unit must be rendered/displayed in the receiver. The T-STD for synchronized data Elementary Streams includes an additional buffer for re-assembling data bytes into synchronized Decoding Access Units. Therefore, as opposed to synchronous Data Access Units, synchronized Data Access Units have been defined for the purpose of presenting data at precise times. Also, as opposed to synchronous Data Access Units, consecutive synchronized Data Access Units may not be of the same size. Consequently, the T-STD for synchronized data Elementary Streams goes one step further by providing a buffer for collecting data bytes pertaining to a Data Access Unit before it is decoded and presented at a precise instant of the receiver System Time Clock. Synchronized data Elementary Streams have stream_type value 0x06 or 0x14.

Bounded Data Blobs

A Data Blob or a Data Module could be a representation of a file, a group of files, a directory of files, a group of directories containing files, a module, a hierarchy of modules, an object, a group of objects, or a bounded data entity whose boundaries are known to the content author [4]. The definition of a Data Module is not defined by the Data Broadcast Standard. However, a Data Module can generally be viewed to represent a finite-sized entity that is to be delivered to a data receiver. Data Modules that have no synchronization requirements associated with them should be delivered using the asynchronous module delivery method of the Download Protocol.

Data Modules that are to be repetitively delivered should use the Data Carousel option of the asynchronous Download Protocol. Otherwise, the non-flow controlled scenario of the Download Protocol should be used.

Synchronized Data Modules should be delivered using the synchronized Download Protocol.

Network Datagrams

Network datagrams, for example Internet Protocol (IP) datagrams or Internetwork Packet Exchange (IPX) datagrams, can be encapsulated in various manners [4]. The different encapsulation methods are intended to meet specific needs of different applications.

ATSC DSM-CC Addressable Sections are the method of delivery for asynchronous network datagrams. Datagrams that do not have any timing or synchronization requirements associated with them should be delivered using the DSM-CC Addressable Sections encapsulation.

Datagrams that are synchronized to a program by using a Presentation Time Stamp (PTS) are conveyed in the synchronized streaming protocol. Examples of synchronized datagrams include datagram triggers, and audio/video associated enhanced data.

Synchronous datagrams are carried using the synchronous streaming protocol.

Streaming Data

Streaming data can be synchronous, synchronized, or asynchronous. Asynchronous streaming data should be carried by the asynchronous Data Download Protocol. The delivery of asynchronous data Elementary Streams is not subject to any timing constraints derived from the presence of MPEG-2 Systems time stamps in the ATSC Transport Stream. In essence, asynchronous data Elementary Streams are not time-sensitive streams and the bytes of such streams are delivered at a fixed "leak rate" from a well-defined Transport System Target Decoder (T-STD) buffer model to the data receiver. Asynchronous data Elementary Streams have stream_type value 0x0B, 0x0D or 0x95.

All data that is synchronous is carried using the synchronous streaming encapsulation where the data is transported in Packetized Elementary Stream (PES) packets. The synchronous data streaming protocol has been harmonized with SCTE DVS132. Synchronous data Elementary Streams are used for applications requiring continuous streaming of data to the receiver at a regular and piece-wise constant data rate. Synchronous data is delivered as a stream of 2-byte synchronous_data_access_units, each synchronous_data_access_unit being associated with a precise delivery time derived directly from the PTS field in the PES packet header and a leak rate specified either in the PES packet header or the synchronous data header structure at the beginning of the PES packet payload.

The Presentation Time Stamps in the PES packets specifies the time of delivery of the first synchronous_data_access_unit in the PES packet relative to the System Time Clock (STC) reconstructed from the PCR fields in the ATSC TS. The leak rate is then used to infer the delivery time for each of the following synchronous_data_access_units. The PCR time stamps may be delivered in the Transport Packets conveying the synchronous data Elementary Stream. The delivery of synchronous_data_access_units is governed by a well-defined T-STD buffer model. The difference between synchronous data and asynchronous data is the fact that synchronous_data_access_units have been defined for the sole purpose of delivering bytes out of the T-STD buffer model according to a strict timing tied to the 27 MHz System Time Clock of the receiver. Synchronous data Elementary Streams have stream_type value 0xC2.

Synchronized data should be carried in the Synchronized Streaming Protocol. (The Synchronized Streaming Protocol has been harmonized with DVB.) Synchronized data Elementary Streams are used for applications requiring presentation of data at precise but not necessarily regular instants and in connection to a reference video, an audio, or another data Elementary Stream. Synchronized data is delivered as a series of Data Access Units spanning one or several PES packets. The time separating two consecutive synchronized Data Access Units is arbitrary. A synchronized data Elementary Stream includes Presentation Time Stamps. The Program Clock Reference time stamps are delivered in the MPEG-2 TS packets of the reference Elementary Stream. The purpose of the PTS time stamps is to specify the instant in time, relative to the STC, at which a Data Access Unit must be rendered/displayed in the receiver. The T-STD for synchronized data Elementary Streams includes an additional buffer for re-assembling data bytes into synchronized Decoding Access Units. Therefore, as opposed to synchronous Data Access Units, synchronized Data Access Units have been defined for the purpose of presenting data concurrently with another media stream. Consequently, the T-STD for synchronized data Elementary Streams goes one step further by providing a buffer for collecting data bytes pertaining to a Data Access Unit before it is decoded and presented at a precise instant of the receiver System Time Clock. Synchronized data Elementary Streams have stream_type value 0x06 or 0x14.

5.9.2c   Data Carousel

The A/90 Standard supports the carriage of data objects using the *non-flow controlled scenario* and the *Data Carousel scenario* of the DSM-CC user-to-network Download Protocol. The ATSC use of the DSM-CC Download Protocol supports the transmission of the following:

- Asynchronous Data Modules

- Asynchronous data streaming

- Non-streaming synchronized data

Transmission errors in data carried by the Download Protocol may be detected, since the DSM-CC sections used include a CRC/checksum field for that purpose.

The Data Carousel sends (and repeats) data at intervals in order to avoid the necessity for a flow control system, as well as repetitions for error protection. The head-end infrastructure is illustrated in Figure 5.9.5.

The non-flow control scenario embodies the unidirectional, one time transmission of a bounded data image to a data receiver. The Data Carousel cyclically repeats the contents of the carousel, one or more times. If a data decoder wants to access a particular module from the Data Carousel, it may simply wait for the next time that the data for the requested module is broadcast. Figure 5.9.6 illustrates a Data Carousel implementation of the Download Protocol.

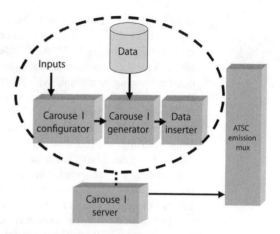

Figure 5.9.5 Head-end infrastructure of the Data Carousel server. (*From* [4]. *Used with permission.*)

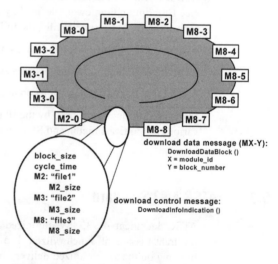

Figure 5.9.6 Cyclic transmission of information in a Data Carousel. (*From* [4]. *Used with permission.*)

5.9.2d   Data Receiver Support

A data receiver may support any combination of the encapsulation protocols [4]. A data receiver that encounters an unsupported encapsulation protocol, or any other unsupported data item delivered via the transport multiplex, should seamlessly discard the unsupported information.

5.9.3   ATSC A/92 Standard

Document A/92, "Delivery of IP Multicast Sessions over ATSC Data Broadcast," specifies the delivery of IP multicast sessions, the delivery of data for describing the characteristics of a session, and usage of the ATSC A/90 Data Broadcast Standard for IP multicast [5]. A/92 defines a standard for the asynchronous transmission of Internet Protocols (IP) specifically including multicast addressing compatible with A/90. This document assumes the use of *Session Description Protocol* (SDP) as an integral part of the IP multicast-based data broadcast service.

Standard A/92 defines a model whereby standard IP network terminology and concepts can be applied to the transmission of IP multicast services in an ATSC Transport Stream. When creating, inserting, and transporting IP packets through a Transport Stream, rules need to be established to define what constitutes an IP network or sub-network and how services on such networks can be discovered. This is critical for all implementations, as there is the strong potential for IP address conflicts without such rules. This standard, thus, establishes a bridge between Internet Protocols and ATSC Protocols. A/92 also documents various cases that may be encountered when an ATSC virtual channel includes one or more IP multicast sessions.

A stand-alone IP multicast service is a Data Service not related to any audio-visual event. This service may include one or several IP multicast sessions. If the service includes multiple sessions, one or several ATSC A/90 applications may be used to signal the sessions.

Document A/92 defines a variety of ways for signaling an IP multicast service within the Service Description Framework defined by A/90. Detailed usage recommendations are provided for each scenario. More specifically, the following scenarios are discussed:

- A single IP multicast session on a single ATSC virtual channel

- Multiple IP multicast sessions on a single ATSC virtual channel

- Multiple IP multicast sessions on multiple ATSC virtual channels

Signaling information is used by the IP multicast receiver to discover the service and to bind the various IP multicast media and SDP announcement streams to particular MPEG-2 Program Elements.

5.9.4   ATSC A/93 Standard

ATSC document A/93, "Synchronized/Asynchronous Trigger Standard," defines a method for the transmission of synchronized data elements, and synchronized and asynchronous events, building on the synchronized delivery mechanisms of A/90. Standard A/93 specifically enables the synchronized delivery of Data Modules through the decoupling of the timing from the delivery of the data element. It also enables the delivery of events to receivers, including application-defined events [6].

A/93 addresses the need to synchronize data with video and to deliver events to receivers. This standard can serve as the transport-layer infrastructure for applications such as the display of a data advertisement (e.g., a web-page that allows purchasing) at specific time points within the video stream.

The key technical issue addressed by the ATSC trigger design was to remedy the lack of a guarantee for a continuous System Time Clock (STC) for accurate frame/field synchronization of data requiring long decoding times. A/93 allows an arbitrarily complex *Data Access Unit*

(DAU) to be activated by an arbitrarily simplified receiver to achieve tight synchronization in the context of a discontinuous time line.

The ATSC Data Broadcast Standard specifies that non-streaming, synchronized Data Access Units are encapsulated with the synchronized Download Protocol. This encapsulation carries a Presentation Time Stamp (PTS) with each DAU, which indicates the time of DAU activation.

The model standardized in A/90 is not guaranteed to achieve the desired synchronization. For example, a problem occurs when DAUs carrying complex data objects having long decoding times are combined with timeline (PCR) discontinuities. In this circumstance, there can be ambiguity in the meaning of the PTS value in the encapsulation. The synchronized trigger is designed to address this situation, as well as other complications.

5.9.4a   Requirements

The goal A/93 is to enable frame-level ("tight") synchronization of arbitrary DAUs on the ATSC transport [6]. To achieve this goal, the design of the synchronized trigger standard was guided by the following requirements:

- The overall trigger design does not preclude long term delays (days or months) between transport of the target and the trigger.

- Triggers are compatible with, and do not adversely impact, synchronization of DAUs conveyed in synchronized MPEG-2 Program Elements.

- Triggers are implemented in such a way as to not be subjected to any problems with MPEG-2 PCR discontinuities.

- Triggers may reference objects and actions in addition to related DAUs (all targets).

- Triggers have a mechanism that allows them to be hidden from unauthorized applications (for example, commercial detection).

- The design is transport-independent, provided it does not negatively impact ATSC transport functionality.

- Targets are signaled with a unique invariant identifier.

- Objects and related DAUs are stored according to certain application buffer model(s) until the trigger is received and processed.

- A trigger references one and only one target.

- The decoding time associated with the trigger is minimal.

- Triggers have no time sequence restrictions with respect to objects and actions.

- The trigger structure permits maintenance of the proper time relationships between the trigger and the target by re-multiplexors.

- Related DAU's precede triggers in time.

- Triggers fit in a single MPEG-2 Transport Stream packet.

- The delivery and processing of triggers is such that there is the capability of it being within N (where N is small) frames of any PCR timeline discontinuity.

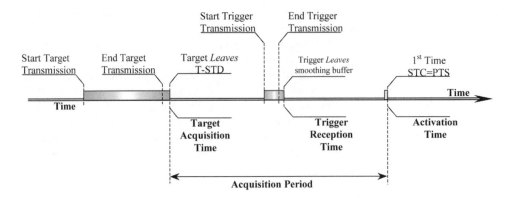

Figure 5.9.7 Trigger time line defined in A/93. (*From* [6]. *Used with permission.*)

5.9.4b   Temporal Concepts

Figure 5.9.7 illustrates the various temporal concepts introduced by A/93 [6]. The *pre-load data*—referred to as the *target*—is considered to be acquired, and the trigger is regarded as received, only after it leaves the smoothing buffer of the *Transport System Target Decoder* (T-STD) associated with the MPEG-2 Program Element that carries it. A synchronized trigger and its target are considered activated when the PTS value matches, for the first time, the 90 kHz portion of the receiver System Time Clock. The acquisition period, during which the receiver is expected to acquire and decode the target, is the period between the time the target leaves its associated T-STD and the time of activation. Unpredictable behavior may occur in the case where the target is not fully decoded and ready at the time of activation.

5.9.4c   Trigger Overview

A key property of trigger design is the ability to decouple the asynchronous delivery (and decoding) from the activation of objects or events (synchronized or asynchronous activation) [6]. ATSC triggers are light-weight constructs that carry pointers to the objects to be activated, to the application that needs to process the pre-load data for purposes of presentation, and some additional self-contained user-data (see Figure 5.9.8). Synchronized triggers also carry a PTS that indicates the point along the content timeline for synchronized presentation.

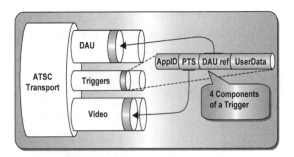

Figure 5.9.8 Overall A/93 trigger structure. (*From* [6]. *Used with permission.*)

While the DAU referenced by a trigger should be emitted *before* the trigger, the PTS of the trigger synchronizes with an instant of the STC as reconstructed by the receiver.

Figure 5.9.9 illustrates a simplified trigger timing diagram. In this example, the STC is discontinuous at time t_0 due to, for example, commercial insertion. In addition, a Data Carousel may be used to deliver the pre-load data (Note: depending upon the application, non-carouseled or other non-Download Protocol-based delivery methods may be used as well). The PTS of the trigger indicates when the pre-load data, referenced by the trigger, needs to be activated; however, standard A/93 does not specify what to do with the data activated, nor whether or how to display it.

Synchronized triggers are activated when the 90 kHz part of the STC in the receiver matches the value of the PTS specified by the trigger. To remove ambiguity regarding activation time, one of the requirements of the standard is that the packet carrying a synchronized trigger must be transmitted after the packet carrying the first PCR value of a new timeline, t_0. The PTS must be sufficiently delayed after t_0 (at least 33 ms) in order to enable the receiver to place the trigger in the buffer, decode it, and activate the pre-load data at the time that the value of the 90 kHz portion of the receiver STC strikes the PTS, denoted t_{pts}.

The pre-load data is transmitted asynchronously, possibly using a Data Carousel. To ensure that the pre-load data arrives and is decoded sufficiently early, consideration must be given to the bandwidth allocated to the carrying carousel as well as its repeat rate. When the receiver fully receives a DAU representing pre-load data, decoding begins immediately, so the object will be in the appropriate state for presentation as directed by the synchronized trigger message.

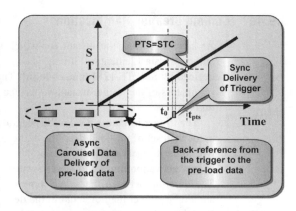

Figure 5.9.9 Simplified ATSC trigger timing diagram. (*From* [6]. *Used with permission.*)

5.9.4d   Activation

Trigger activation is the process of enabling the target referenced by a trigger, which may cause rendering [6]. For asynchronous triggers, activation occurs as soon as the trigger decoding is complete; the trigger decoding may start as early as when it has been fully received. For synchronized triggers, activation occurs at the time that the value of the 90 kHz portion of the receiver STC strikes the PTS carried in the trigger; as before, the trigger decoding starts as soon as it is fully received.

The A/93 Standard does not present any normative receiver behavior requirements with regard to the presentation of the data. In particular, there are no normative display (or un-display) specifications.

Triggers may be transmitted repeatedly to ensure capture through random tuning. Upon repeated transmission, however, it is critical to ensure that the identities of the triggers are generated so as to ensure their uniqueness.

5.9.4e   Content Creation Ramifications

At authoring time, the instant for which synchronization occurs for data relative to a video, audio, or data Elementary Stream is captured relative to a story timeline; for example such as defined by SMPTE 12M time code [6]. These *synchronization time instants* are regarded as the originators of the trigger instances defined in A/93, with their activation times in story timeline units.

The trigger instances need to be generated before emission. Upon generating a trigger and an associated set of pre-load data, the authoring system may need to compute an arrival time relative to the trigger activation time point. Furthermore, trigger generation requires conversion of trigger activation times and target acquisition times from the story timeline units to STC units. The duration between the arrival time of the target and the first trigger activation time point should be greater than or equal to the decoding time needed by any receivers to decode the pre-load data. This duration may be derived by the authoring system based on a set of metadata associated with the pre-load data to be synchronized.

Because the pre-load data can be delivered asynchronously (e.g., via a Data Carousel), the authoring system needs to compute a maximum target acquisition time period during which the receiver should not purge the pre-load data from the cache. This maximum target acquisition period is metadata-associated to the pre-load data via an acquisition descriptor. The maximum target acquisition period is the period between the earliest target acquisition time and the latest activation time for a specific target using all emitted triggers.

Note that the calculation of time delays needs to be robust so as to avoid synchronization errors being introduced by re-multiplexers that will not likely parse the trigger content to analyze activation times. Unpredictable behavior may occur in the event that the pre-load data is not available (or accessible or decoded) at the activation time.

5.9.5   ATSC A/94 Standard

ATSC standard A/94 defines an Application Reference Model (ARM), including a binding of Application Environment facilities onto the ATSC Data Broadcast Standard [7]. This standard includes a system-wide resource naming scheme, a state model, Data Models, and constraints and extensions to A/53, A/65, and A/90 to implement Application Environments.

The architecture and facilities of this standard are intended to apply to terrestrial (over-the-air) broadcast systems and receivers. In addition, the same architecture and facilities may be applied to other transport systems (such as cable or satellite).

5.9.5a   System Assumptions

Standard A/94 builds on ATSC DTV standards for the general transport and carriage of data [7]. Specifically, it is built on A/53 (including A/52), A/65, A/90, A/91, A/92, A/93, and the Transport Stream File System Standard (A/95). The relationship of this standard to other ATSC standards is shown in Figure 5.9.10. Note that the location of these blocks do not designate subset relationships.

5.9.5b   Functional Overview

The A/94 standard defines an architecture and framework for ATSC DTV A/90 Data Service applications and the environments in which those applications are processed [7]. This architecture is specified in terms of the following models:

- Data Model

- Application Model

- Application Environment Model

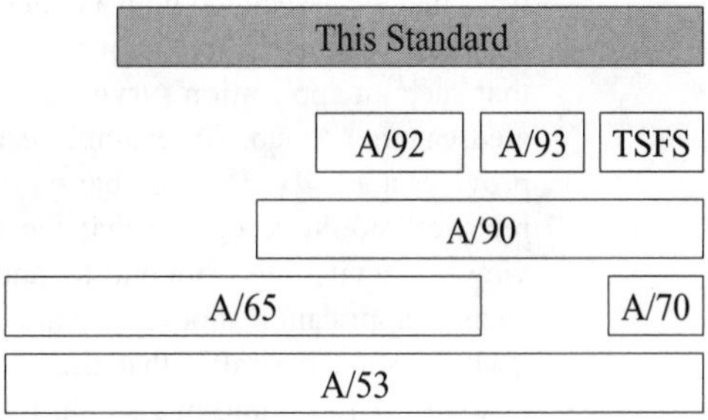

Data Models

Figure 5.9.10 Relationship among ATSC Standards. (*From* [7]. *Used with permission.*)

The following basic Data Models are supported by A/94 [7]:

- **Internet Protocol (IP) packets**—the standard carriage of internet protocol packets, just as on any other IP network transport, as defined in A/90.

- **Modules**—bounded sequences of bytes that use the DSM-CC data download name space as defined in A/90.

- **Files**—bounded sequences of bytes with a text-based, hierarchical namespace, and attributes found for traditional files in common operating systems such as defined in ATSC A/95.

- **Streams**—unbounded sequences of bytes.

- **Triggers**—structures used to provide synchronization services containing a reference to an application that is intended to process it, a reference to a target, or opaque data bytes. It may optionally contain a Presentation Time Stamp.

Application Model

An *Application Model* provides a means by which applications are organized, delivered (interchanged), and processed. [7]

A *Resource* is an embodiment of one of the Data Models described previously. A resource is equivalent to SMPTE data essence material.

An *Application* is a collection of resources and their related metadata, along with application metadata.

An *Application Environment* is the receiver environment that processes an application.

Application Categories

An *Application Category* is its relationship to the virtual channel [7]. The categories supported by A/94 are:

- Bound

- Shared

A *Bound Application* is one that is associated with a single virtual channel and can be activated only by tuning to that channel, and is stopped by tuning away. In contrast, some applications may be associated with a collection of related virtual channels, such as a set of ATSC minor channels within a single major channel (e.g., all XBC channels: 10.1, 10.2, etc). It is desirable that such an application survive channel changes as long as the receiver stays within the associated channel group. An example is a broadcaster-specific electronic program guide (EPG) that provides a small navigation bar on the bottom of the screen for a group of related channels. The receiver should recognize that the same application is signaled on the multiple channels the viewer is switching from and to, and not stop, then reload and restart the application. However, such an application should suspend during this channel transition. This is a *Shared Application*.

It is often desirable that the execution of certain applications be independent of channel changes. An example of such an Unbound Application is a non-broadcaster-specific EPG. The application resides in the receiver after being downloaded through a virtual channel or other means. A viewer can activate and stop the application regardless which virtual channel is being watched. The states and transitions needed for the category of *Unbound Application* are outside the scope of Standard A/94.

Application Announcement

The future availability of an application for possible processing and presentation by an Application Environment may be announced in the transport [7]. The intention of application announcement is to notify the receiver of the future availability of the application so that the receiver can take appropriate action to schedule use of the application and/or inform the viewer.

The announcement of an application is characterized by its metadata:

- Profile and Level

- Availability time and duration

- Rating information

- Language

and the following optional metadata:

- Title

- Description

Note that A/90 provides for announcement of events within Data Services only, and not individual applications.

Application Signaling

The availability of an application for processing and presentation by an Application Environment is signaled in the transport [7]. The signaling of an application is accompanied by metadata that describes all the resources that embody the application. This Application Signaling Metadata is as follows:

- Application Identifier

- Profile and Level

- Category

- Invocation Directive (bootstrap or runtime)

- Identifier of the root resource

It is not expected that this metadata will be presented to a viewer, but will instead be used by the Application Environment for processing the application.

Resource Signaling

Each resource listed in the application can include the following metadata [7]:

- Identifier

- Content type

In addition, for *Module Data Models*, there is the additional metadata:

- Acquisition Directive

and, for *file Data Models*:

- Miscellaneous attributes (as defined in A/95)

A *resource identifier* takes the form of an absolute *Universal Resource Identifier* (URI); this identifier is globally unique. A *resource content type* takes the form of a MIME media type; note that MIME media types may take parameters.

An optional Acquisition Directive is conveyed by means of an application descriptor, as defined in A/93, which indicates to the Application Environment that it should promptly acquire the resource, with the expectation that a trigger referencing the resource will be arriving soon.

Security and Policy

There is no requirement for added support at the transport level for security models or conditional access [7]. This topic is not covered in the A/94 standard; readers are instead referred to ATSC document A/70.

However, content authors and broadcasters may wish to assert their policy positions with respect to control of certain actions that a Data Service may take. This takes the form of asserting a denial of certain "policies," where each "policy" gives permission for a certain specific type of action in the Application Environment. In the absence of any explicit assertion to deny a policy, any permissions affected by the policy remain grantable. To assert denial of policies a *Broadcaster Policy Descriptor* is transmitted. In general, this will result in one or more permissions being denied.

5.9.6   ATSC A/95 Standard

Document A/95 defines the ATSC Transport Stream File System (TSFS) standard for delivery of hierarchical name-spaces, directories, and files [8]. This standard builds upon the Data Service delivery mechanism defined in A/90, and other documents in the data broadcast suite of standards described in this chapter (Figure 5.9.11). The TSFS is based on the *Object Carousel* design specified in ISO/IEC 13818-6:1998[3].

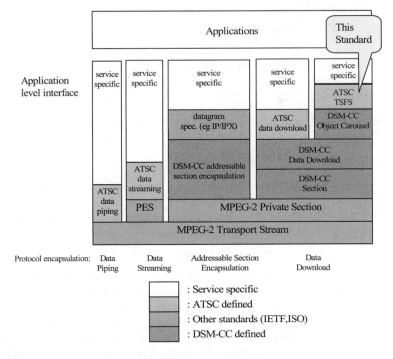

Figure 5.9.11 Graphical encapsulation overview and relation to other standards. (*From* [8]. *Used with permission.*)

Figure 5.9.12 illustrates the general structure of a digital storage media (DSM) directory object and a DSM file object. A DSM directory object includes an object information (objectInfo) structure and a binding structure referencing zero or more "child" objects (DSM::Directory or DSM::File objects). The binding information includes an objectInfo field for the purpose of providing early information about a referenced object.

Each TSFS has one *Service Gateway* object, which serves as the top level directory of the TSFS. The objects in the TSFS can all be reached by following a directory path (sequence of directory links) starting from the Service Gateway. An object in a TSFS can be referenced from multiple directories in the TSFS, so there can be multiple paths from the Service Gateway to an object in the TSFS. Such paths cannot create circular routes in the directory reference structure. For example, the situation where directory A contains a reference to directory B, which in turn contains a reference to directory A, cannot occur.

The names that appear in the binding structures of the Service Gateway object all conform to the syntax of an absolute *Uniform Resource Identifier* (URI) and the names that appear in the binding structures of lower-level directories of a TSFS all conform to the syntax of a relative

3. "Information technology — Generic coding of moving pictures and associated audio information — Part 6: Extensions for DSM-CC," September 1, 1998.

(a)

(b)

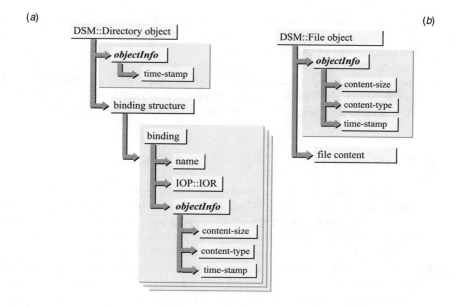

Figure 5.9.12 File system conventions: (*a*) directory object, (*b*) file object.. (*From* [8]. *Used with permission.*)

path. The effect of this structure is that each file or directory object in a TSFS, with the exception of the Service Gateway object itself, has an associated absolute URI that can be used to reference the object, obtained by concatenating the names in a sequence of directory references leading from the Service Gateway to the object, with slashes (/) as delimiters between them. If there are multiple such sequences of directory references leading to the object, then it has multiple URIs associated with it.

An *Interoperable Object Reference* (IOR) referencing an object in the same TSFS contains the object key for the referenced object, the carousel ID and module ID of the Data Carousel module containing the object, a "tap" identifying the Program Element containing the message describing the delivery parameters for the module, and optionally a tap identifying the Program Element containing the messages carrying the module itself. These messages gives such information as the block size, module size, module version, and delivery timeout intervals for the module.

All objects of a TSFS are carried in a single virtual channel. However, it is also possible to have bindings that contain references to objects in other TSFSs (often called "soft links"). These other TSFSs may be in the same virtual channel or a different virtual channel—even a different virtual channel in a different Transport Stream. Thus, the logical name space of a TSFS may span multiple virtual channels and even multiple Transport Streams.

An IOR referencing an object in a different TSFS identifies the TSFS containing the object and gives the directory path (sequence of directory links) in that TSFS which leads from the Service Gateway of that TSFS to the object. Organization of the ATSC TSFS into the Data Modules of an ATSC Data Carousel is not specified by A/95.

To allow unambiguous identification of a TSFS in an ATSC Transport Stream, a new protocol_encapsulation value is defined for the Data Service Table of the Data Broadcast Standard to signal the Object Carousel encapsulation. For each TSFS used by an application in a Data Service, the DST contains a tap referencing the Program Element containing the DSI (*Download Server Initiate*) message that signals the location of the Service Gateway object for the TSFS. In the case where this Program Element resides in a remote ATSC virtual channel or a remote ATSC Transport Stream, it is signaled in the Network Resources Table (NRT).

5.9.7   ATSC A/96 Standard

Interactive content has long been assumed to hold an important place in the future of digital television. ATSC Standard A/96, "ATSC Interaction Channel Protocols," represents a significant step toward this goal. Remote interactivity spans a wide range of possible applications and markets, including E-commerce transactions during commercials, electronic banking, polling, email services, and a whole spectrum of other services yet to be defined.

Standard A/96 defines a core suite of protocols to enable remote interactivity in television environments. Remote interactivity requires the use of a two-way *Interaction Channel* that provides for communications between the *Client Device*, such as a DTV receiver or set-top box, and one or more *Remote Servers* [9].

A typical topology for interactive television involves a client using a modem to exchange data with an Interactive Content Service Provider (ICSP). In order to communicate with the ICSP, the client may need to establish contact first with an Interaction Channel Provider (ICP), a role that may be filled by a conventional Internet Service Provider (ISP).

In a more complex access scheme, the interactive television client could be part of a home local area network (LAN) incorporating other elements such as hubs, switches, routers, servers, firewalls, and related devices. The LAN communicates with the Interaction Channel Provider typically through a broadband access line such as a DSL or cable modem.

Data exchanged between the two ends of the system (the consumer and the service provider) might include, for example:

* Application data related to particular television programs

* News on-demand extracted from a Web server

* Credit card information uploaded during an e-commerce transaction

Interactive television services, by definition, allow a certain degree of interaction between a user and an application running on a DTV receiving device and/or on a Remote Server. One way of providing this interactivity is by using *Carousel Data* that can be stored locally. In this case, a user will run the application software based on data and code stored on the client machine. A second method involves user access of data or software from a Remote Server using an Interaction Channel.

A/96 relies on established Internet protocols. Figure 5.9.13 illustrates the parallel infrastructure defined by the standard and the manner in which it relates to other ATSC standards.

An application in the form of files carrying code and data could be delivered using *Forward Broadcast* Protocols. For ATSC, the broadcast files could include ACAP content carried using the multiple layers of ATSC data transport protocols, and announcement protocols provided for by PSIP. Upon execution, the application may invoke Interaction Channel Protocols to communi-

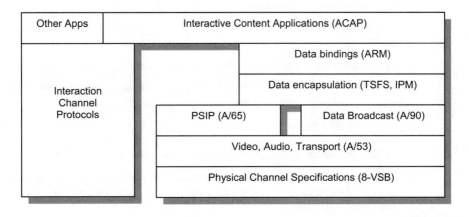

Figure 5.9.13 Relation of Interaction Channel Protocols and other ATSC standards. (*From* [9]. *Used with permission.*)

cate with ICSP Remote Servers and provide the remote interactivity experience. Figure 5.9.13 illustrates this case by showing an ACAP protocol layer on top of both forward broadcast and Interaction Channel Protocols.

Additional applications such as e-mail, games with remote interactivity, or Web services could be deployed separately from broadcast facilities using the standardized Interaction Channel Protocols. The figure illustrates this case by showing the block 'Other Apps' independently structured on top of the Interaction Channel Protocols.

5.9.8 ATSC A/97 Standard

ATSC standard A/97, "Software Download Data Service," specifies a Data Service that may be used to download software to a terminal device using a Transport Stream via an appropriate physical layer [10]. This service may be used to effect updates or upgrades of firmware, operating system software, device driver software, native application software, middleware, and other types of software that reside in a terminal device. A/97 specifies the announcement, signaling, and encapsulation process for the delivery of this download Data Service.

The content and format of the software download data is not defined by A/97. Rather, the formats and interpretations of the software download payload are defined by each user.

The behavior and facilities of A/97 are intended to apply primarily to terrestrial television broadcast systems and receivers. In addition, the same behavior and facilities may be specified and/or applied to other transport systems, such as cable or satellite.

In order to better understand the design rationale, consider the following typical application scenarios.

Scenario 1: A carousel broadcasts modules targeting a single device.:

- Typically used in an environment where there is no aggregator and a single manufacturer is creating a carousel to support only one hardware/software model.
- The carousel targets a single hardware/software version at a time.
- Additional hardware/software versions would be supported by terminating the carousel and restarting with new announcement, signaling, and data.

This is the simplest model.

Scenario 2: A carousel broadcasts modules targeting multiple devices.:

- Typically used in an environment where there is an aggregator supporting multiple manufacturers or where a single manufacturer would like to support multiple hardware/software versions on a single carousel.
- The carousel targets multiple hardware/software versions at a time.
- Announcements are a critical part of the functionality of this scenario.

Scenario 3: Multiple carousels.:

- Typically used in an environment where there is a combination of aggregators and/or individual manufacturers all creating carousels for a single transport.
- Each virtual channel may contain carousel(s) of Scenario 1 or Scenario 2.
- Multiple Virtual Channels containing carousels for software downloads may exist on a single transport.

  Note that Scenario 1 is a simplification of 2, and Scenario 2 is a simplification of Scenario 3. The software download Data Service is based on the 2-layer carousel scenario of ATSC A/90, including the DSI (Download Server Initiate), DII (Download Information Indication), and DDB (Download Data Block) messages as describe in A/90. Top-level signaling is accomplished via a new Virtual Channel Table service_type. Announcement is accomplished via schedule information added to the DSI. Thus, the DSI is used for both signaling and announcement; and the DII is used for some parts of the module signaling.

  The application scenarios described above are supported by this design as follows:

- Scenario 1 is a single carousel in a single Virtual Channel and contains only a single group. Multiple downloads are managed serially through complete version changes of all DSI and DII messages.
- Scenario 2 is also a single carousel in a single Virtual Channel, but it makes use of multiple groups. Multiple downloads are managed concurrently through changes to the DSI and DII messages as needed.
- Scenario 3 is a combination of scenarios 1 and/or 2 in multiple Virtual Channels. Multiple downloads are managed as in those scenarios on a per channel basis.

5.9.9   ATSC A/101 Standard

The Advanced Common Application Platform (ACAP) specification, developed as the result of a harmonization effort between the ATSC DTV Application Software Environment (DASE) and CableLabs' Open Cable Application Platform (OCAP) specifications, is the culmination of extensive effort by a dedicated team of specialists from dozens of organizations representing diverse industry segments. ACAP provides consumers with advanced interactive services while providing content providers, broadcasters, cable and satellite operators, and consumer electronics manufacturers with the technical details necessary to develop interoperable services and products.

5.9.9a   About Interactive Television

The rollout of the digital television infrastructure facilitates a new era in service to the consumer built around two-way interactive technologies. Two worlds that were once barely connected—television and the Internet—are now on the verge of combining into an entirely new service: namely, interactive television. Thanks to the ongoing transition of television from analog to digital, it is now possible to efficiently combine video, audio, and data within the same signal.

This combination leads to powerful new applications. For example, computers can be turned into traditional TV receivers and digital set-top boxes can host applications such as interactive TV, e-commerce, and customized programming.

The term *interactive television* (ITV) is broad and not entirely well defined. However, it certainly includes the following general categories:

- Customized news, weather and traffic

- Stock market data, including personal investment portfolio performance in real-time

- Enhanced sports scores and statistics on a selective basis

- Games associated with program

- On-line real-time purchase of everything from groceries to software without leaving home

- Video on demand (VOD)

There is no shortage of reasons why ITV is viewed with considerable interest around the world. The backdrop for ITV growth comes from both the market strength of the Internet and the technical foundation that supports it. With the rapid adoption of digital video technology in the cable, satellite, and terrestrial broadcasting industries, the stage is set for the creation of an ITV segment that introduces to a mass consumer market a whole new range of possibilities.

For example, technologies are available that support interactive features for game shows, sports and other programs, interactive advertising, e-mail, and Internet access. Rather than concentrating just on Web services, the goal is to deliver a better television experience.

5.9.9b   About ACAP

In essence, ACAP makes it appear to interactive programming content that it is running on a so-called *common receiver*. This common receiver contains a well-defined architecture, execution model, syntax, and semantics. As a "middleware" specification for interactive applications,

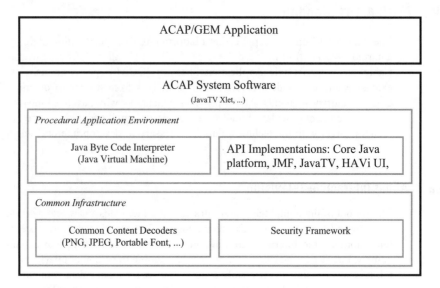

Figure 5.9.14 ACAP-J system architecture. (*From* [11]. *Used with permission.*)

ACAP gives content and application authors assurance that their programs and data will be received and run uniformly on all brands and models of receivers.

An *ACAP Application* is a collection of information that is processed by an *Application Environment* in order to interact with an end-user or otherwise alter the state of the Application Environment.

ACAP Applications are classified into two categories depending upon whether the initial application content processed is of a *Procedural* or a *Declarative* nature. These categories of applications are referred to as procedural (ACAP-J) and declarative (ACAP-X) applications, respectively. An example of an ACAP-J application is a Java TV™ Xlet composed of compiled Java byte code in conjunction with other multimedia content such as graphics, video, and audio. An example of an ACAP-X application is a multimedia document composed of XHTML markup, style rules, scripts, and embedded graphics, video, and audio.

An ACAP application need not be purely procedural or declarative. In particular, an ACAP-J application may reference declarative content, such as graphics, or may construct and cause the presentation of markup content. Similarly, ACAP-X applications may use script content, which is procedural in nature. Furthermore, an ACAP-X application may reference an embedded Java TV Xlet.

The architecture and facilities of ACAP are intended to apply to broadcast systems and receivers for terrestrial broadcast and cable TV systems. In addition, the same architecture and facilities may be applied to other transport systems, such as satellite.

ACAP is primarily based on the GEM (Globally Executable Multimedia Home Platform) specification developed by the DVB consortium and DASE (DTV Software Application Environment) developed by the ATSC. ACAP includes additional functionality from Cable Labs' OCAP specification. ACAP builds upon GEM by adding specification elements in order to offer

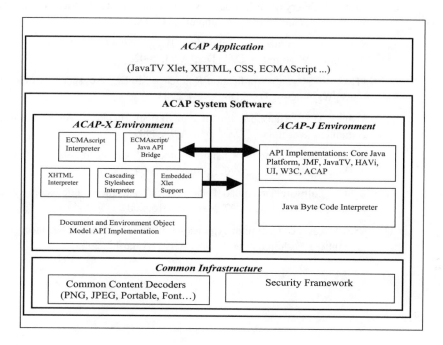

Figure 5.9.15 ACAP Application and system software. (*From* [11]. *Used with permission.*)

a high degree of interoperability among different environments based on digital TV specifications from ATSC and SCTE.

Where only ACAP-J applications are supported, the application and system software are as shown in Figure 5.9.14.

Where ACAP-X applications are supported, the application and systems software are as illustrated in Figure 5.9.15.

5.9.10 References

1. NAB: "An Introduction to DTV Data Broadcasting," *NAB TV TechCheck*, National Association of Broadcasters, Washington, D.C., August 2, 1999.

2. "Navigation of DTV Data Broadcasting Services," *NAB TV TechCheck*, National Association of Broadcasters, Washington, D.C., November 1, 1999.

3. ATSC: "ATSC Data Broadcast Standard," Advanced Television Systems Committee, Washington, D.C., Doc. A/90, July 26, 2000.

4. ATSC: "Recommended Practice—Implementation Guidelines for the ATSC Data Broadcast Standard," Advanced Television Systems Committee, Washington, D.C., Doc. A/91, May 10, 2001.

5. ATSC: "Delivery of IP Multicast Sessions over ATSC Data Broadcast," Advanced Television Systems Committee, Washington, D.C., Doc. A/92, 2002.

6. ATSC: "Synchronized/Asynchronous Trigger Standard," Advanced Television Systems Committee, Washington, D.C., Doc. A/93, 2002.

7. ATSC: "ATSC Data Application Reference Model," Advanced Television Systems Committee, Washington, D.C., Doc. A/94, 2003.

8. ATSC: "ATSC Transport Stream File System Standard," Advanced Television Systems Committee, Washington, D.C., Doc. A/95, 2003.

9. ATSC: "ATSC Interaction Channel Protocols," Advanced Television Systems Committee, Washington, D.C., Doc. A/96, February 3, 2004.

10. ATSC: "Software Download Data Service," Advanced Television Systems Committee, Washington, D.C., Doc. A/97, November 16, 2004

11. ATSC: "Advanced Common Application Platform (ACAP)," Advanced Television Systems Committee, Washington, D.C., Doc. A/101, August 2, 2005.

DTV Implementation Considerations

As with any new technology, the implementation of various aspects of digital television have experienced growing pains. Certain requirements and trade-offs—not anticipated during initial design of a given device, system, or standard—inevitably must be resolved while the technology is being implemented.

The distinction between a digital video system intended for production applications and a system intended for the transmission of image and sounds is an important one. Because of their closed-loop characteristics, production systems can be of any practical design. The process of developing a production system can focus simply on those who will directly use the system. It is not necessary to consider the larger issues of compatibility and public policy, which drive the design and implementation of over-the-air broadcast systems.

Although the foregoing is certainly correct, in the abstract, it is obvious that the economies of scale argue in favor of the development of a production system—even if only closed-loop—that meets multiple applications. The benefits of expanded markets and interoperability between systems are well documented. It was into this environment that DTV was born.

This section examines some of the implementation issues involved in the transition from analog to digital broadcasting. Many more issues will be identified—and resolved—before the DTV transition is completed. The chapters that follow are intended as a starting point.

In This Section:

6.1

Video Production Issues

Laurence J. Thorpe, Jerry C. Whitaker[1]

6.1.1 Introduction

The resolution of the displayed picture is the most basic attribute of any video production system. Generally speaking, an HDTV image has approximately twice as much luminance definition horizontally and vertically as the 525-line NTSC system or the 625-line PAL and SECAM systems. The total number of luminance picture elements (*pixels*) in the image, therefore, is 4 times as great. The wider aspect ratio of the HDTV system adds even more visual information. The HDTV image is 25 percent wider than the conventional video image for a given image height; the ratio of image width to height in HDTV systems is 16:9, or 1.777. The conventional video image has a 4:3 aspect ratio.

As a result of these attributes, the HDTV image may be viewed more closely than is customary in conventional television systems. Full visual resolution of the detail of conventional television is available when the image is viewed at a distance equal to about 6 or 7 times the height of the display. The HDTV image may be viewed from a distance of about 3 times picture height for the full detail of the scene to be resolved.

6.1.1a HDTV Defined

Although there is no single, universal definition for HDTV, it is generally accepted to encompass several elements which are described by various consumer, broadcast, and regulatory groups. HDTV offers the potential for approximately twice the horizontal and twice the vertical resolution of current (NTSC) television. When combined with a wide screen format (16:9 aspect ratio), this can result in considerably more visual information than conventional television. When referring to consumer products, HDTV sets:

- Can have 720 or 1080 active vertical scanning lines.

---

1. From *Standard Handbook of Video and Television Engineering*, 4th. ed., Jerry C. Whitaker (ed.), McGraw-Hill, New York, N.Y., 2003. Used with permission.

- Are capable of decoding the transmitted 720 × 1280 and 1080 × 1920 ATSC formats and displaying them as a 16:9 aspect ratio image. These two HD formats can potentially provide over eight times as much picture information as delivered over broadcast NTSC.

While these high-definition transmission formats will be supported by such sets, the actual delivered resolution may vary by broadcaster, by product, and by program. HDTV is normally accompanied by digital surround-sound capability.

6.1.1b   Production Systems vs. Transmission Systems

Bandwidth is perhaps the most basic factor that separates production HDTV systems from transmission-oriented systems for broadcasting. A closed-circuit system does not suffer the same restraints imposed upon a video image that must be transported by radio frequency means from an origination center to consumers. It is this distinction that has led to the development of varied systems for production and transmission applications. Terrestrial broadcasting of NTSC video, for example, is restricted to a video baseband that is 4.2 MHz wide. The required bandwidth for full resolution HDTV, however, is on the order of 30 MHz. Fortunately, video-compression algorithms are available that can reduce the required bandwidth without noticeable degradation and still fit within the restraints of a standard NTSC, PAL, or SECAM RF channel. The development of efficient compression systems was, in fact, the breakthrough that made the all-digital HDTV system possible.

Video compression involves a number of compromises. In one case, a tradeoff is made between higher definition and precise rendition of moving objects. It is possible, for example, to defer the transmission of image detail, spreading the signal over a longer time period, thus reducing the required bandwidth. If motion is present in the scene over this longer interval, however, the deferred detail may not occupy its proper place. Smearing, ragged edges, and other types of distortion can occur.

6.1.1c   Business and Industrial Applications

Although the primary goal of the early thrust in HDTV equipment development centered on those system elements essential to support program production, a considerably broader view of HDTV anticipated important advances in its wider use. The application of video imaging has branched out in many directions from the original exclusive over-the-air broadcast system intended to bring entertainment programming to the home. Throughout the past three decades, television has been applied increasingly to a vast array of teaching, training, scientific, corporate, and industrial applications. The same era has seen the emergence of an extensive worldwide infrastructure of independent production and postproduction facilities to support these needs.

As the Hollywood film industry became increasingly involved in supplying prime-time programming for television (on an international basis) via 35 mm film origination, video technology was harnessed to support off-line editing of these film originals, and to provide creative special effects. Meanwhile, as the world of computer-generated imaging grew at an explosive pace, it too began penetrating countless industrial, scientific, and corporate applications, including film production. The video industry has, in essence, splintered into disparate (although at certain levels, overlapping) specialized industries, as illustrated in Figure 6.1.1. Any of these video application sectors is gigantic in itself. It was into this environment that HDTV was born.

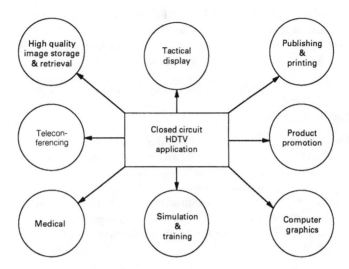

Figure 6.1.1 Applications for high-definition imaging in business and industrial facilities.

Apart from the issues of terrestrial (or cable and/or direct-broadcast satellite) distribution of HDTV entertainment programs, there exists another crucial issue: electronic imaging as a whole. Video technology is being applied to a vast diversity of applications. These include medical, teleconferencing, science, art, corporate communication, industrial process control, and surveillance. For some of these applications, 525 NTSC has been adequate, for some barely adequate, and for others woefully inadequate.

6.1.1d   Broadcast Applications

As with many things technical, the plans of the designers and those of the end users do not always coincide. It was assumed from the beginning of the standardization process for HDTV that the result would be a system specifically for the delivery of pictures and sound of superb quality—a quantum leap beyond NTSC. The flexible nature of the ATSC DTV system permits broadcasters to decide whether they would like to send to viewers one superquality HDTV program or several "NTSC-equivalent" programs.

Obviously, the choice of HDTV programming or multiple-stream standard-definition programming has a considerable impact on facility design and budget requirements. Once that decision has been made, the focus moves to implementation issues:

- Studio-to-transmitter (STL), intercity relay (IRC), and satellite links

- Master control switching and routing

- Production equipment (cameras, switchers, special effects systems, recorders, and related hardware)

- Studios and sets for wide-screen presentations

The conversion from NTSC to DTV often is compared with the long-ago conversion from black and white to color. Many of the lessons learned in the late 1950s and early 1960s, however, are of little use today because the broadcast paradigm has shifted. But the most important lesson from the past still is valid: build the technology around the programming, not the other way around.

6.1.1e   Computer Applications

One of the characteristics that set the ATSC effort in the U.S. and DVB in Europe apart from the NTSC/PAL/SECAM efforts of the past was the inclusion of a broad range of industries—not just broadcasters and receiver manufacturers, but all industries that had an interest in imaging systems. It is, of course, fair to point out that during work on the analog black-and-white and color standards, there were no other industries involved in imaging. Be that as it may, the broad-based effort encompassed by the ATSC and DVB systems has ensured that the standards will have applications in far more industries than simply broadcast television.

6.1.2   HDTV and Related Media

The core of HDTV production is the creation of high-quality images. As HDTV was emerging in the early 1980s, a quite separate and initially unrelated explosion in electronic imaging was also under way in the form of high-resolution computer graphics. This development was propelled by broad requirements within a great variety of business and industrial applications, including:

- Computer-aided design (CAD)

- Computer-aided manufacturing (CAM)

- Printing and publishing

- Creative design (such as textiles and decorative arts)

- Scientific research

- Medical diagnosis

A natural convergence soon began between the real-time imagery of HDTV and the non-real-time high-resolution graphic systems. A wide range of high-resolution graphic display systems is commonly available today. This range addresses quite different needs for resolution within a broad spectrum of industries. Some of the more common systems are listed in Table 6.1.1.

The ATSC DTV system thus enjoys a good fit within an expanding hierarchy of computer graphics. This hierarchy has the range of resolutions that it does because of the varied needs of countless disparate applications. HDTV further offers an important wide-screen display organization that is eminently suited to certain critical

Table 6.1.1 Common Computer System Display Formats

Horizontal Pixels	Vertical Pixels (Active Lines)
640	480
800	600
1024	768
1280	1024
1280	1536
2048	1536
2048	2048

Table 6.1.2 Basic Characteristics of Common Computer Graphics Displays (*After* [1])

Parameter	Value				
Horizontal Resolution	640	800	800	1024	1280
Vertical Resolution	480	600	600	768	1024
Active Lines/Frame	480	600	600	768	1024
Total Lines/Frame	525	628	666	806	1068
Active Line Duration, μs	20.317	20.00	16.00	13.653	10.119
f_h (Hz)	37.8	37.879	48.077	56.476	76.02
f_v (MHz)	72.2	60.316	72.188	70.069	71.18
Pixel Clock, MHz	31.5	40	50	75	126.5
Video Bandwidth, MHz	15.75	20	25	37.5	63.24

demands. The 16:9 display, for example, can encompass two side-by-side 8 × 11-in pages, important in many print applications. The horizontal form factor also lends itself to many industrial design displays that favor a horizontally oriented rectangle, such as automobile and aircraft portrayal.

6.1.2a   Resolution Considerations

With few exceptions, computers were developed as stand-alone systems using proprietary display formats [1]. Until recently, computers remained isolated with little need to exchange video information with other systems. As a consequence, a variety of specific display formats were developed to meet computer industry needs that are quite different from the broadcast needs.

Among the specific computer industry needs are bright and flickerless displays of highly detailed pictures. To achieve this, computers use progressive vertical scanning with rates varying from 56 Hz to 75 Hz (or higher), referred to as *refresh rates*, and an increasingly high number of lines per picture. High vertical refresh rates and number of lines per picture result in short line durations and associated wide video bandwidths.

Because the video signals are digitally generated, certain analog resolution concepts do not directly apply to computer displays. To begin with, all *analog-to-digital* (A/D) conversion concerns related to sampling frequencies and associated anti-aliasing filters are nonexistent. Furthermore, there is no vertical resolution ambiguity, and the vertical resolution is equal to the number of active lines. The only limiting factor affecting the displayed picture resolution is the capabilities of the display system.

The computer industry uses the term *vertical resolution* when referring to the number of active lines per picture and *horizontal resolution* when referring to the number of active pixels per line. This resolution concept has no direct relationship to the classic television resolution concept as developed for analog TV and can be misleading (or at least confusing). Table 6.1.2 summarizes the basic characteristics of common computer graphics displays.

6.1.2b   Film Production and Media Assets

In today's distribution environment, feature films are usually shot on 35 mm film [2]. After the film is distributed theatrically, using 35 mm prints, it is usually converted by telecine to standard NTSC, PAL, and/or other video standards and distributed on broadcast quality digital and/or analog tape for broadcast, satellite, and cable delivery. It is subsequently duplicated to DVD for distribution in the home video rental and sell-through markets. Other forms of video entertainment and information, which are not intended for film release, are often produced on video. The exception is when productions, intended for television or video, are shot on film in anticipation of a possible future re-release in a higher resolution format. This method is commonly employed to protect the future value of unique and/or valuable content, or *media assets*.

The concept of capturing, editing, and archiving on a higher resolution format than the current distribution system is, of course, nothing new; it dates back to the classic television sitcom "I Love Lucy." The financial rewards of such an approach are not lost on program producers. As delivery systems improve, the original media asset can be copied into better formats, retaining their original value.

6.1.3   Characteristics of the Video Signal

High-definition television has improved on earlier techniques primarily by calling more fully upon the resources of human vision [3]. The primary objective of HDTV has been to enlarge the visual field occupied by the video image, an attribute that can be used to benefit interactive TV applications. This attribute has called for larger, wider pictures that are intended to be viewed more closely than conventional video. To satisfy the viewer upon this closer inspection, the HDTV image must possess proportionately finer detail and sharper outlines.

6.1.3a   Critical Implications for the Viewer and Program Producer

In its search for a "new viewing experience," early experiments conducted an extensive psychophysical research program in which a large number of attributes were studied. Viewers with non-technical backgrounds were exposed to a variety of electronic images, whose many parameters were then varied over a wide range. A definition of those imaging parameters was being sought, the aggregate of which would satisfy the average viewer that the TV image portrayal produced an emotional stimulation similar to that of the large-screen film cinema experience.

Central to this effort was the pivotal fact that the image portrayed would be large—considerably larger than current NTSC television receivers. Some of the key definitions being sought by researchers were precisely how large, how wide, how much resolution, and the optimum viewing distance of this new video image.

A substantial body of research gathered over the years has established that the average U.S. consumer views the TV receiver from a distance of approximately seven picture heights. This translates to perhaps a 27-in NTSC screen viewed from a distance of about 10 ft. At this viewing distance, most of the NTSC artifacts are essentially invisible, with perhaps the exception of cross color. Certainly the scanning lines are invisible. The luminance resolution is satisfactory on camera close-ups. A facial close-up on a modern high-performance 525-line NTSC receiver, viewed from a distance of 10 ft, is quite a realistic and pleasing portrayal. But the system quickly fails on many counts when dealing with more complex scene content.

Wide-angle shots (such as jersey numbers on football players) are one simple and familiar example. TV camera operators have adapted to the inherent restrictions of 525-line NTSC, as witnessed by the continual zooming in for close-ups during most sporting events. The camera operator accommodates for the technical shortcomings of the conventional television system and

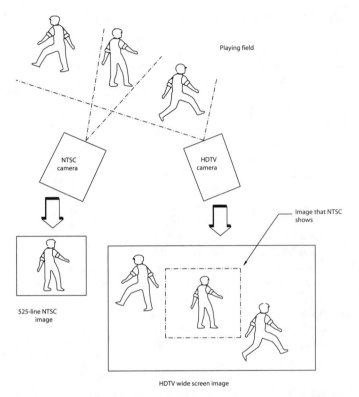

Figure 6.1.2 An illustration of the differences in the scene capture capabilities of conventional video and HDTV. (*After* [3].)

delivers an image that meets the capabilities of NTSC, PAL, and SECAM quite reasonably. There is a penalty, however, as illustrated in Figure 6.1.2. The average home viewer is presented with a very narrow angle of view—on the order of 10°. The video image has been rendered "clean" of many inherent disturbances by the 10-ft viewing distance and made adequate in resolution by the action of the camera operator and director; but, in the process, the scene has become a small "window". The now "acceptable" television image pales in comparison with the sometimes awesome visual stimulation of the cinema. The primary limitation of conventional TV systems is, therefore, image size. A direct consequence is further limitation of image content; the angle of view constantly is constricted by the need to provide adequate resolution. There is significant, necessary, and unseen intervention by the TV program director in the establishment of image content that can be passed on to the home viewer with acceptable resolution.

Compared with the 525-line NTSC signal (or the marginally better PAL and SECAM systems), HDTV offers a vast increase in total information contained within the visual image. If all this information is portrayed on an appropriate HDTV studio monitor, the dramatic technical superiority of HDTV over conventional technology easily can be seen. The additional visual information, coupled with the elimination of composite video artifacts, portrays an image almost totally free (subjectively) of visible distortions, even when viewed at a close distance.

On a high-quality monitor, HDTV displays a technically superb picture. The *information density* is high; the picture has a startling clarity. However, when viewed from a distance of approximately seven picture heights, it may be indistinguishable from a good NTSC portrayal. The wider aspect ratio is the most dramatic change in the viewing experience at normal viewing distances.

6.1.3b   Image Size

It is clear that HDTV is demanded by consumers. If consumers are to retain the average viewing distance of 10 ft, then the minimum image size required for an HDTV screen for the average definition of a totally new viewing experience is about a 75-in diagonal. This represents an image area considerably in excess of present "large" 27-in NTSC (and PAL/SECAM) TV receivers. In fact, as indicated in Figure 6.1.3, the viewing geometry translates into a viewing angle close to 30° and a distance of only three picture heights between the viewer and the HDTV screen. Compare this with the viewing angle for conventional systems at 10°, as shown in Figure 6.1.4.

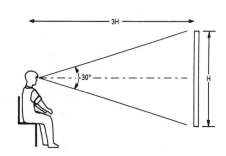

Figure 6.1.3 Viewing angle as a function of screen distance for HDTV. (*After* [3].)

HDTV Image Content

There is more to the enhanced viewing experience than merely increasing picture size [3]. The larger artifact-free imaging capability of HDTV allows a new image portrayal that capitalizes on the attributes of the larger screen. As mentioned previously, as long as the camera operator appropriately fills the 525 (or 625) scanning system, the resulting image (from a resolution viewpoint) is actually quite satisfactory on conventional systems. If, however, the

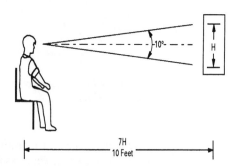

Figure 6.1.4 Viewing angle as a function of screen distance for conventional video systems. (*After* [3].)

same football game is shot with an HDTV camera and the angle of view of the lens is adjusted to portray the same resolution (in the picture center) as the 525 camera when capturing a close-up of a player on its 525 screen, a vital difference between the two pictures emerges: the larger HDTV image contains considerably more information, as illustrated in Figure 6.1.2.

The HDTV picture shows more of the football field—more players, more of the total action. Thus, the HDTV image is radically different from the NTSC portrayal. The individual players are portrayed with the same resolution on the retina—at the same viewing distance—but a totally different viewing experience is provided for the consumer. The essence of HDTV imaging is this greater sensation of reality.

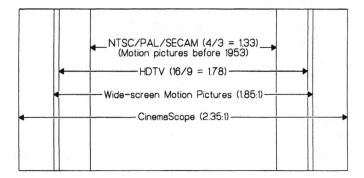

Figure 6.1.5 Comparison of the aspect ratios of television and motion pictures. (*After* [5].)

6.1.3c   Format Development

Established procedures in the program production community provide for the 4:3 aspect ratio of video productions and motion picture films shot specifically for video distribution. This format convention has, by and large, been adopted by the computer industry for desktop computer systems.

In the staging of motion picture films intended for theatrical distribution, no provision generally is made for the limitations of conventional video displays. Instead, the full screen, in wide aspect ratios—such as CinemaScope—is used by directors for maximum dramatic and sensory impact. Consequently, cropping of essential information may be encountered more often than not on the video screen This problem is particularly acute in wide-screen features where cropping of the sides of the film frame is necessary to produce a print for video transmission. This objective is met in one of the following ways:

- *Letter-box* transmission with blank areas above and below the wide-screen frame.

- Printing the full frame height and cropping equal portions of the left and right sides to provide a 4:3 aspect ratio. This process frequently is less than ideal because, depending upon the scene, important visual elements may be eliminated.

- Programming the horizontal placement of a 4:3 aperture to follow the essential picture information. Called *pan and scan*, this process is used in producing a print or in making a film-to-tape transfer for video viewing. Editorial judgment is required for determining the scanning cues for horizontal positioning and, if panning is used, the rate of horizontal movement. This is an expensive and laborious procedure and, at best, it compromises the artistic judgments made by the director and the cinematographer in staging and shooting, and by the film editor in postproduction.

One of the reasons for moving to a 16:9 format is to take advantage of consumer acceptance of the 16:9 aspect ratio commonly found in motion picture films. Actually, however, motion pictures are produced in several formats, including:

- 4:3 (1.33)

- 2.35, used for 35 mm anamorphic CinemaScope film

- 2.2 in a 70 mm format

Still, the 16:9 aspect ratio generally is supported by the motion picture industry. Figure 6.1.5 illustrates some of the more common aspect ratios.

6.1.4 References

1. Robin, Michael: "Digital Resolution," *Broadcast Engineering*, Intertec Publishing, Overland Park, Kan., pp. 44–48, April 1998.

2. Leathers, David: "Production Considerations for DTV," in *NAB Engineering Handbook*, 9th ed., Jerry C. Whitaker (ed.), National Association of Broadcasters, Washington, D.C., pp. 1067–1072, 1999.

3. Thorpe, Laurence J.: "Applying High-Definition Television," *Television Engineering Handbook*, rev. ed., K. B. Benson and Jerry C. Whitaker (eds.), McGraw-Hill, New York, N.Y., pg. 23.4, 1991.

6.1.5 Bibliography

Baldwin, M. Jr.: "The Subjective Sharpness of Simulated Television Images," *Proceedings of the IRE*, vol. 28, July 1940.

Belton, J.: "The Development of the CinemaScope by Twentieth Century Fox," *SMPTE Journal*, vol. 97, SMPTE, White Plains, N.Y., September 1988.

Benson, K. B., and D. G. Fink: *HDTV: Advanced Television for the 1990s*, McGraw-Hill, New York, N.Y., 1990.

Fink, D. G.: "Perspectives on Television: The Role Played by the Two NTSCs in Preparing Television Service for the American Public," *Proceedings of the IEEE*, vol. 64, IEEE, New York, September 1976.

Fink, D. G: *Color Television Standards*, McGraw-Hill, New York, N.Y., 1986.

Fink, D. G, et. al.: "The Future of High Definition Television," *SMPTE Journal*, vol. 9, SMPTE, White Plains, N.Y., February/March 1980.

Fujio, T., J. Ishida, T. Komoto, and T. Nishizawa: "High-Definition Television Systems—Signal Standards and Transmission," *SMPTE Journal*, vol. 89, SMPTE, White Plains, N.Y., August 1980.

Hubel, David H.: *Eye, Brain and Vision*, Scientific American Library, New York, N.Y., 1988.

Judd, D. B.: "The 1931 C.I.E. Standard Observer and Coordinate System for Colorimetry," *Journal of the Optical Society of America*, vol. 23, 1933.

Kelly, R. D., A. V. Bedford, and M. Trainer: "Scanning Sequence and Repetition of Television Images," *Proceedings of the IRE*, vol. 24, April 1936.

Kelly, K. L.: "Color Designation of Lights," *Journal of the Optical Society of America*, vol. 33, 1943.

Miller, Howard: "Options in Advanced Television Broadcasting in North America," *Proceedings of the ITS*, International Television Symposium, Montreux, Switzerland, 1991.

Pitts, K. and N. Hurst: "How Much Do People Prefer Widescreen (16 × 9) to Standard NTSC (4 × 3)?," *IEEE Transactions on Consumer Electronics*, IEEE, New York, N.Y., August 1989.

Pointer, R. M.: "The Gamut of Real Surface Colors, *Color Res. App.*, vol. 5, 1945.

SMPTE Recommended Practice RP 199-1999, "Mapping of Pictures in Wide-Screen (16:9) Scanning Structure to Retain Original Aspect Ratio of the Work," SMPTE, White Plains, N.Y., 1999.

6.2

Audio Production Issues

6.2.1 Introduction

The realism of a video presentation depends to a great degree on the realism of the accompanying sounds. This important point has not be lost on the designers of HDTV systems. Particularly in the close viewing of HDTV images, if the audio system is monophonic, the sounds seem to be confined to the center of the screen. The result is that the visual and aural senses convey conflicting information. From the beginning of HDTV system design, it was clear that stereophonic sound—at minimum—must be used. The generally accepted quality standard for high-fidelity audio was set by the digital compact disc (CD). This medium covers audio frequencies from below 30 Hz to above 20 kHz, with a dynamic range of 90 dB or greater.

Sound is an important element in the viewing environment. To provide the greatest realism for the viewer, the picture and the sound should be complementary, both technically and editorially. The sound system should match the picture in terms of positional information and offer the producer the opportunity to use the spatial field creatively. The sound field can be used effectively to enlarge the picture. A *surround sound* system can further enhance the viewing experience.

6.2.2 The Aural Image

There is a large body of scientific knowledge focused on how humans localize sound. Most of the research has been conducted with subjects using earphones to listen to monophonic signals to study *lateralization*. *Localization* in stereophonic listening with loudspeakers is not as well understood, but the research shows the dominant influence of two factors: *interaural amplitude* differences and *interaural time delay*. Of these two properties, time delay is the more influential factor. Over intervals related to the time it takes for a sound wave to travel around the head from one ear to the other, interaural time clues determine where a listener will perceive the location of sounds. Interaural amplitude differences have a lesser influence. An amplitude effect is simulated in stereo music systems by the action of the stereo balance control, which adjusts the relative gain of the left and right channels. It is also possible to implement stereo balance controls based on time delays, but the required circuitry is more complex.

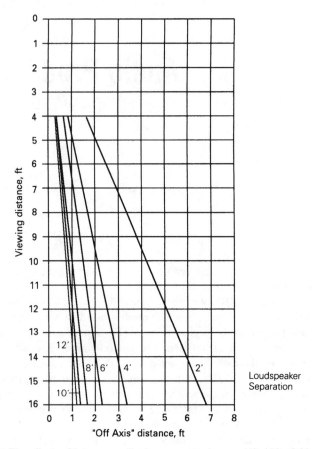

Figure 6.2.1 The effects of listener positioning on center image shift. (*After* [1].)

A listener positioned along the line of symmetry between two loudspeakers will hear the center audio as a phantom or *virtual image* at the center of the stereo stage. Under such conditions, sounds—dialogue, for example—will be spatially coincident with the on-screen image. Unfortunately, this coincidence is lost if the listener is not positioned properly with respect to the loudspeakers. Figure 6.2.1 illustrates the sensitivity of listener positioning to aural image shift. As illustrated, if the loudspeakers are placed 6 ft apart with the listener positioned 10 ft back from the speakers, an image shift will occur if the listener changes position (relative to the centerline of the speakers) by just 16 in. The data shown in the figure is approximate and will yield different results for different types and sizes of speakers. Also, the effects of room reverberation are not factored into the data. Still, the sensitivity of listener positioning can be seen clearly. Listener positioning is most critical when the loudspeakers are spaced widely, and less critical when they are spaced closely. To limit loudspeaker spacing, however, runs counter to the purpose of wide-screen displays. The best solution is to add a third audio channel dedicated exclusively to the transmission of center-channel signals for reproduction by a center loudspeaker positioned at the video display, and to place left and right speakers apart from the display to emphasize the wide-

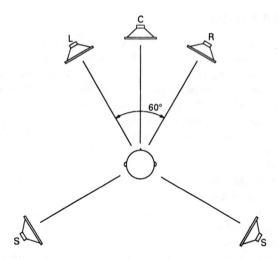

Figure 6.2.2 Optimum speaker placement for HDTV viewing. (*After* [2].)

screen effect. The addition of *surround sound* speakers further improves the realism of the aural component of the production.

6.2.2a Matching Audio to Video

It has been demonstrated that even with no picture to provide visual cues, the ear/brain combination is sensitive to the direction of sound, particularly in an arc in front of and immediately in back of the listener. Even at the sides, listeners are able to locate direction cues with reasonable accuracy. With a large-screen display, visual cues make the accuracy of sound positioning even more important.

If the number of frontal loudspeakers and the associated channels is increased, the acceptable viewing/listening area can be enlarged. Three-channel frontal sound using three loudspeakers provides good stereo listening for three or four viewers. The addition of two or more rear channels permits surround sound effects.

Surround sound presentations, when done correctly, significantly improve the viewing experience. For example, consider the presentation of a concert or similar performance in a public hall. Members of the audience, in addition to hearing the direct performance sound from the stage, also receive reflected sound, usually delayed slightly and perhaps diffused, from the building surfaces. These acoustic elements give a hall its tonal quality. If the spatial quality of the reflected sound can be made available to the home viewer, the experience will be enhanced greatly. The home viewer will see the stage performance in high-definition and hear both the direct and indirect sound, all of which will add to the feeling of being present at the performance.

In sports coverage, much use can be made of positional information. In a tennis match, for example, the umpire's voice would be located in the center sound field—in line with his or her observed position—and crowd and ambient sounds would emanate from left and right. See Figure 6.2.2.

6.2.2b   Making the Most of Audio

In any video production, there is a great deal of sensitivity to the power of the visual image portrayed through elements such as special effects, acting, and directing that build the scene. There are a number of subtleties in the visual area, understood and manipulated by video specialists. By the same token, there are psychoacoustic subtleties relating to how humans hear and experience the world around them that audio specialists can manipulate to their advantage.

Reverb, for example, is poorly understood; it is more than just echo. This tool can be used creatively to trigger certain psychoacoustic responses in an audience. The brain will perceive a voice containing some reverb to be louder. Echo has been used for years to effectively change positions and dimensions in audio mixes.

To use such psychoacoustic tools is to work in a delicate and specialized area, and audio is a subjective discipline that is short on absolute answers. One of the reasons it is difficult to achieve good quality sound is because it is hard to define what that is. It is usually easier to quantify video than audio. Most people, given the same video image, come away with the same perception of it. With audio, however, accord is not so easy to come by. Musical instruments, for example, are harmonically rich and distinctive devices. A violin is not a pure tone; it is a complex balance of textures and harmonics. Audio offers an incredible palette, and it is acceptable to be different. Most video images have any number of absolute references by which images can be judged. These references, by and large, do not exist in audio.

When an audience is experiencing a program—be it a television show or an aircraft simulator computer game—there is a balance of aural and visual cues. If the production is done right, the audience will be drawn into the program, putting themselves into the events occurring on the screen. This *suspension of disbelief* is the key to effectively reaching the audience.

6.2.3   References

1.   Torick, Emil L.: "HDTV: High Definition Video—Low Definition Audio?," *1991 HDTV World Conference Proceedings*, National Association of Broadcasters, Washington, D.C., April 1991.

2.   Holman, Tomlinson: "The Impact of Multi-Channel Sound on Conversion to ATV," *Perspectives on Wide Screen and HDTV Production*, National Association of Broadcasters, Washington, D.C., 1995.

6.2.4   Bibliography

Hamasaki, Kimio: "How to Handle Sound with Large Screen," *Proceedings of the ITS*, International Television Symposium, Montreux, Switzerland, 1991.

Holman, Tomlinson: "Psychoacoustics of Multi-Channel Sound Systems for Television," *Proceedings of HDTV World*, National Association of Broadcasters, Washington, D.C., 1992.

Keller, Thomas B.: "Proposal for Advanced HDTV Audio," *1991 HDTV World Conference Proceedings*, National Association of Broadcasters, Washington, D.C., April 1991.

Lagadec, Roger, Ph.D.: "Audio for Television: Digital Sound in Production and Transmission," *Proceedings of the ITS*, International Television Symposium, Montreux, Switzerland, 1991.

Miller, Howard: "Options in Advanced Television Broadcasting in North America," *Proceedings of the ITS*, International Television Symposium, Montreux, Switzerland, 1991.

Slamin, Brendan: "Sound for High Definition Television," *Proceedings of the ITS*, International Television Symposium, Montreux, Switzerland, 1991.

6.3
Digital System Architectures

6.3.1 Introduction

In a discussion of facility design, there are many commonalities between a conventional analog plant and one built around digital systems for operation as a DTV station. There are a number of areas, however, where the special requirements of digital video, audio, and ancillary data must be considered. Furthermore, the changing nature of the digital video production/transmission plant demands a level of operational flexibility never before required.

6.3.1a DTV Implementation Scenarios

The transition from a conventional NTSC plant to DTV is dictated by any number of factors, depending upon the dynamics of a particular facility. The choices, however, usually can be divided into four overall groups [1]:

- Simple pass-through

- Pass-through with limited local insert

- Local HDTV production

- Multicasting

Clearly the first two scenarios are appropriate for the transition phase from analog to digital broadcasting. With the eventual sunset of NTSC, however, the final two scenarios will become the norm.

Local production in high-definition suggests the following basic capabilities:

- Pass-through of network DTV programming

- Local insertion of interstitials and promos

- Unconstrained switch-point selection

- Local insertion of full-screen or keyed logos

- Local insertion of weather crawls or alerts

- Full range of surround sound audio processing

- Enhanced video capabilities

- High-definition cameras in the studio and in the field

- High-definition video production editing capabilities

The most flexible of the four basic scenarios, multicasting permits the station to adjust its program schedule as a function of day parts and/or specific events. The multicasting environment is configured based on marketplace demands and corporate strategy. To preserve the desired flexibility, the facility must be designed to permit dynamic reconfiguration.

6.3.2 DTV Plant Latency and Timing Issues

Programs starting and finishing at unpredictable times can cause great inconvenience to viewers, particularly if they are trying to watch or record successive programs on different stations. A study group of the ATSC Implementation Subcommittee concluded that this can be a serious problem, which to avoid upsetting the viewing public, broadcasters would have to address at some point [2]. An illustration of the problem appears in Figure 6.3.1. From the figure, it follows that

$$D = d_d + d_r + d_c + d_s + d_t + d_m \hspace{3cm} (6.3.1)$$

Where:
D = total delay
d_d = distribution delay
d_r = routing delay
d_c = conversion delay
d_s = switching delay
d_t = transmission and multiplexer delay
d_m = makeup delay to equalize the NTSC and DTV paths

This signal latency problem results from the processing of standard- and high-definition digital television formats for intra-plant signal routing, inter-plant distribution, and delivery to the home. With the deployment of digital technology, the issue of latency can be problematic. Compressed formats are present in modern television plants in a variety of locations; from tape machine to transmitted stream, from file server to format converter. As the operators of these plants convert signal formats from analog to digital, from one digital format to another, and/or encode or decode the compressed streams, delays are introduced. This delay can vary tremendously depending on the path taken from the originating signal point to the final emissions point. Thus, a local station connected to the network by a short fiber connection might be delayed by 7 s or less while a cable system sending QAM over the cable fed by a local station connected, in turn, to the network by satellite might take 20 s.

In a single program thread, latencies can be ignored for the most part, and the total input/output delay allowed to fall where it may. However, when the need arises for multiple feeds or synchronicity in one form or another, or when the time that the program "airs" is determined to be critical, these various delays must be accounted for and an appropriate engineering solution applied.

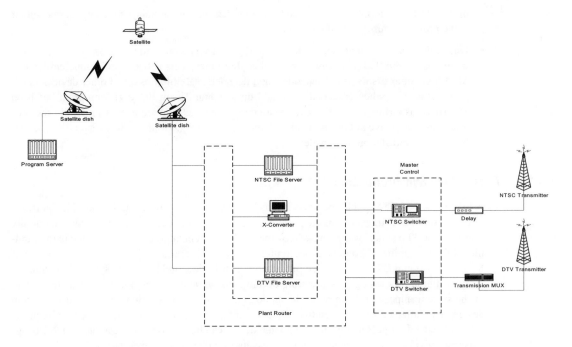

Figure 6.3.1 A typical television distribution chain. Latencies can be experienced at each link in the process. (*From* [2[. *Used with permission.*)

For example, a plant may take in a high-definition feed, convert it to NTSC (or vice-versa), and plan to broadcast both in a simulcast scenario. If no corrections are applied, the feeds will be out of "time" with respect to each other. Depending on the complexity of the operation, timing differences could be in the order of between 5 and 25 s. One option is to ignore the differential, and allow one broadcast to lag the other. However, if it is desired that commercial insertions be accomplished by one control facility, some provision must be made to deal with the delay, either by bringing the programs back into sync, or by calculating a fixed delay for the insertion of material into the lagging program. There may also be a desire to lock the timing of the two programs together for a variety of operational or programming reasons. This would require the insertion of a fixed delay into the leading program.

Such operations may be needed even in the case where two program streams, one NTSC and one digital, are provided to a station, since the processing of the two signals may be through paths and equipment of significantly different latencies.

The following fundamental issues are apparent:

- Because some stations want to use one master control facility to insert interstitial material into both the NTSC and DTV program streams, the station must ensure that appropriate delays are inserted into the various paths within the station. This assumes that the program source (network or syndicator) supplies one feed and the station converts to the other feed prior to transmission. Should two feeds be provided, additional delays may be required, either at the program source or at the station.

- One of the minor but irritating problems for digital TV stations and networks is how to ensure programs are emitted precisely at their scheduled times. This is because the inherent latency of MPEG compressors, ATSC encoders and decoders, satellite links, and other devices in the program chain—when cascaded—can add up to a significant delay. If a program has been advertised as airing at $T = X$, then, when latencies are determined to be significant, the program needs to arrive at the station at no later than $T = X - D$, where D is the sum of the distribution path and station latency delays.

6.3.2a   Audio/Video Synchronization

A related area of DTV plant operation is synchronization between the audio and video portions of the program as it travels from the point of origination to the consumer. This issue, commonly referred to as "lip sync," was addressed by the ATSC Implementation Subcommittee in the Finding "Relative Timing of Sound and Vision for Broadcast Operations" [3].

The end-to-end DTV audio/video production, distribution, and broadcast system is a complex array of digital processing, compression, decompression, and storage devices. Each component in the system imposes a latency on the audio and/or video signals flowing through it. System design goals often call for the relative audio/video latency through each component to be in the sub-millisecond range. Operationally, unequal delays can be imposed on the audio and video signals respectively, and these delays compromise audio/video synchronization.

Overview

One of the overarching goals of the DTV broadcasting system is to deliver audio and video in proper synchronization to the viewer [3]. Because each digital audio and video component in the chain from production to reception imposes some degree of latency on the signals passing through it, and the delays imposed on the audio and video signals are typically unequal, each component harbors the potential to cause an audio/video synchronization error at its output. The overall audio-video synchronization error is the algebraic sum of the individual synchronization errors encountered in the chain. While a given synchronization error may cause either a positive or negative differential shift in audio/video timing, the video signal is typically subjected to greater delay than the audio signal, and the tendency is therefore toward video lagging behind audio. Thus, there is a requirement to monitor audio/video synchronization at various points within in the system and to make corrections, where required, in order to deliver to the viewer audio/video synchronization within the required tolerance.

In addition, there are points within the end-to-end chain that require A/V synchronization to be maintained, such as switching points, monitoring points, and transmission/encoding points.

For the purposes of analysis, the end-to-end DTV system may be divided into four segments:

- Acquisition and production/post production (the contribution system)

- Release facility and distribution system

- Local broadcast station

- Home receiver

It follows that steps must be taken to ensure that the audio and video signals delivered at the output stage of each of the four segments are synchronized within a tight tolerance.

Contribution System

At the production/post-production stage, audio/video synchronization errors can occur in the capture stage, in film-to-video transfer, and in editing [3]. The product may be delivered on video tape or by various electronic means, but whatever the delivery medium, steps must be taken to ensure that audio/video synchronization in the delivered product falls within the required tolerance.

Release Facility

The release facility segment contains a number of devices through which the DTV audio and video signals are passed, which variously impose compression and de-compression, processing, and storage and their attendant differential delays on the signals [3]. The process of distributing the signals to affiliate stations typically requires compression and decompression steps. It is incumbent on the release facility to correct the differential audio/video delays that the signals experience within the plant so that the initial timing relationship is restored to a tight tolerance before the signals reach the distribution encoder. Synchronization to a tight tolerance should be maintained in any encode/decode process that is involved in delivering the signals to the affiliate station, so that the tight A/V synchronization can be monitored at switching and other points.

Broadcast Station

The affiliate station segment contains a number of devices that are similar to those encountered in the release facility segment and that generate the same types of differential audio-video delays, including switching and monitoring points [3]. Audio/video synchronization should be restored to a tight tolerance before the signals are input to the broadcast station audio and video encoding devices to assure that the presentation time stamps placed on the audio and video access units by the encoder faithfully represent correct synchronization.

Timing Recommendations

The ATSC Implementation Subcommittee Finding suggests that under all operational situations, at the inputs to the DTV encoding devices, the sound program should be tightly synchronized to the video program. Specifically, the sound program should never lead the video program by more than 15 ms, and should never lag the video program by more than 45 ms [3].

MPEG-2 models the end-to-end delay from an encoder's signal input to a decoder's signal output as constant. This end-to-end delay is the sum of the delays from encoding, encoder buffering, multiplexing, transmission, demultiplexing, decoder buffering, decoding, and presentation. Presentation Time Stamps are required in the MPEG bit stream at intervals not exceeding 700 ms. The MPEG System Target Decoder model allows a maximum decoder buffer delay of 1 s. Audio and video Presentation Units that represent sound and pictures that are to be presented simultaneously may be separated in time within the transport stream by as much as 1 s. In order to produce synchronized output, the IS Finding suggests that the receiver must recover the encoder's System Time Clock (STC) and use the Presentation Time Stamps (PTS) to present the audio-video content to the viewer with a tolerance of ±15 ms of the time indicated by PTS.

Although practical aural and visual presentation devices typically have finite and different inherent delays, and may have additional delays imposed by post-processing or output functions, the system target decoder models these delays as zero. Such delays must be corrected before the audio and video signals are presented to the viewer.

6.3.3   Program Interchange Identification Requirements

Another study item from the ATSC IS focused on program interchange identification requirements and solutions [4]. Common practice in the identification of programs during the production and distribution processes, including program interchange between organizations, involves human-readable labels and, sometimes, machine-readable labels attached to physical media, with no widely accepted standards for such labeling. Any identification information embedded within the content consists of opening slates recorded in the video that must be read by humans. Programs are often accompanied by format sheets that provide the timing of segments within the programs, but the timings so provided are often suspect and require verification if tight control of program integration is to be maintained.

The change to digital operation offers the opportunity to provide the means for communication of identification, segment timing, and similar characteristics that can support the automated integration of content and the mechanized confirmation of what actually airs. This capability becomes progressively more important as the number of channels multiplies and as operators must simultaneously monitor, but perhaps not control, an increasing number of services. Dependence upon operators up and down a distribution chain manually intervening to insert commercials and interstitials in response to program content and cues is rapidly becoming an anachronism.

To achieve the greatest benefit to the industry in terms both of accuracy of program release and integration, and of labor savings through use of automation, it is necessary to establish standard methods for carrying the required information together with program content. Fortunately, a number of essential underlying techniques that enable meeting these requirements have been developed.

6.3.3a   Identification Solutions

There are two basic categories of requirements that must be addressed if the desired improvements in operating practices are to be achieved [4]:

- The identification of the content itself

- Carriage of information about the internal structure of the content sufficient to allow accurate integration of the content into complete services having seamless continuity

Identification of the content must account for both individual productions and series, properly identifying episodes that are members of series. It must also account for a wide variety of versions that are possible for each individual production and for episodes in series. For the sake of efficiency, it should be possible to uniquely identify content by series, episode, and version using a single number. That number should appear frequently embedded within the data of the content so as to be quickly available throughout the duration of the content. This will allow rapid confirmation of the identity of the content when it first appears and when it is joined in progress.

Use of just a number, however, would require that anyone needing to know more about the content look up the number in a database. To avoid the necessity for database lookup, it should also be possible to carry detailed information about the identity of the series, episode, and version similarly embedded within the content data. It should be possible to carry with the content at least the information that would be available from the database that registers the unique number assigned to the series, episode, and version. It should also be possible to carry other information that further identifies the content or describes some of its characteristics.

6.3.3b Media Management for Digital Television

ATSC document A/57, "Content Identification and Labeling for ATSC Transport," defines a means to uniquely identify content (an *Audiovisual Work*). The identifier may be one of several ID systems, including the International Standard Audiovisual Number (ISAN, ISO 15706[1]) as extended to include version information (V-ISAN). The V-ISAN identifier allows unique identification of a root program, episode, and version [5]. The document also defines an MPEG-2 content identification descriptor for the carriage of this identifier. Furthermore, A/57 defines the semantics of the use of the MPEG-2 Content Identification Descriptor for labeling streaming Audiovisual Content in the ATSC Transport.

V-ISAN

Future television operations must be more efficient than current operations because they will be required to carry far more content, of far greater variety, from a far greater number of sources than they do now, and they must do so with fewer personnel [5].

Within the television and film industry there are many places where it is critically important that a particular movie, series episode, commercial, or other material (and their many versions) be succinctly and uniquely identified. These places include just about all aspects of production, post-production, network and broadcast operations:

Different versions of programs can be created by performing minor edits on the original work, or by combining different language audio and/or subtitling. Each such version must be properly identified. Currently, different aspects of operations and different types of content each may have their own identification schemes. Converting between these schemes is tedious and potentially error-prone. A V-ISAN scheme meets this need and enables the integration of operations.

The core function of V-ISAN is to provide a unique identifier for a single selected piece of completed content. For example, each combination of episode and each version of that episode of a specific program would have an associated V-ISAN identifier. The V-ISAN identifier is intended to label and reference finished, distributed works, which differentiates it from a Universal Material Identifier (UMID, SMPTE 330M)—a numbering system developed specifically for production and post-production use.

A V-ISAN identifier may be used for a variety of purposes and transported in a number of ways. It serves as the key to index a variety of databases that reference the content that it identifies. Another example of its use is for linking program guide information to programs.

A V-ISAN identifier can be associated with any complete program, commercial, promotional announcement, or other finished content. Preferably, a V-ISAN identifier is assigned shortly

---

1. More information on ISAN and its registrar, ISANIA, can be found at www.agicoa.org/

before creation of the content, but after it is certain that the creation will take place. This timing allows all related systems and parties to prepare for the distribution of the new content. However, a V-ISAN identifier may also be assigned during or after creation.

V-ISAN Elements

The three elements described below are collectively referred to as a V-ISAN identifier, which together identify a unique piece of content [5]. Each element is a binary number. When presented in human readable form, they are be presented in a compact, consistent, and easily readable manner.

- **Root Identifier:** A binary number assigned to a piece of content. Closely related content is distinguished by version and episode. For example, an episodic program will have a common root identifier for all episodes and versions. The Root Identifier is assigned by the registration system after a system user provides the mandatory descriptive data.

- **Episode Identifier**: A binary number assigned by the registration system after the content owner provides the registration system with adequate episode descriptive information. All episodes of a series should have the same Root Identifier. Programs that do not have episodes (e.g., movies), use an episode identifier of zero.

- **Version Identifier**: A binary number assigned by the registration system after the content owner provides adequate version descriptive information. All versions that are editorially related to the original content have the same Root Identifier and if episodic have the same episode identifier. All versions are uniquely distinguished by their having different version numbers. Assignment of different version identifiers may be based on language, aspect ratio, subtitling, or other similar characteristics. Each editorially-significant variation in the content receives a different version identifier.

6.3.4   PSIP Implementation Considerations

PSIP information is generated through a process illustrated in Figure 6.3.2 for the general case. Figure 6.3.3 places the PSIP generator in perspective relative to the ATSC transmission system, and Figure 6.3.4 describes the reception and decoding process [6].

Having illustrated the overall program and data flow, the specifics of key elements can be described. To review, the primary tables are:

- System Time Table, STT, which provides the time reference

- Master Guide Table, MGT, which provides version, size, and *program identification* (PID) of all other tables (except the STT)

- Virtual Channel Table, VCT, which provides attributes for all virtual channels in the Transport Stream

- Rating Region Table, RRT, which provides rating information for multiple geographic regions

- Event Information Table, EIT, which provides information relating to events on the virtual channels

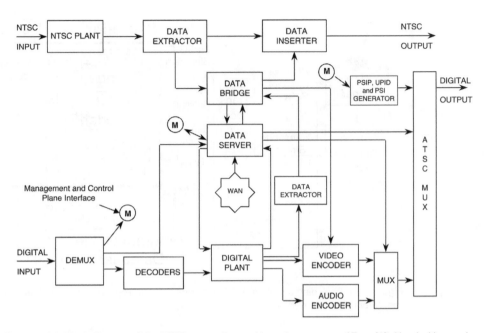

Figure 6.3.2 Block diagram of the PSIP generation and insertion process. (*From* [6]. *Used with permission.*)

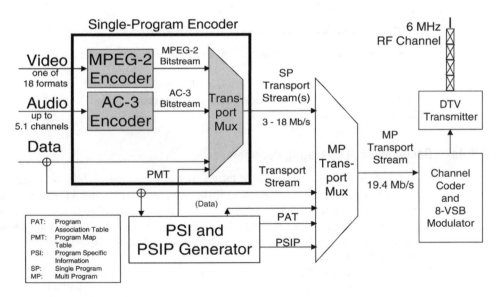

Figure 6.3.3 Block diagram of the ATSC transmission system, including PSIP generation and insertion. (*From* [6]. *Used with permission.*)

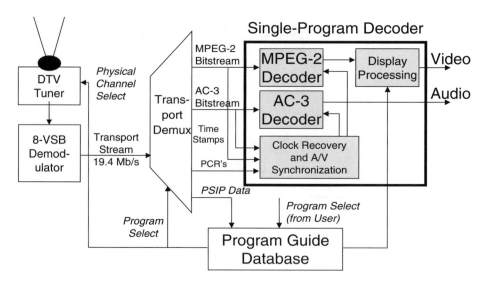

Figure 6.3.4 Bock diagram of an ATSC decoder, including PSIP extraction and program guide generation. (*From* [6]. *Used with permission.*)

- Extended Text Table, ETT, which provides detailed descriptions of virtual channels and events

- Directed Channel Change Table, DCCT

- Directed Channel Change Selection Code Table, DCCST

Not all of these tables are required for terrestrial and cable applications, as detailed in Table 6.3.1.

The STT is a short table that is sent in packets with the base PID. It contains the current time, sent once every second, and is expected to be very accurate. Note that the PSIP standard has an implementation tolerance of ± 1 s.

6.3.4a   Recommended Operating Parameters

Although proper implementation of PSIP at the television station level is not particularly complex, neither is it always straightforward. To address this issue, the ATSC developed a Recommended Practice to explain the operator-oriented elements of PSIP and to provide practical examples of typical station operation [6]. The document was also intended to provide guidelines for designers of PSIP-related hardware and software to optimize user interface information for such equipment.

There are certain "must have" items and "must do" rules of operation. If the PSIP elements are missing or wrong, there may be severe consequences, which will vary depending on the receiver design. The following are key elements that must be set and/or checked by each station:

- **Transport Stream Identification**. The pre-assigned TSIDs must be set correctly in all three locations (PAT, VCT common information, and virtual channel-specific information).

Table 6.3.1 PSIP Tables Required for Transmission in the Broadcast and Cable Modes

Table	Required for Broadcast?	Required for Cable?
STT	Yes	Yes
MGT	Yes	Yes
VCT	Yes (TVCT)	Yes (CVCT)
RRT	Conditional[1]	Conditional[1]
EIT	Yes (EIT-0, -1, -2, -3). All others optional.	Optional, except when required.
ETT	Optional	Optional

[1] If a Content Advisory Descriptor is transmitted, the associated RRT, except for RRT 01, is transmitted.

- **System Time Table**. The STT time should be checked daily and locked to house time. The tolerance in the standard is to allow for unusual conditions and ideally should be inserted into the TS within a few milliseconds after each seconds-count increment of the house time.

- **Short Channel Name**. This is a seven-character name that can be set to any desired name indicating the virtual channel name. For example, a station's call letters followed by SD1, SD2, SD3, and SD4 to indicated various SDTV virtual channels or anything else to represent the station's identity (e.g., WNABSD1, KNABSD2, WNAB-HD, KIDS, etc.).

- **Major Channel**. In most cases, the previously assigned, paired NTSC channel is the major channel number.

- **Service Type**. The service type selects DTV, NTSC, audio only, data, etc., and must be set as operating modes require.

- **Modulation Mode**. A code for the RF modulation of the virtual channel.

- **Source ID**. The Source ID is a number that associates virtual channels to events on those channels. It typically is automatically updated by PSIP equipment or updated from an outside vendor.

- **Service Location Descriptor** (SLD). Contains the MPEG references to the contents of each component of the programs plus a language code for audio. The PID values for the components identified here and in the PMT must be the same for the elements of an event/program. Some deployed systems require separate manual setup, but PID values assigned to a VC should seldom change.

The receiver requires this information in order to properly function and to permit the construction of a basic program guide. Figure 6.3.5 shows what a typical electronic program guide (EPG) might look like.

The maximum cycle time/repetition rate of the tables should be set or confirmed to conform with the suggested guidelines given in Table 6.3.2 for mandatory PSIP tables and Table 6.3.3 for optional PSIP tables.

It is recommended that broadcasters set up a minimal set of three days of tables. The primary cycle time guidelines are illustrated in Figure 6.3.6. The recommended table cycle times given result in a minimal demand on overall system bandwidth. Considering the importance of the information that these PSIP tables provide to the receiver, the bandwidth penalty is trivial.

Chan	Name	6:00 PM	6:30 PM	7:00 PM	7:30 PM	8:00 PM	8:30 PM
6-0	XYZ	City Scene		Travel Log		Movie: *Speed II*	
6-1	XYZ	City Scene		Travel Log		Movie: *Speed II* (HD)	
6-2	XYZ	Movie: *Star Trek—The Voyage Home*				Tune 6-1 for Movie: *Speed II* (HD)	
6-3	LNC	Local News		Airport Info		HD Program on 6-1	

Figure 6.3.5 Example of what an electronic program guide might look like. Extensions can enable thematic browsing and sorting. (*From* [6]. *Used with permission.*)

Table 6.3.2 Mandatory PSIP Table Suggested Repetition Rates (*After* [6].)

PSIP Table	Transmission Cycle
MGT	Once every 150 ms
TVCT	Once every 400 ms
EIT-0	Once every 0.5 seconds
EIT-1	Once every three seconds
EIT-2 and EIT-3	Once every minute
STT	Once every second

Table 6.3.3 Suggested Repetition Rates for Optional PSIP Tables (*After* [6].)

PSIP Table		Transmission Cycle
	DCC request in progress	150 msec
DCCT[1]	2 seconds prior to DCC request	400 msec
	No DCC	n/a
DCCSCT		Once per hour
ETT		Once every minute
EIT-4 and higher		Once every minute
DET		A later version of the Recommended Practice will address data services.

[1] A/65A specifies the following repetition rates for DCC per specified conditions.

Most Common Mistakes

Experience has shown that certain errors are common in many PSIP implementations [6]. These problems typically include the following:

• Missing tables, particularly the STT and EIT.

• Major channel number set to the DTV RF channel number, rather than the associated (legacy) NTSC channel number.

• TSID set to 0 or 1, the NTSC TSID, or another station's TSID; or not set the same in the three required places.

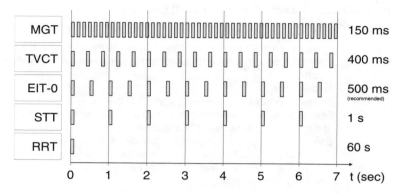

Figure 6.3.6 Maximum cycle times for PSIP tables. (*From* [6]. *Used with permission.*)

- System time missing or set to 00:00:00 on 1/6/1980

Some receivers react to these errors by not tuning to that station.

6.3.4b   About the VCT

There is *essential* station-specific VCT information that the broadcaster must input for viewers to be able to properly tune programs [6]. This information is given in Table 6.3.4.

Because the VCT allows each minor channel to also be assigned a permanent short name and channel name, and since each minor channel will keep the same TSID, carrier frequency (zero or not), and modulation mode over time, the PSIP encoder system software should allow the user to create a local look-up table that associates each minor channel number with these fixed values so the user can then create new VCTs simply by entering the minor channel number of each desired minor channel to be put in the new VCT. The PIDs for each component in a minor channel should not be changed as any changes are expected to increase the time it takes for the receiver to tune the station.

Updating the VCT

Even though the TSID and other parameters for each virtual channel can be a permanent assignment, the minor channels that the station is using may change over time [6]. When a program on a new virtual channel is announced in the EIT, the PSIP Standard requires that the VCT contain EIT VCT information, and vice versa. Because of this, it is recommended that broadcasters update the VCT first to reflect a change in the channel lineup and then use the appropriate source_id in constructing the EITs.

A new VCT containing updated information can be transmitted at any time with the version_number increased by one. It is required that the virtual channel be in the VCT as soon as an EIT that will use that virtual channel is sent. This gives receivers the opportunity to scan the frequencies and detect the channel presence. The system design assumes the receivers scan all RF channels at least once just after being turned off. Filling three days worth of EITs once a day

Table 6.3.4 Station-Specific PSIP Data (*After* [6].)

Data	Action by Broadcaster	Example
major channel number	Entered once. (Use the same channel number as the NTSC channel number assignment. If no paired NTSC channel, use the assigned DTV channel number.)	2
minor channel numbers	Entered once.	1
analog TSID	Entered once. One of a pair (see digital TSID below).	0D4A
digital TSID	Entered once. Assigned with analog TSID in consecutive pairs by the FCC during licensing of the new DTV assignment.	0D4B
service location descriptor	Entered once as pointers to each video, audio, and data stream.	2
source id	Entered once for each virtual channel or automatically generated.	7
service type	Entered once. (Tells the receiver whether the associated minor channel is providing digital or analog service.)	2
short name	Input once.	NABDT
modulation mode	Entered once for each virtual-minor number	0x04
carrier frequency	Recommend zero.	0
MPEG program number	Entered once for each virtual channel. The MPEG program number must be unique within the Transport Stream and shall not be zero.	1
ETM location	None	N/A
access controlled	yes/no	0
hidden	yes/no	0
hide guide	yes/no	1

should reduce the risk of not having information at the time of tuning to just those sets which are never turned off or experience extended power outages.

6.3.4c   Event Information Table

The Event Information Table (EIT) is the PSIP table that carries program schedule information for each virtual channel [6]. MPEG-2 has a construct called a *Program*; in ATSC standards, TV programs are called *Events*. Each instance of an EIT covers a three-hour time span, and provides the following information for each programming source:

- Event start time

- Event duration

- Event title

- A pointer to optional descriptive text for the event

- Program content advisory data (optional, but if present it must go here)

- Caption service data (sometimes optional, but when present must go here)

- Audio service descriptor (required if audio is present)

Most of this data is provided 'under the covers' from other systems to the PSIP generator. The user just needs to enter what program is on when, and select the proper operating parameters.

Each EIT covers a period of three hours. The PSIP generator should automatically convert from local time to the universal time used inside the system. (The receiver converts back to that receiver's local time.) EIT-0 represents the "current" three hours of programming. For terrestrial PSIP, the first four EITs (EIT-0, EIT-1, EIT-2, and EIT-3) are required by PSIP Standard A/65 [7].

Daily updates of the EITs can be done for all programming, but the current EIT has some additional needs. It is strongly recommended that a daily update be done at or near the normal close of business and that this update have at least three days worth of station-correct programming announcements (24 EITs). As most receivers will be acquiring future EITs in the early morning hours while "off", this enables receivers to miss one day of acquisition and still have the next days' events. Adding one days' worth of EITs will add about 1 kbits/s of data to the transport overhead. It is desirable to send all 16 days worth of EITs to cover consumers' setting of recorders during a vacation.

The EIT-0 has some special needs as it contains the closed caption, ratings information, and other essential data about the current program. The connection from master control to the PSIP generator should enable direct updates of current program parameters in EIT-0. By contrast, the EITs for the future are primarily informational, and less critical to system performance as long as the station virtual channel line-up is not changed.

EITs should not be sent describing test signal occurrences in a virtual channel.

Each EIT has space for event titles. The receiver recommendation is to display the first 30 characters of the title, and it is recommended that the first 30 characters be chosen carefully to maximize the chance of meaningful display by receivers. If it is desired to send additional information about the entire event, this is sent in another structure—the ETT. Such information would optionally be presented to consumers, usually after an action. Receivers may have limited support for descriptive text so there may be a trade-off between covering more events and more data about each event. Also, the rate this information is sent can be adjusted by setting the time interval between ETTs to make more efficient use of bandwidth. If however, these were set longer than one minute apart, receiver "off" search time would be increased.

All this information is carried in PSIP data packets called *descriptors*. It is recommended that at a minimum three descriptors be sent when needed:

- The Content Advisory Descriptor (EIT)

- AC-3 Audio Descriptor (EIT and PMT)

- Caption Service Descriptor (EIT)

5.1.1 PMCP

In an effort toward automated generation of PSIP data, the ATSC developed standard A/76, "Programming and Metadata Communication Protocol" [8]. This standard makes it possible to integrate the various information sources that are needed to compile the key PSIP tables. PMCP is designed to permit broadcasters, professional equipment manufacturers, and program service providers to interconnect and transfer data among systems that eventually must be communicated to the PSIP generator. These systems include:

- Traffic

- Program management

- Listing services

- Automation

- MPEG encoder

Because PSIP and other DTV metadata is typically developed and/or processed by several separate systems, communicating the appropriate metadata to the PSIP generator can be problematic.

PMCP is intended to solve this problem by defining a method for communicating metadata that the PSIP generator requires. The overall goal is to ensure proper PSIP implementation while requiring minimum manual intervention by the broadcaster. Equipment manufacturers, system designers, and broadcasters can use the tools provided in PMCP to help achieve that goal. While targeted primarily at PSIP, the schema is extensible for other types of metadata, and can be used for the carriage of private information within the current structure.

PMCP is based on a protocol utilizing XML (*Extensible Markup Language*) message documents and the heart of the standard is an XML schema that defines the message structure, the elements allowed, their relationships, and attributes. XML is a W3C standard that allows structuring of information in a text document that is both human- and machine-readable.

Because PSIP and other DTV metadata is originated or processed by several separate systems and equipment, up to now there have been difficulties in communicating the metadata to the PSIP generator. Implementing PMCP will help ensure that transmitted PSIP information is complete and correct, with minimum manual intervention.

PMCP references and is complementary to other existing ATSC standards. It also supports the ISO V-ISAN standard for unique identification of program content and carries all the information needed in one message structure for:

- Virtual channels

- PSIP events

- Programs

- System Time Table

- Regional Ratings Table

6.3.1   References

1.   Course notes, "DTV Express," PBS/Harris, Alexandria, Va., 1998.

2.   ATSC: "Implementation Subcommittee Finding: Systems Evaluation Working Group Report on Latency and Timing Issues," Advanced Television Systems Committee, Washington, D.C., Doc. IS/266, June 13, 2002.

3.   ATSC: "Implementation Subcommittee Finding: Relative Timing of Sound and Vision for Broadcast Operations," Advanced Television Systems Committee, Washington, D.C., IS/191, June 25, 2003.

4.   ATSC: "Implementation Subcommittee Finding: Program Interchange Identification Requirements and Solutions," Advanced Television Systems Committee, Washington, D.C., Doc. IS/214, July 28, 2001.

5.   ATSC: "Content Identification and Labeling for ATSC Transport," Doc. A/57A, Advanced Television Systems Committee, Washington, D.C., July 1, 2003.

6. ATSC: "Recommended Practice: Program and System Information Protocol Implementation Guidelines for Broadcasters," Advanced Television Systems Committee, Washington, D.C., Doc. A/69, June 25, 2002.

7. ATSC: "Program and System Information Protocol for Terrestrial Broadcast and Cable—Revision C," Advanced Television Systems Committee, Washington, D.C., Doc. A/65C, January 2, 2006.

8. ATSC: "Programming Metadata Communication Protocol," Advanced Television Systems Committee, Washington, D.C., Doc. A/76, November 10, 2004.

6.3.1   Bibliography

ATSC: "Implementation Subcommittee Report on Findings of the Top Down Meetings," Advanced Television Systems Committee, Washington, D.C., Doc. IS/095, October 30, 1998.

ATSC: "Implementation of Data Broadcasting in a DTV Station," Advanced Television Systems Committee, Washington, D.C., Doc. IS/151, November 1999.

Extensible Markup Language (XML) 1.0, World Wide Web Consortium Recommendation, http://www.w3.org.

Sheth, Amit, and Wolfgang Klas (eds.): *Multimedia Data Management*, McGraw-Hill, New York, N.Y., 1996.

SMPTE Metadata Dictionary, http://www.smpte.org/.

SMPTE: "System Overview—Advanced System Control Architecture, S22.02, Revision 2.0," S22.02 Advanced System Control Architectures Working Group, SMPTE, White Plains, N.Y., March 27, 2000.

mental component in the waveform of Figure 3.1.5. Note that frequency is expressed as a multiple of the fundamental frequency f1. The numerals are the harmonic numbers. Only the fundamental f1 and the third and fifth harmonics (f3 and f5) are present. 3-12

Figure 3.1.8 Characteristics of speech. (*a*) Waveforms showing the varying area between vibrating vocal cords and the corresponding airflow during vocalized speech as a function of time. (*b*) The corresponding amplitude-frequency spectrum, showing the 100-Hz fundamental frequency for this male speaker. 3-13

Figure 3.2.1 (*a*) The relationship between the incident sound, the reflected sound, and a flat reflecting surface, illustrating the law of reflection. (*b*) A more elaborate version of (*a*), showing the progression of wavefronts (the curved lines) in addition to the sound rays (arrowed lines). (*c*) The reflection of sound having a frequency of 100 Hz (wavelength 3.45 m) from a surface with irregularities that are small compared with the wavelength. (*d*) When the wavelength of the sound is similar to the dimensions of the irregularities, the sound is scattered in all directions. (*e*) When the wavelength of the sound is small compared with the dimensions of the irregularities, the law of reflection applies to the detailed interactions with the surface features. 3-21

Figure 3.2.2 (*a*) Differing direct and reflected path lengths as a function of receiver location. (*b*) The interference pattern resulting when two sounds, each at the same sound level (0 dB) are summed with a time delay of just over 5 ms (a path length difference of approximately 1.7 m). (*c*) The reflection signal has been attenuated by 6 dB (it is now at a relative level of –6 dB, while the direct sounds remains at 0 dB); the maximum sound level is reduced, and perfect nulls are no longer possible. The familiar comb-filtering pattern remains. 3-23

Figure 3.2.3 Behavior of a point monopole sound source in full space (4π) and in close proximity to reflecting surfaces that constrain the sound radiation to progressively smaller solid angeles. 3-25

Figure 3.2.4 Stylized illustration of the diffraction of sound waves passing through openings and around obstacles. (*a*) The case where the wavelength is large compared with the size of the opening and the obstacle. (*b*) The case where the wavelength is small compared with the size of the opening and the obstacle. 3-26

Figure 3.2.5 A simplified display of the main sound radiation directions at selected frequencies for: (*a*) a trumpet, (*b*) a cello. 3-27

Figure 3.2.6 The refraction of sound by wind and by temperature gradients: (*a*) downwind or in a temperature inversion, (*b*) upwind or in a temperature lapse. 3-28

Figure 3.3.1 The human ear: (*a*) cross-sectional view showing the major anatomical elements, (*b*) a simplified functional representation. 3-32

Figure 3.3.2 Contours of equal loudness showing the sound pressure level required for pure tones at different frequencies to sound as loud as a reference tone of 1000 Hz. 3-34

Figure 3.3.3 The standard frequency-weighting networks used in sound-level meters. 3-36

Figure 3.3.4 Detection threshold for pure tones of various frequencies: (*a*) in insolation, (*b*) in the presence of a narrow band (365 to 455 Hz) of masking noise centered on 400 Hz at a sound level of 80 dB, (*c*) in the presence of a making tone of 400 Hz at 80 dB. 3-37

Jerry C. Whitaker is Vice President of Standards Development at the Advanced Television Systems Committee (ATSC), Washington, D.C. He was previously President of Technical Press, a consulting company based in the San Jose (CA) area. Mr. Whitaker has been involved in various aspects of the electronics industry for over 25 years, with specialization in communications. Current book titles include the following:

- Editor-in-chief, *Standard Handbook of Video and Television Engineering*, 4th ed., McGraw-Hill, 2003

- Editor-in-chief, *Standard Handbook of Audio and Radio Engineering*, 2nd ed., McGraw-Hill, 2001

- Editor-in-Chief, *Standard Handbook of Broadcast Engineering*, McGraw-Hill, 2005

- Editor, *Television Receivers*, McGraw-Hill, 2001

- Editor, *Audio/Video Professional's Field Manual*, McGraw-Hill, 2001

- Editor, *Master Handbook of Video Production*, McGraw-Hill, 2002

- Editor, *Master Handbook of Audio Production*, McGraw-Hill, 2002

- Editor, *Audio/Video Protocol Handbook*, McGraw-Hill, 2001

Mr. Whitaker has lectured extensively on the topic of electronic systems design, installation, and maintenance. He is the former editorial director and associate publisher of *Broadcast Engineering* and *Video Systems* magazines, and a former radio station chief engineer and television news producer.

Mr. Whitaker is a Fellow of the Society of Broadcast Engineers and an SBE-certified Professional Broadcast Engineer. He is also a Fellow of the Society of Motion Picture and Television Engineers.